New Records of the Great Dying in South China

New Record of Great Dying in South China series systematically documents the recently discovered Permian-Triassic biota, including fossil plants, brachiopods, bivalves, radiolarians, conodonts, foraminifers, ostracods, and stratigraphy across the PTB interval. The series focuses the taxonomy, phylogeny, biodiversity, biostratigraphy, evolution of biota, paleoecology and paleogeography. The series summarizes the processes of mass extinction and environmental crisis.

More information about this series at https://link.springer.com/bookseries/15421

Jianxin Yu · Jean Broutin · Zongsheng Lu
Editors

Plants and Palynomorphs around the Permian-Triassic Boundary of South China

Springer

Editors
Jianxin Yu
School of Earth Sciences, State Key Laboratory of Biogeology and Environmental Geology
China University of Geosciences
Wuhan, China

Jean Broutin
Faculty of Sciences and Engineering (CR2P, MNHN, CNRS)
Sorbonne Université
Paris, France

Zongsheng Lu
School of Earth Sciences
China University of Geosciences
Wuhan, China

ISSN 2524-4574 ISSN 2524-4582 (electronic)
New Records of the Great Dying in South China
ISBN 978-981-19-1491-1 ISBN 978-981-19-1492-8 (eBook)
https://doi.org/10.1007/978-981-19-1492-8

This Springer imprint is published by the registered company Springer Nature Singapore Pte Ltd.
The registered company address is: 152 Beach Road, #21-01/04 Gateway East, Singapore 189721, Singapore

Preface

Permian-Triassic is an important stage for the earth to evolving from Paleozoic to Mesozoic. Hence, geologists in the world have been working hard on the origin, scale, content and characteristics of this specific period of development. Presently, profound achievements have been made in the study of the changes of marine biome and geologic events. The establishment of marine Permian-Triassic GSSP-global boundary stratotype sections and points in Meishan, Changxing County, Zhejiang Province, China, has marked the great progress in the research of large-scale strata-types (Yin et al. 2001). Consequently, a great interest arose for an accurate determination and identification of an accessory GSSP in terrestrial environments. This causes induced international competition in locating a global stratotype section for such no typical section which could be acknowledged by most geologists. The previous studies on a Permian-Triassic boundary in terrestrial facies in China have been concentrated on its northern area, specially in Urumqi-Fukang-Jimusar on the southwestern margin of the Junggar Basin, where the P/T transition deposits are terrestrial. Detailed study carried out on biostratigraphy, event-stratigraphy, sedimentation and climate changes in this area facilitates this section to be competitive for a "terrestrial" GSSP of the terrestrial section. Lozovsky (1998) once suggested that the Junggar Basin in Xinjiang, China and the Moscow Basin in Russia should be the ideal candidates for the global Permian-Triassic stratotype section with their completeness and abundance in vertebrate faunas, which was interrupted by the Dalongkou event. At present, other basins, such as Karoo Basin in South Africa, Russian Terrane, Bowen Basin and Sydney Basin in Australia, are all impressive and competitive for the global stratotype. The well-conducted biostratigraphic studies on these basins have revealed the abnormal section boundary which can be correlated with that of the marine facies and thus proved the existence of various events of global significance, for instance, bacteria events, volcanism, celestial collision, and environmental collapse. Therefore, it is still necessary to find and recommend another better candidate section for the terrestrial GSSP in China.

The turn from Permian to Triassic is one of the most crucial periods of the dramatic earth evolution, which especially led to the extinction of 90% marine organisms and

70% terrestrial organisms. However, opinions vary about the origin for such "catastrophic" extinction. Hence, many hypotheses crop up to account for the principle of this global change, such as volcanic events, celestial collision, anoxic deposition, environmental change, magnetic overturn, and sea level changes. On such account, the research on the biostratigraphy and paleoclimatology of this period is of great significance for the precise determination of the accessory stratotype section and points of the terrestrial Permian-Triassic boundary and the solution to causes of the biologic extinction.

During the Permian to Triassic transition, the Western Guizhou-Eastern Yunnan of the upper Yangtze Platform in South China was located on the eastern margin of the Kangdian Massif. In this period, the area was generally in an offshore environment, receiving successively sediments of terrestrial facies-marine/terrestrial transitional facies and marine facies. The whole P/T section is well preserved with well-developed biome, continuous strata, and complete sequence, favorable for the study of the terrestrial stratigraphy and accessory stratotype sections. Hence, it can be an important area for both marine and terrestrial facies Permian-Triassic boundary division and correlation in South China and even all over the world. Studies carried out before 1949 were concentrated on the preliminary geological understanding of the macro-outlook of stratigraphic biology. Since then, more understandings and stratigraphic researches have led to the study of the sedimentary environment and facies as well as the lithological paleogeography. For example, Nanjing Institute of Geology and Paleontology of Chinese Academy of Science (1980) has made a considerable summary of the stratigraphy in the area, based on the detailed biogeologic study. They believe that the Late Permian terrestrial coal-bearing strata or with marine intercalations in the border area of Guizhou and Yunnan contain a complete sequence and are continuous with the overlying "Kayitou Sandshale Formation", which is now proved to be the very beginning of Early Triassic based on identified bivalves, ostracods, insects, and palynomorphs and also the facies alteration of the bottom strata of Feixianguan Formation. Yang (1987) and Yin et al. (1995) worked in Panxian on the border area on the transition section from marine to terrestrial facies and on the biostratigraphy. Wang and Yin (2001) discovered clay layers of terrestrial Permian-Triassic boundary with event stratigraphic index during their research on the Permian-Triassic stratigraphy of terrestrial facies and first approach to the sedimentary facies and sequence stratigraphy in eastern Yunnan and western Guizhou. The discovery of this boundary clay layer has laid to a foundation for the further analysis of the terrestrial boundary stratigraphy in the area and also provided favorable condition for the correlation of marine and terrestrial facies boundary strata as well as for the determination of Permian-Triassic boundary. Peng (2002) has further investigated the terrestrial Permian-Triassic boundary and discussed the terrestrial biological event process. All these studies indicate that the study of the sedimentation environment and sequence stratigraphy is of great implication to locate a candidate for being the terrestrial Permian-Triassic accessory section and point stratotype section in China.

On the basis of the previous studies, this new approach covers many Permian-Triassic sections of marine-terrestrial alternation facies and terrestrial facies. It studies paleontology, stratigraphy, and climatology in the area by the analysis of the large plant fossils collected from the outcrops and microfloras obtained by chemical process with multidisciplinary approaches so as to present a systematic account of the Permian-Triassic flora. It will expound on the genera and species composition and distribution of the flora as well as their interrelations to point out the evolution of the Permian-Triassic flora in western Guizhou and eastern Yunnan. This will be, on the one hand, to establish the terrestrial, marine-terrestrial alternation and marine biostratigraphic boundary, their correlation and, on the other hand, to enrich the studies of the Late Permian gigantopterids flora and Early Triassic Lycophyta in South China, which may fill the blank of knowledge on the Earliest Triassic flora assemblage in this area. It is expected that this study will help the further probing of the plant composition, evolution, and migration of the late Cathaysia flora and their relationship with the flora in other areas and with the paleogeographic environment.

This dissertation is based on the field observation and collecting of the practical sections and laboratory analysis, testing, and processing of various samples replaced in a literature review as exhaustive as possible of related papers, published in China and other countries. The content and the innovations of this dissertation can be found as follows.

1 Content of the Study

The focus of this dissertation is the study of the paleofloras and their evolution dynamics during the critical geological period of the Permian-Triassic transition, based on abundant very plant fossils discovered in the marine-terrestrial alternation and terrestrial strata in western Guizhou and eastern Yunnan, South china. This study intends to analyze the characteristics of the flora in this area and their patterns of extinction and survival as well as their migration and evolution dynamics. It actually consists of the following aspects.

Field Geological Investigation This involved observations of the rock types in different areas, their structure and tectonics, thickness change, spatial distribution, and paleogeographic location on the one hand. These observations show that there is no obvious lithologic difference in the different facies, which are a set of continuous deposition section dominated by siltstone and silty mudstone. The difference is that the Late Permian Xuanwei Formation includes interbedded coal beds or coal seams. On the other hand, it deals with the ancient soils with abundant root systems. Then, samples of each type in the different areas mentioned above have been collected and prepared for laboratory analysis to propose accurate correlations.

Study of Fossil Macrofloras It is necessary to work out the composition of the plant fossil assemblages before the study of the flora composition, their extinction and survival as well as their evolution dynamics in this critical geological period. As indicated above, the plant assemblage of the Latest Permian Xuanwei Formation in South China was studied in 1970s, but there was a lack of analysis of its extinction mode and survival processes owing to the strategy available for these latest Paleozoic plants. Meanwhile, the number of the Earliest Triassic plant fossils is very rare which were found, until the 1980s, only in Siberia. Subsequently, Wang and Wang (1989) found plant fossils in Liujiagou Formation from Shanxi, North China. In recent years, with our deeply research, "miniaturized"? fossil plants, with special shape, were found in the Earliest Triassic most basal beds of the Kayitou Formation. The study of these valuable fossils has provided evidence for a more precise location of the terrestrial Permian-Triassic boundary.

Study of Fossil microfloras Although the four sections of the above-mentioned Permian-Triassic time, all contain rich plant fossils, it is necessary to choose one as a terrestrial accessory stratotype section and point for the correlation with the marine GSSP, Meishan section (GSSP) in Changxing, Zhejiang Province. Biologically, only the palynological assemblages can be the key for this choice. In the meantime, the pollen data may compensate for the diversity distortion caused by the limited preservation of macro-plant fossils and help to an accurate evaluation of the floral extinction mode. In addition, this study also deals with the analysis of spores in situ of some plants.

Study of Fossil Plants of Specific Implications Biostratigraphic and palaeogeographic analyses are made on the typical elements of plant assemblages of different ages, such as *Lobatannularia* and Gigantopterids of Latest Permian Xuanwei Formation and *Annalepis* and *Peltspermum* of Early Triassic Kayitou Formation (Induan Age).

2 Innovations Arising From This Study

It has been bothering paleobotanists and stratigraphers whether plant fossils can be used to determine accurately the continental Permian-Triassic boundary. Hence, this dissertation intends to find the index of terrestrial Permian-Triassic biostratigraphic boundary in South China that should be of practical implication.

The plant fossils found in the Kayitou Formation in western Guizhou and eastern Yunnan, reported for the first time in this dissertation, should filled up the blank of the plant assemblages successions during the Permian-Triassic transition in South China. Additionally, this assemblage is a totally new flora for its difference in outlook from those known in other areas. Therefore, it is quite necessary and edge-cutting for this dissertation to study in detail the composition of this flora.

This study is quite interdisciplinary, as it covers aspects in paleobotany, palynology, phytogeography, and plant evolutionary dynamics in order, to work out the terrestrial plant evolution, extinction, migration, and recovery mechanism across Permian and Triassic. See Fig. 1 for various symbols used in the illustrations

During the research, the authors have made excursions to observe and collect first-hand data many times until now. The Permian-Triassic boundary sections observed are varying from marine to terrestrial facies, respectively deep-water Xinmin section

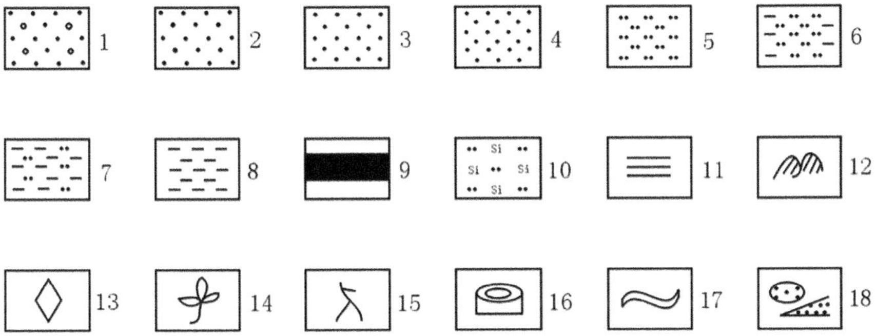

Fig. 1 Illustrates the legends used in the monograph. 1. Pebbly coarse sandstone; 2. Coarse sandstone; 3. Medium-grained sandstone; 4. Fine-grained sandstone; 5. Siltstone; 6. muddy siltstone; 7. Silty mudstone; 8. Mudstone or claystone; 9. Coal seam (line); 10. Siliceous siltstone; 11. Horizontal bedding; 12. Cross bedding; 13. Pyrite crystal; 14. Fossil leave; 15. Fossil root; 16. Petrified wood; 17. Footprint fossil; 18. Lenticular rock

in Anshun and Duanshan Section in Huishui, shallow-water Zhongzai section in Liuzhi, marine-terrestrial alternation Tucheng section in Panxian, Kele section in Hezhang and Mide section in Xuanwei, terrestrial Lubei in Huize, Chahe and Zhejue sections in Weining, and so on.

Wuhan, China — Jianxin Yu
Paris, France — Jean Broutin
Wuhan, China — Zongsheng Lu

References

Lozovsky VR (1998) The Permian-Triassic boundary in the continental series of Eurasia. Palaeogeogr Palaeoclimatol Palaeoecol 143:273–283

Nanjing Institute of Geology and Palaeontology, Academia Sinica (1980) Late permian coal-bearing strata and palaeontological Fauna in western Guizhou and eastern Yunnan. Science Press, Beijing, pp 1–277

Peng YQ, Wang SY, Wang YF, Yang FQ (2002) A proposed area for the study of the Accessory Section and Point of the Terrestrial Permian-Triassic boundary. J China Univ Geosci. 13:157–162

Wang SY, Yin HF (2001) Study on terrestrial Permian-Triassic boundary in eastern Yunnan and western Guizhou. China Univ Geosci Press. Wuhan. 88 pp (in Chinese with English abstract)

Wang ZQ, Wang LX (1989) Earlier Early Triassic fossil plants in the Shiqianfeng Group in North China. Shanxi Geology, 14(1):23–40 (in Chinese with English abstract)

Yang ZY, Yin HF, Wu SB, Yang FQ, Ding MH, Xu GR (1987) Permian-Triassic boundary stratigraphy and fauna of China. Geological Memoirs, People's Republic of China Ministry of Geology and Mineral Resources, Series 2, 6. Geological Publishing House, Beijing, pp 1–379

Yin HF, Ding MH, Zhang KX, Tong JN, Yang FQ, Lai XL (1995) Dongwuan-Indosinian (Late Permian-Middle Triassic) ecostratigraphy of the Yangtze region and its margins. Science Press, Beijing, pp 1–338

Acknowledgments

We are grateful to the senior adviser of the editorial committee, Professors Hong-Fu Yin and Wei-Hong He for their organization, push, and encouragement during the preparation of this book. One of the authors, Jian-Xin Yu, especially wishes to express her sincere thanks to Profs. Qing-Lai Feng and Jean Broutin for their supervision in the junior stage of her research career. We also wish to thank Professor Qi-Sheng Huang for his great contribution to the identifications on plants fossils in this book and help in supervising the students. We are also grateful to Zhuo Feng, Sheng-Hui Deng, and Grauvogel-Stamm Lea for their helpful discussions on some of the genera described in this book. We thank graduate students Qian-Qian Song, Wen-Jin Pan, Wei Zhang, Dao-Liang Chu, and so on from China University of Geosciences (Wuhan) for Participating in the workfield and collecting the specimens of plant fossils for us. This work has been supported by NSFC (Grant Nos. 92055201, 41930322, 41661134047, 41572005), the Ministry of Education of China (BP0820004 of 111 Project). We sincerely thank several publishers for their kind permissions for reprinting some published fossil images, including Elsevier, Springer, and China University of Geosciences (Wuhan) Press.

Contents

Chapter 1
Literature Review

Jianxin Yu, Xiao Shi, and Hui Li

A good study has been made of the marine facies Permian–Triassic boundary stratigraphy sequence, biome change and geological events. Meishan D section in Changxing of Zhejiang, China and its base of 27c bed has been chosen as the GSSP (Yin et al. 2001) due to the work of the Chinese geologists for decades since Yin et al. suggested that the first appearance of the Conodont species *Hindeodus parvus* should mark the base boundary of the Triassic. This definite has put a good end to the important boundary stratigraphic study in marine environment.

However, the determination of GSSP for the marine Permian–Triassic boundary arouses people's interests to work out an accessory section and point of terrestrial facies for such a boundary. Attention is focused on the interrelationship between the environmental change and biologic extinction when the continent was more sensitive and critical in preserving the fossils. It will be more complicated to determine a stratotype boundary because of particularity of the biologic preservation on land. Therefore, some specialists working for the boundary stratigraphy suggest to make a continental boundary stratigraphy division and correlation to those of the marine facies (Lozovsky 1991; Lucas et al. 1992). While some geologists tend to find terrestrial accessory section and point (ASP) in South Africa (Lucas et al. 1996) as the Permian–Triassic Beaufort Group of Karoo Basin yields abundant vertebrates and

J. Yu (✉)
China University of Geosciences (Wuhan), Wuhan, Hubei, P.R. China
e-mail: yujianxin@cug.edu.cn

X. Shi
College of Earth Sciences, Jilin University, Changchun, Jilin, P.R. China
e-mail: xiaoshi@jlu.edu.cn

H. Li
Jiangxi Key Laboratory for Mass Spectrometry and Instrumentation, East China University of Technology, Nanchang, P.R. China
e-mail: huili_201860188@ecut.edu.cn

J. Yu et al. (eds.), *Plants and Palynomorphs around the Permian-Triassic Boundary of South China*, New Records of the Great Dying in South China,
https://doi.org/10.1007/978-981-19-1492-8_1

is thus an important area for the study of terrestrial Permian–Triassic boundary. In contrast, China, located in the eastern Tethys area, is not only one of the best studied areas for marine facies Permian–Triassic stratigraphy boundary in the world, but also an area with well-developed terrestrial Permian–Triassic strata. Particularly, the Junggar Basin in Xinjiang consists of a complete series of continental Permian–Triassic sequence with continuous depositions and fine outcrops and also contains abundant fossil fauna (e.g. ostracods, bivalves, vertebrates) and fossil plant near Permian–Triassic boundary. It is one of the ideal areas to study and establish a terrestrial Permian–Triassic stratotype Section and Point. In addition, similar areas for the ASP are found in Moscow Terrane of Russia, Ural areas of Kazakhstan, Sydney Basin and Bowen Basin of Australia, and Germany and its adjacent areas.

1.1 International Studies

1.1.1 Research of Terrestrial Permian–Triassic Boundary

It is known in the study of the terrestrial stratigraphic sections that terrestrial vertebrates diversify apparently above and below the Permian–Triassic boundary. Angara flora in Russia and Cathaysian flora in China, prosperous in Late Permian, disappeared at the end of the epoch and thus fossil plants could rarely be found in early Early Triassic. It was only in basalmost Triassic that some simple new plant assemblages began to appear. In Sydney Basin, southeastern Australia, there is obvious change of terrestrial plants, with the sudden decrease in flora diversity and local species diversity and the total disappearance of Glossopterids. It was not until Middle Triassic that the flora began to recovery in the area. While in Israel, *Lueckisporites virkkiae* assemblage, the gymnosperm pollen, predominant in Late Permian disappeared totally in the claystone below the boundary but marine acritarch and large amount of Lycopsida spore occurred instead. This indicates an important break in the palynomorph assemblages. Ochev (1973) attributes the change of vertebrates at terrestrial Permian–Triassic boundary to the change of sedimentary facies and suggests that the real change should happen between Early and Middle Triassic. On the contrary, Smith (1995) and Ward (2000) believe that the biologic extinction must have influenced the environment during their study of the Permian–Triassic boundary in Karoo Basin, South Africa. It is generally thought that the change of bended river into braided river reflects the increase of the river slope, and that this is resulted from the orogeny, which is the idea of some early researchers of the Karoo Basin. However, Smith (1995) and Ward (2000) maintained that no tectonic orogeny has happened near the Permian–Triassic boundary. Therefore, the change of the river type, from the highly bended river to lowly bended river (anomaly layer) in Late Permian to braided river deposition in Early Triassic, should have been caused by the biologic extinction rather than the change of the slope. They stated that the increase of the deposition load caused the bended river to change into braided river when the ground surface, especially the river bank, lost its protection from the plants because of the biologic extinction, in particular the death of plants of large root systems in big numbers. This kind of biologic extinction and environment change of global characteristics should lead to the ^{13}C deviation near the Permian–Triassic boundary.

The paleobiologists (Visscher et al. 1996; Peng et al. 2006) has found a fungi events in the terrestrial Permian–Triassic boundary strata from Europe, North America, Africa, Australia, Madagascar, Israel, Greenland and South China, characterized by the widespread *Reduviasporonites* (*Tympanicysta*) and the equivalent *Chordecystia*, which can be correlated in paleontology and geochemistry with other records. They believe that this fungi event could have been brought about by the serious air pollution caused by the large-scale volcanic eruption in Siberia. While Krassilov (1999) assumes that this could be resulted from the ground water rise and wide distribution of rivers due to the global transgression. Wang Ziqiang (1997) has also found a large number of fungi and algae traces in the remains of gymnosperm in North China and also a lot of small tumors on the seeds owing to the parasitic fungi. He proposes that the terrestrial ecosystem evolution falls into three steps: progressive declining of the ecosystems in Permian, collapse of the ecosystems (cluster extinction) in Late Permian and continual revival and development of the ecosystem in Triassic. He suggests that the entire biologic evolution is closely related to the environment change-the global desertification (the change from semi-arid climate to arid climate).

1.1.2 Research to Terrestrial Permian–Triassic Transition Flora

Before 1920s, almost all the paleobotanic research was completed by international specialists. Even at the present time, the Chinese research level in the field of paleobotany may be 10 years behind that of other countries. For the research on the terrestrial Permian–Triassic plants is now mainly conducted in Europe, America, Australia and India. The following aspects are focused on the study of the biologic evolution between Paleozoic and Mesozoic.

1. International Cooperative Research on the response of the equatorial subtropical vegetation to the Ice Age in Late Paleozoic, reorganization of Terrestrial Plant Ecosystem during the end of an Ice age: "A Geosystems Approach to the Intert-errestrial Comparison of Early and Mid-Permian floras". This group makes up of many paleobotanists from china, France, Holand, Germany, Brazil and South Africa and so on. The scientific implication of this collaborative research lies in the fact that the global greenhouse effect triggered by the glacier melting in Late Paleozoic is more correlative than other elements in other ages and/or since Late Cenozoic in the geologic history and so can be used to infer the climate change in modern times. It is assumed that the rapid glacier melting resulted in the obvious change of the floras, including extinction, spatial distribution

changes of the equatorial and temperate zone vegetation, and even the vegetation structural change on a global scale. Although the Late Paleozoic floras and their geographic distribution are different from those in the present earth, their change mechanism, the geographic distribution features and changing ratio may be the pattern for the correlation of the similar modern greenhouse effect in the world.

2. One of the international paleobotanic research focuses is on the ecologic characteristics of the floras for different floral Provinces in the critical P/T geologic period, from the angle of relationship between the fossil plant taphonomy and cuticles anatomy and the environment change and the response of the vegetation to the global climate change in Late Paleozoic to make a prediction of future CO2 change in atmosphere.
3. International paleobotanists are now more interested in the sudden (unexpected) mass extinction of the floras in the end of Permian and the new Lycophyta flora in Early Triassic. Some of them believe that the former happened in a long time with only variations of time in different areas, while others tend to agree that the catastrophic event is related with the vascular plant evolution (Retallack 1995; Eshet et al. 1995; Steiner et al. 2003; Nowak et al. 2019, 2020).

1.2 Studies in China

The terrestrial Permian–Triassic strata in China is mainly distributed in its northern area, while they are outcropped in a few numbers on the massif margins in the southern area, with typical area found in western Guizhou and eastern Yunnan.

1.2.1 Establishinga Fairly Complete Framework of Stratigraphic Units in Western Guizhou and Eastern Yunnan

Many geologists have done great jobs on the strata formed at the turn of Permian and Triassic in their previous researches. Yang Ruiji, Li Peixian, Zhou Tongshun and Qu Lifan have made a systematic study of the terrestrial Permian–Triassic strata in Dalongkou, Jimsar of Xinjiang, which is presented in their monograph "Terrestrial Permian–Triassic Stratigraphy and Faunas, Dalongkou, Jimsar of Xinjiang". Li Zishun, Zhan Lipei and Dai Jinye have made a detailed study of the marine strata near Permian–Triassic boundary in Sichuan and Shaanxi, represented by their book "Permian–Triassic Biostratigraphy and eventostratigraphy in Northern Sichuan and South Shaanxi". The research groups headed by the two academicians, Prof. Yang Zunyi and Prof. Yin Hongfu, have made world-known achievements in the study of the Permain-Triassic stratigraphy. They have already finished many projects funded by IGCP (IGCP-106,203,359) and China Natural Science Fund, represented by their books "Permian–Triassic Boundary Strata and Faunas in South China", "Permian–Triassic Transition Geologic Events in South China", "Permo-Triassic Events in the Eastern Tethys" and "Proceedings of the International Conference on Pangea and the Paleozoic–Mesozoic Transition".

Terrestrial Permian–Triassic stratigraphic study in South China started in 1930s, when some famous geologists, Yue Senxun, Huangjiqing, Wang Zhuquan et al. made investigations of the Late Permian strata in the area. In 1970s, Nanjing Institute of Geology and Paleontology of the Chinese Academy of Science made a special study of Permian–Triassic biostratigraphy, represented by their books "Late Permian Coal-bearing Strata and Palaeontological Fauna in western Guizhou and eastern Yunnan" and "Palynology of Upper Permian and Lower Triassic Strata of Fuyuan District, eastern Yunnan".

On the base of the previous worker, the further study (including paleontology, event stratigraphy, sequence stratigraphy and the analysis of biomarker and geochemistry (Wang and Yin 2001; Peng et al. 2002, 2005, 2006; Yang et al. 2005; Yin et al. 2007; Yu et al. 2007, 2010, 2015; Chu et al. 2016, 2020; Feng et al. 2019) in western Guizhou and eastern Yunnan has developed one after the other in recent years. These studies had laid a good foundation for further research on the Permian–Triassic boundary correlation with the marine section, patterns of extinction and survival of the flora and so on in South China.

1.2.2 Research on the Paleofloras

The study of Permian floras in China can be said to be closely related with the coal exploration and the discovery of gigantopterids. However, such study made with modern method was mainly started by some European scholars roughly in mid nineteenth century. Schenk (1883), Halle (1927), and Stockmans and Mathieu (1939) studied Permian plant fossils and made their publications one after the other. Although literature on this aspect is rather rare and the record of the fossils beds is not clear, the particularity of the Permian floras dominated by gigantopterids is quite outstanding. "Palaeozoic Plants from Central Shansi", published by the famous Sweden paleobotanist Halle in 1927 is considered as the classic for the Carboniferous and Permian flora studies, thus a milestone for the Chinese paleobotanic development indicating a new stage for the Permian flora study in China.

It was in 1930s that the Chinese geologists and paleobotanists began their studies on Permian flora. Among whom Sze (1933, 1942, 1947) made greater contributions. The birth of New China brought the spring of science to the study of the paleoflora. Some new and important achievements written by the Chinese paleobotanists were made either in depth or on scale of the study. The studies include chiefly (1) the classification and evolution of gigantopterides by some scholars (Li and Yao 1983a, b; Yao and Crane 1986) based on their morphologic classification, propagation organs, wood anatomy and alteration; (2) the discovery and investigation of the Permian

Glossopterids flora from Qinghai-Tibet plateau, Cathaysian flora and the Cathaysian-Gondwana mixed flora(?), represented by the achievements made by Li and Yao, Li Xingxue (1986), Xu et al. and Li et al. (1991), which have filled up the blank of the Permian flora research in the area and provided practical example for the approach to the forming and evolution of the plateau and the relationship between Cathaysian flora and Gondwana flora; (3) study of the geographic system of the Permian plants: represented by the achievements made by Li et al. (1980), who is the first to divide the Cathaysian flora into North China flora subregion and South China subregion and by Xu Ren and Li Xingxue (1986), who have discussed respectively the relationship between the Gondwana flora and Cathaysian flora in terms of plate tectonics; and (4) some new understandings in the study of the Permian–Triassic plant alteration and extinction, illustrated by the publications of Yao Zhaoqi et al. (1980) and Ouyang Shu (1980), Wang Ziqiang (1993 and 1997), Fang Zongjie (2004) and Yu et al. (2007).

The Triassic flora research in China began in the late nineteenth century, mainly represented by the study of the plant fossils from Xujiahe of Sichuan made by German paleobotanist Schenk (1884). The first Chinese paleobotanist on the Triassic flora should be late Si Xingjian (Sze 1933, 1935, 1936), who studied the plant fossils from the Late Triassic coal-bearing strata. After this study was the study of the Yanchang flora in north Shaanxi made by Pan Zhongxiang (Pan 1936). Great achievements have been made in this area since 1949, especially in the discovery and detailed investigation of the Late Triassic flora fossils.

However, the Early and Middle Triassic floras were first formally reported in China only at the end of 1970s. The first discovery of *Pleuromeia* from Early Triassic Heshanggou Formation in Qinshui Basin, Shanxi made by Wang Ziqiang (Wang Lixin et al. 1978) aroused the great attention of the paleobontanists both from China and other countries. At the same time, Zhou et al. (1979) reported the discovery of late Early Triassic flora in Jiuqujiang of Qionghai, Hainan, which laid the foundation for the Early Triassic flora research in South China. More research made by Meng Fansong (1992) in the area has enriched the study of the Early Triassic flora. Since 1980s, new advances have been made in the Early Triassic research in North China, especially the flora study of the upper part of Shihchienfeng Group., Liujiaogou Formation and Heshanggou Formation (Wang Ziqiang 1982; Wang Ziqiang et al. 1990). This not only clarifies the basically same characteristics of the Early Triassic flora in North China as those from the West European Buntsandstein flora, but also provides important evidence for the formal classification of the Shihchienfeng Group in the area. Moreover, the discovery of *Annularia shirakii* Kaw., *Gigantopteris dictyophylloides* Gu et Zhi, etc. in the basement of the Lower Triassic Kayitou Formation on the border of Yunnan and Guizhou (Yao et al. 1978, 1980) and their occurrence in the same layers or overlying layers of the Early Triassic bivalves should be of great significance for the understanding of the floral assemblages in the beginning of Triassic in South China and for the evidences of continuous deposition between Paleozoic and Mesozoic.

Presently, the Chinese Triassic flora studies are characterized by the noticeable progress in the entire reconstruction of the ancient plants, reproductive biology, plant anatomy and systematic evolution. More achievements are also made in the comprehensive study of the geographic provinces based on the data of plate tectonics, paleomagnetism and paleozoology (Wu Shunqing 1983; Wang Ziqiang 1985; Yin Hongfu et al. 1986; Sun Ge 1987, 1993; Yu et al. 2010; Feng et al. 2019).

To sum up, the Chinese studies of Permian–Triassic plants have undergone over one hundred years of history. They have developed from the morphologic description to the probing of basic theories, integrating rich practical information with the study of the global geologic system. This indicates that paleobotany parallels with other basic geologic theories in studying the whole earth.

1.3 Studies in the Research Area for the Monograph

The research area situated on the border of western Guizhou and eastern Yunnan, belongs to the South China subregion of Cathaysian flora. In this area, coal-bearing strata of terrestrial and marine-terrestrial facies are developed with abundant gigantopterids. The earliest study was started by Xie Jiarong, who studied the Permian–Triassic fossil plants when investigating Xuanwei coal series. Then, Yunnan Geological Bureau and Guizhou Geological Bureau made some simple collection and identifications of plant fossils to determine the dating of coal beds when prospecting coal in the area. While the real systematic study of the Permian–Triassic flora in the area began in 1970 and 1980s. During this time, paleobotanists and palynologists, represented by Yao Zhaoqi (1978), Zhao Xiugu (1980) and Ouyang Shu (1986), clarified the Permian–Triassic flora assemblage outlook and main characteristics with the study of the preserved fossil plants. Their results suggest that the genera and species of the assemblage in the bottom of the Lower Triassic are very monotonous, containing relict elements of Late Permian Cathaysian flora, and thus should represent the declining period of the flora.

Recently, many scholars have been making further research of the Permian–Triassic biologic extinction in the area. For instance, Wang and Yin (2001), in their study of the boundary strata in the area, proposes that an extinction event happened in the terrestrial biome in the turn from Permian and Triassic and the plants alternated in sudden change while showing a progressive alternation in the process. Yang et al. (2005), Yin et al. (2007) and Yu et al. (2007, 2015) point out that the composition of the flora near the boundary does not have substantial change, showing only a gradual reduction of the genera and species number, and so they should be of an extinction mode of multi-stages. Whereas the finding of many new plant fossils in Kayitou Formation of marine-terrestrial transition strata in the last ten years has certainly provided new information for the boundary determination and the study of the plant revival after the extinction (Yu et al. 2010; Feng et al. 2019).

It should be noted that arguments are now heated on the geological time of Kayitou Formation, for it has a lack of standard fossils. Yao et al. (1980) assumes that it should be Early Triassic. Wang and Yin (2001) suggest that this formation should be a time-crossing lithostratigraphic unit in the terrestrial strata such as those in Zhejue of Weining and Mide and Laibin of Xuanwei and should be the deposition of late Late

Permian and early Early Triassic and that the marine facies deposition in Panxian-Shuicheng area should be Early Triassic. Liu and Yao (2002) think that it should be the earliest Triassic. Chen Jinhua, on the other hand, believes that the marine strata of the Kayitou Formation should be the product of transgression in Late Permian and so should be dated Latest Permian. Until now, the geological age of Kayitou Formation is divergent in the field of paleonotology (Shen et al. 2011; Yu et al. 2015; Chu et al. 2016; Feng et al. 2019).

References

Chu DL, Yu JX, Tong JN, Benton MJ, Song HY, Huang YF, Song T, Tian L (2016) Biostratigraphic correlation and mass extinction during the Permian-Triassic transition in terrestrial-marine siliciclastic settings of South China. Glob Planet Change 146:67–88

Chu DL, Grasby SE, Song HJ, Corso JD, Wang Y, Mather TA, Wu Y, Song HY, Shu WC, Tong JN, Wignall PB (2020) Ecological disturbance in tropical peatlands prior to marine Permian-Triassic mass extinction. Geology 48:288–292

Eshet Y, Rampino MR, Visscher H (1995) Fungal event and palynological record of ecological crisis and recovery across the Permian-Triassic boundary. Geology 23:967–970

Fang ZJ (2004) The Permian-Triassic boundary crisis: patterns of extinction, collapse of various ecosystems, and their causes. In: Rong JY, Fang ZJ (eds) Mass extinction and recovery, evidences from the Palaeozoic and Triassic of South China. University of Science and Technology of China Press, Hefei, pp 785–928 (in Chinese with English summary)

Feng Z, Bertling M, Noll R, Ślipiński A, Roessler R (2019) Beetle borings in wood with host response in early Permian conifers from Germany. PalZ 93(3):409–421

Halle TG (1927) Fossil plants from southwestern China. Palaeontologia Sinica Series A 1(2):1–26

Krassilov VA, Afonin SA, Barinova SS (1999) Tympanicysta and the terminal Permian events. Permophiles 35:16–17

Li XX (1986) The mixed Permian Cathaysia-Gondwana flora. Palaeobotanist 35(2):211–222

Li XX, Yao ZQ (1983a) Fructifications of gigantopterids from South China. Palaeontographica Abstract B 85(13):11–26

Li XX, Yao ZQ (1983b) Current Studies of Gigantopterids. Palaeontologia Cathayana 1:319–326

Li XX, Rigby J, Wu YM (1991) A biostratigraphical comparison between the Lower Gondwana flora of south Xizang (Tibet) and Kashmir. In. Utrid HR, Campos AC (eds) Gondwana Seven Proceedings. Univ. Sao Paulo, Brazil, pp 533–538

Li XX (1980) New advances of studies in the Cathaysia flora. In. Scientific paper on geology for international exchange. Prepared for the 26th International Geological Congress (4), Stratigraphy and Palaeontology. Beijing: Geological Publishing House, pp 73–81 (in Chinese with English summary)

Liu LJ, Yao ZQ (2002) Geological age of the Kayitou Formation. J Stratigr 26(3):235–237

Lozovsky VR (1991) Proposal for a new working group on continental beds at the Permian/Triassic boundary in the continental series. Permophyles 19:7–11

Lucas SG, Lozovsky VR, Cheng ZW (1992) Proposal to STS for formation of a working group on the nonmarine Triassic timescale. Albertiana 10:41

Lucas SG, Kozur H, Molina-Garza R (1996) Reconsideration of Dalongkou as an auxiliary GSSP for the Permian-Triassic boundary. Albertiana 18:10–11

Meng FS (1992) Fossil Plants. In: Yichang Institute of Geology and Mineral Resources et al., eds. Geology of Hainan Island, 1. Stratigraphy and Palaeology. Beijing: Geological Publishing House: 166–170, 175–183 (in Chinese)

Nowak H, Vérard C, Kustatscher E (2020) Palaeophytogeographical patterns across the Permian-Triassic boundary. Front Earth Sci 8:609

Nowak H, Schneebeli-Hermann E, Kustatscher E (2019) No mass extinction for land plants at the Permian-Triassic transition. Nature Communications 10(1)

Ochev VG (1973) About the character of the changes in fauna of terrestrial vertebrates at the boundary of the Permian and Triassic. Bull Moip Ser Geol 48:1

Ouyang S, Li ZP (1980) The micro-floral in the Kayitou Bed of Fuyuan, Yunnan and its stratigraphic and paleobotanic significance. In: Nanjing Institute of Geology and Palaeontology, Academica Sinica, (ed), Late Permian coal-bearing strata and palaeontological fauna in western Guizhou and eastern Yunnan. Science Press, Beijing, pp 123–194 (in Chinese)

Ouyang S (1986) Palynology of upper Permian and lower Triassic strata of Fuyuan district, eastern Yunnan. Science Press, Beijing, pp 122 (in Chinese with English Summary)

Peng YQ, Wang SY, Wang YF, Yang FQ (2002) A proposed area for the study of the Accessory Section and Point of the Terrestrial Permian-Triassic boundary. J China Univ Geosci 13:157–162

Peng YQ, Zhang SX, Yu JX, Yang FQ, Gao YQ, Shi GR (2005) High-resolution terrestrial Permian-Triassic eventostratigraphic boundary in western Guizhou and eastern Yunnan, southwestern China. Palaeogeogr Palaeoclimatol Palaeoecol 215:285–295

Peng YQ, Yu JX, Gao YQ, Yang FQ (2006) Palynological assemblages of non-marine rocks at the Permian-Triassic boundary, western Guizhou and eastern Yunnan, South China. J Asian Earth Sci 28:291–305

Retallack GJ (1995) Permian-Triassic life crisis on land. Science 267(5194):77–80

Schenk A (1884) Die wahrend der Reise des Grafen Bela Szechenyi in China gesamlten fossilien Pflanzen. Palaeontographica 31(3):163–182

Schenk A., 1883. Pflanzen aus der Steinkohlen Formation. In: Richthofen F von. China, 4. Berlin: 238–239.

Shen WJ, Sun YG, Lin YT, Liu DH, Chai PX (2011) Evidence for wildfire in the Meishan Section and implications for Permian-Triassic events. Geochim Cosmochim Acta 27:1992–2006

Smith RMH (1995) Changing fluvial environments across the Permian-Triassic boundary in the Karoo Basin, South Africa and possible causes of tetrapod extinction. Palaeogeogr Palaeoclimatol Palaeoecol 117:81–104

Steiner MB, Eshet Y, Rampino M, Schwindt DM (2003) Fungal abundance spike and the Permian-Triassic boundary in the Karoo Supergroup (South Africa). Palaeogeogr Palaeoclimatol Palaeoecol 194:405–414

Stockmans F, Mathieu FF (1939) La Flore paleozoique du Bassin houiller de Kaiping (China). Mus Roy Hist Nat, Belgique:49–165

Sun G (1987) On late Triassic geofloras in China and principles for palaeophytogeographic regionalization. Acta Geol Sinica 61(1):1–9

Sze HC (1933) Fossil Pflanzen aus Shensi, Szechuan und Kueichow. Palaeont Sinica Ser A 1(3):1–32

Sze HC (1936) Ueber einen Vorl/iufer von Pterophyllum in der Unteren Shihhotze Serie in Shansi. Bull Geol Soc China 15(4):467–476

Sze HC (1942) Ueber ein neues Exemplar von Psaronius aus dem Omeishan Basalt in Weining (Kueichou) mit besonderer Berticksichtigung des Alters des Basaltes in Stidwestchina. Bull Geol Soc China 22(1–2):105–131

Sze HC (1947) On the structure of Psaronius sinensis from the Omeishan Basalt Series in Southwestern China. J Geol Amer 55(3):160–166

Sze HC (1935) Ueber das Vorkommen von Callipteris conferta Sternb. in der Unteren Shihhotze Serie in central Shansi. Bull Geol Soc China 14(4):569–578

Visscher H, Brinkhuis H, Dilcher DL, Elsik WC, Eshet Y, Looy CV, Rampino MR, Traverse A (1996) The terminal paleozoic fungal event: evidence of terrestrial ecosystem destabilization and collapse. Proc Natl Acad Sci USA 93(5):2155–2158

Wang ZQ (1997) Permian Supaia fronds and an associated Autunia fructification from Shanxi, China. Palaeontology 40:245–277

Wang ZQ, Wang LX (1982) A new species of the lycopsid Pleuromeia from the Early Triassic of Shanxi, China and its ecology. Palaeontology 25(1):215–225

Wang ZQ, Wang LX (1990) Late Early Triassic fossil plants from upper part of the Shiqianfeng Group in North China. Shanxi Geol 14(1):23–40 (in Chinese with English abstract)

Wang LX, Xie ZM, Wang ZQ (1978) On the occurrence of *Pleuromeia* from the Qinshui Basin in Shanxi Province. Acta Palaeont Sinica 17(2):195–212 (in Chinese wtih English abstract)

Wang SY, Yin HF (2001) Study on terrestrial Permian-Triassic boundary in eastern Yunnan and western Guizhou. China University of Geosciences Press, Wuhan, pp 88. (in Chinese with English abstract)

Wang ZQ (1993) Evolutionary ecosystem of Permian–Triassic redbeds in North China: a historical record of global desertification. In: Lucas SG and Morales M. (eds)

Ward PD, Montgomery DR, Smith RMH (2000) Altered river morphology in South Africa related to the Permian-Triassic extinction. Science 289:1740–1743

Wu SQ (1983) On the Late Triassic, Lower and Middle Jurassic floras and phytogeographic provinces of China. In: Editorial Committee of Fundamental Theory of Palaeontological Book Series (ed) Palaeobiogeographic provinces of China. Beijing: Science Press, pp 121–130 (in Chinese)

Yang FQ, Yin HF, Yu JX, Zhang SX, Huang JH, Peng YQ, Huang QS, Zhao QM (2005) Study of the terrestrial Permian-Triassic boundary in Chahe, Weining, Guizhou province. Sci China (series D) 35:519–529 (in Chinese)

Yao ZQ (1978) On the age of "Gigantopteris coal series" and Gigantopteris-flora in South China. Acta Palaeontol Sin 17:81–89 (in Chinese with English abstract)

Yao Z, Crane PR (1986) Gigantopterid leaves with cuticles from the Lower Permian of China

Yao ZQ, Xu JT, Zhen ZG, Mo ZG (1980) Late Permian biostratigraphy and the Permian-Triassic boundary in Western Guizhou and Eastern Yunnan. In: Nanjing Institute of Geology and Palaeontology (ed.), Stratigraphy and palaeontology of Late Permain coal-bearing formations in Western Guizhou and Eastern Yunnan. Science Press, Beijing, pp 1–69 (in Chinese)

Yin HF, Feng QL, Lai XL, Baud A, Tong JN (2007) The protracted Permo-Triassic crisis and multi-episode extinction around the Permian-Triassic boundary. Glob Planet Change 55:1–20

Yu JX, Peng YQ, Zhang SX, Yang FQ, Zhao QM, Huang QS (2007) Terrestrial events across the Permian- Triassic boundary along the Yunnan-Guizhou border, SW China. Global Planet Change 55:193–208

Yu JX, Broutin J, Huang QS, Grauvogel-Stamm L (2010) *Annalepis*, a pioneering lycopsid genus in the recovery of the Triassic land flora in South China. C. R. Palevol. 9:479–486

Yu JX, Broutin J, Chen ZQ, Shi X, Li H, Chu DL, Huang QS (2015) Vegetation changeover across the Permian-Triassic boundary in Southwest China extinction, survival, recovery and palaeoclimate: a critical review. Earth-Sci Rev 149:203–224

Zhao XG, Mo ZG, Zhang SZ, Yao ZQ (1980) Late Permian fossil floras of western Guizhou and eastern Yunnan. In: Nanjing Institute of Geology and Palaeontology, Academica Sinica, (Ed.), Late Permian coal-bearing strata and palaeontological fauna in western Guizhou and eastern Yunnan. Science Press, Beijing, pp 70–122 (in Chinese)

Zhou ZY, Li BX (1979) A preliminary study of the Early Triassic plants from the Qionghai District Hainan Island. Acta Palaeotol Sin 18(5):444–466 (In Chinese with English abstract)

Chapter 2
Stratigraphy of Studied Area Across the Permian–Triassic Boundary, South China

Jianxin Yu, Hui Li, Jean Broutin, and Zongsheng Lu

The research area lies in eastern Yunnan–western Guizhou, southwest of China (Fig. 2.1), in the eastern part of the Yunnan–Guizhou plateau, where highland accounts to over 90% of the land with an average altitude of more than 1,000 m. The area is rather complicated in geology with widespread karsts with unique mountainous relief and climate, as described by the Chinese saying, "There is never a sunny day for three days and there is never a level land for three meters." However, its particular relief of natural beauty holds over 10 minority groups, namely Miao, Buyi, Tujia, Yi, Hui, Zhuang, Yao, etc., who have developed their own ethnic culture and institutions. It hence provides rich and colorful tourism resources groups with many national-level scenic sites. However, it was just this relief that made the area most inaccessible in history and hence its economy most underdeveloped. With the improvement of its economy in recent years, highways are building in the area and so make it more convenient for section observation and description and also for sample collection for all the surveyed sections can be found on the new ways.

J. Yu (✉) · Z. Lu
China University of Geosciences (Wuhan), Wuhan, Hubei, P. R. China
e-mail: yujianxin@cug.edu.cn

Z. Lu
e-mail: zshlu@cug.edu.cn

H. Li
Jiangxi Key Laboratory for Mass Spectrometry and Instrumentation, East China University of Technology, Nanchang, P. R. China
e-mail: huili_201860188@ecut.edu.cn

J. Broutin
Faculty of Sciences and Engineering (CR2P, MNHN, CNRS), Sorbonne Université, Paris, France
e-mail: jean.broutin@upmc.fr

J. Yu et al. (eds.), *Plants and Palynomorphs around the Permian-Triassic Boundary of South China*, New Records of the Great Dying in South China,
https://doi.org/10.1007/978-981-19-1492-8_2

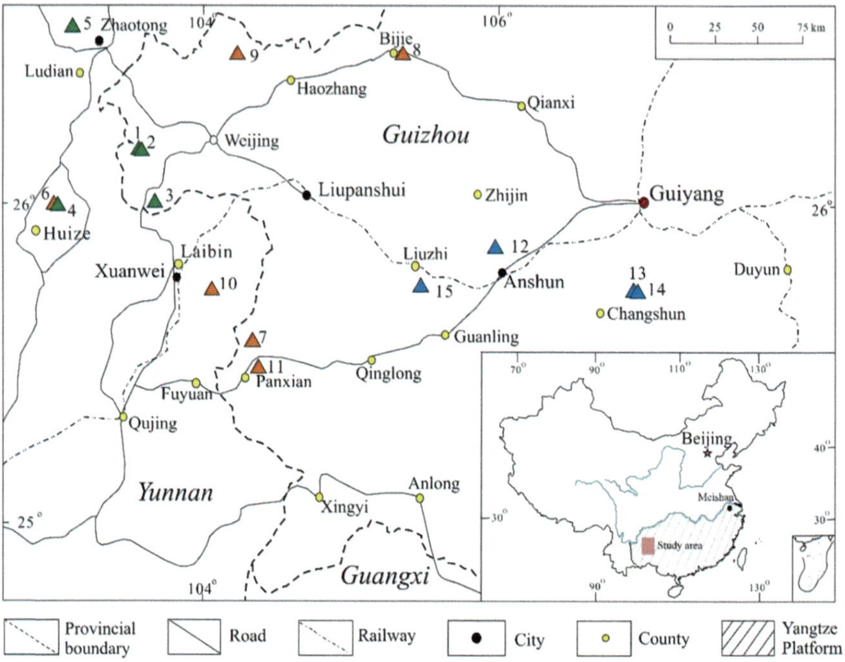

▲ Terrestrial facies sections (1-Chahe Section; 2-Jiucaichong Section; 3-Zhejue Section; 4-Lubei Section; 5-Guanbachong Section)

▲ Terrestrial-marine facies sections (6-Mide Section; 7-Tucheng Section; 8-Jinzhong Section; 9-kele Section; 10-Chinahe Section; 11-Wadu Section)

▲ Marine facies sections (12-Xinming Section; 13-Duanshan Section; 14-Kejiao Section; 15-Zhongzhai Section)

Fig. 2.1 Distribution of studied sections in Western Yunnan and Eastern Guizhou

2.1 Lithostratigraphy Units Across Permian–Triassic Boundary in Research Area

The research area is situated on the western margin of the Yangtze Platform, i.e. the border area between western Guizhou and eastern Yunnan on the eastern side of the Kangdian Massif (Oldland) (Fig. 2.2). As the stratigraphic units in the area are named, modified and even supplemented by different researchers, the terms are confused in the definition of the layer or even name of the same units (Table 2.1). The lithostratigraphy across the non-marine Permian–Triassic boundary, dealt with in this research, include Late Permian coal-bearing Xuanwei Formation, Early Triassic variegated clastic rocks of the Kayitou Formation (Table 2.1).

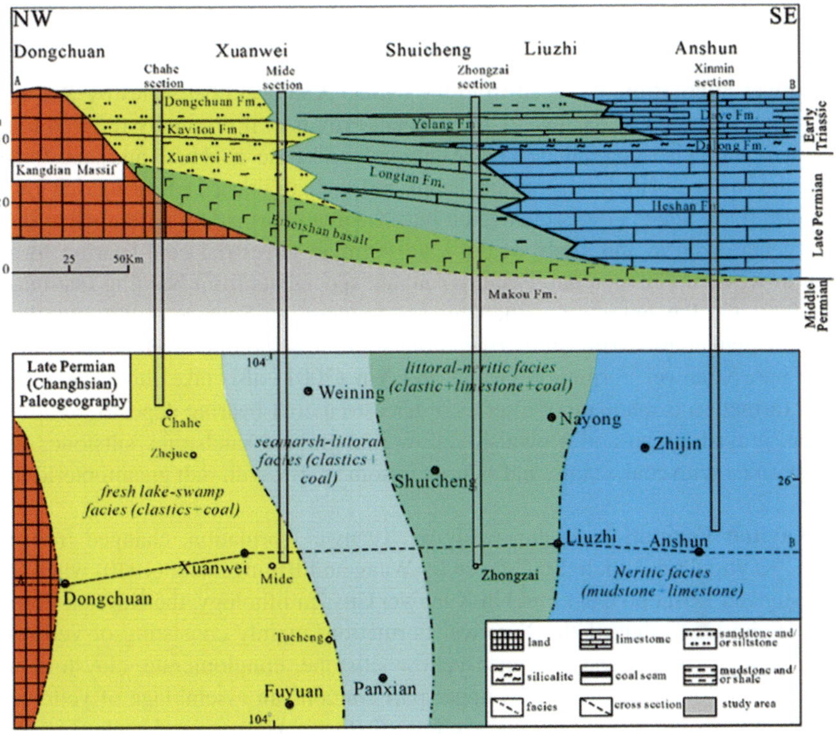

Fig. 2.2 Late Permian Changhsingian paleogeography and lithostratigraphy of western Guizhou and eastern Yunnan (modified after Wang and Yin 2001; Yao et al. 1980)

Table 2.1 Stratigraphic classification near the non-marine Permian–Triassic boundary in western Guizhou–eastern Yunnan (words in Italic is indicate stratigraphic period in the book)

Stage		Deep-water carbonate and silicolite facies	Shallow sea carbonate facies	Neritic clastic facies	Nearshore facies	Terrestrial facies
T 1	*Induan*	*Luolou Formation*	*Daye Formation*	*Yelang Formation*	Feixianguan Formation	Dongchuan Formation
					Kayitou Formation	*Kayitou Formation*
P3	*Changhsingian*	*Dalung Formation*	*Wujiaping Formation*	*Lungtan Formation*	*Xuanwei Formation*	*Xuanwei Formation*
	Wuchiapingian	Lungtan Formation	Lungtan Formation			
P2	Capitanian	Dongwu Movement			Emeishan Basalt	Emeishan Basalt
	Wordian	Maokou Formation	Maokou Formation	Maokou Formation	Maokou Formation	Maokou Formation

2.1.1 Non-Marine Facies Depositional Area

Xuanwei Formation: changed from the named Xuanwei coal series by Xie et al. (1941) in Dasuopo of Xuanwei City, Yunnan Province. It originally referred to the series of yellowish-green shale, coarse-grained sandstone, coal beds and sandy shale, over the basalt and under the red sandstone and shale (Table 2.1), bearing rich gigantopterid flora. Sheng (1962) changed it into Xuanwei Formation. After their systematic generalization and study of the Late Permian terrestrial coal-bearing formation in western Guizhou and eastern Yunnan, specialists from Nanjing Institute of Geology and Paleontology of the Chinese Academy of Science (1980) put all the late Permian coal-bearing strata (including some sea-flooding layers locally) in the area into Xuanwei Formation. Wang and Yin (2001) also take this terminology. This formation is made up of a series of terrestrial coal-bearing deposits, yellowish green, greyish-green and brownish-yellow fine-grained sandstone, siltstone, claystone, shale with coal-seams, and siderite in some parts, with rich gigantopterid flora (Fig. 2.2).

Kayitou Formation: is the overlying Xuanwei Formation, changed from the term "Kayitou Sand-shale Bed" given by Wang and Biqingchang (1940) when they investigated Dasuopo Coal Field in Xuanwei City. In lithology, the Kayitou Formation is similar to that of the Xuanwei Formation, mainly consisting of yellowish green, greyish-green and brownish-yellow siltstone, conglomerate, claystone and shale in the lower part, while its upper part contains an assemblage of yellowish-green, grayish-green and brownish-yellow siltstone, claystone and shale (Fig. 2.3). This formation is in a progressive transition with the overlying Dongchuan Formation consisting of purple-red clastics and underlying Xuanwei Formation in conformity. The widespread appearance in south China of the lycopod *Tomiostrobus* (=*Annalepis*), accompanied with bivalves, ostracodas and conchostraca has been suggested to coincide with the end of the EPPC and the beginning of Triassic (Yu et al. 2015).

Generally, the two series of the deposits across the non-marine Permian–Triassic boundary in western Guizhou and eastern Yunnan are easy to distinguish by their colors. The Xuanwei Formation is composed of a series of gray or grayish-white clastic rocks intercalated with coal beds and seams, the upper Kayitou Formation, variegated (gray, yellowish-green and purple), no coal beds or seams.

2.1.2 Shallow-Water Facies (Neritic Clastic Facies) Depositional Area

The Permian–Triassic interval in the facies area includes the Lungtan and the Yelang Formations (Fig. 2.2). The Changhsingian Lungtan Formation is dominated by fine sandstones, muddy siltstones and calcareous mudstones, intercalated with coal seam,

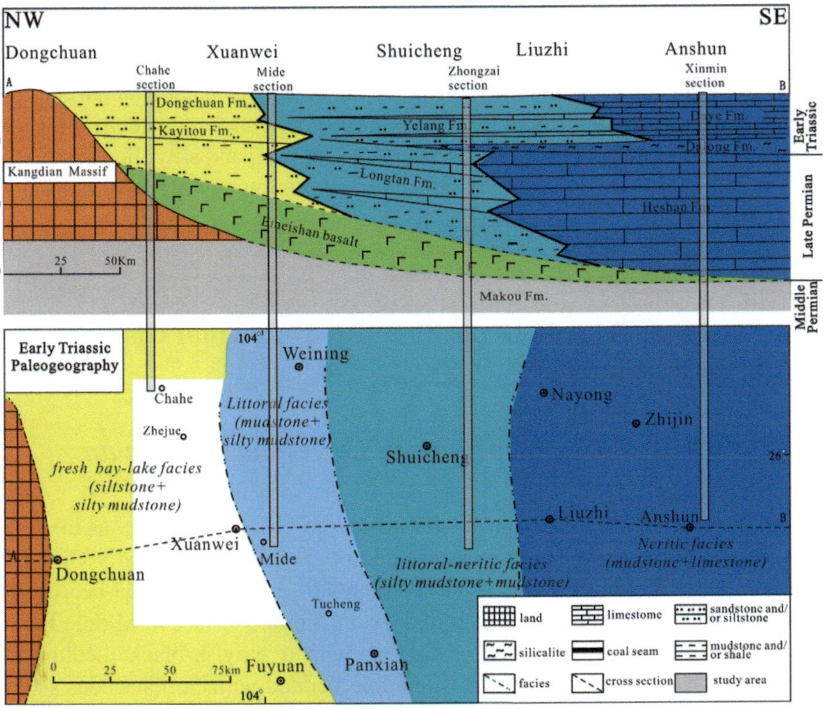

Fig. 2.3 Early Triassic Induan paleogeography and lithostratigraphy of western Guizhou and eastern Yunnan (modified after Wang and Yin 2001; Yao et al. 1980)

limestones or lens-shaped limestones, with fine laminations in calcareous mudstones (Zhang et al. 2014). It yields abundant marine fauna and plant fossils.

The Induan Yelang Formation comprises light gray thin-bedded argillaceous limestones, yellowish green calcareous mudstones and silty mudstones, intercalated with pale volcanic ash, with conodonts, bivalves, gastropods and brachiopods (Fig. 2.3) (Peng et al. 2007; He et al. 2008a; Gao et al. 2009; Zhang et al. 2013), no plant fossils.

2.1.3 *Deep-Water Facies (Shallow Marine Carbonate Facies) Depositional Area*

The Changhsingian Talung Formation is distributed in the western part of the Yangtze sea and is laterally equivalent to the Heshan Formation in eastern areas (Fig. 2.2). The Talung Formation mainly comprises dark grey siliceous mudstones, siliceous limestones and calcareous mudstones, intercalated with volcanic ash (Wu et al. 2018). It has abundantly yielded ammonoids and small brachiopods, and a few bivalves, foraminifers, radiolarians (Xiang et al. 2013) and Plant fossils, which mainly comprise *Lepidostrobophyllum*, *Paracalamites*, *Pecopteris*, *Gigantopteris*, *Cordaites*, *Sphenobaiera* and large conifer with good cuticle (Li et al. 2019).

The Induan Daye Formation is composed of thin-bedded calcareous mudstones, argillaceous limestones (namely marls in the basal part) and silty limestones (overlying marls) (Fig. 2.3), with graded bedding and hummocky cross bedding in silty limestones.

2.2 Permian–Triassic Deposition Environmental Changes in Research Area

2.2.1 Carbonates Terrace Stage in Early-Middle Permian

In Late Devonian-Early Permian, western Guizhou-eastern Yunnan was in an environment of marine carbonates terrace. While in Middle Permian early Qixian, the area was a tidal flat-lagoon, when the sediments form terrestrial coal-bearing clastics (Liangshan Formation). It turned into a semi-limited terrace in Middle Permian middle-late Qixian, when series of dark gray and grayish-black medium-thick micrites and micritic bio-clastic limestone interbedded with carbonaceous shale. In Late-middle Middle Permian Maokouan, the paleogeographic pattern and deposition characteristics of the Qixian were inherited. However, the terrace was considerably wider in this stage, where medium-thick to thick massive carbonates assemblages were developed, dominated by bio-clastic limestone and limestone bearing dolomite masses.

2.2.2 Continent and Littoral-Neritic Sea Stage in Late Permian-Early Late Triassic

The Early Permian paleogeographic framework was changed by Dongwu movement (see Table 2.1). With the large-scale eruption of Emeishan basalt in the bordering area of Sichuan–Guizhou–Yunnan, most of the area was uplifted to lead to disconformable contact between Middle and Upper Permian, and thus formed Kangdian Oldland with its long axis of the main body extending approximately South to North in Yunnan Province. The research area is just located on the eastern side of this oldland (Figs. 2.2 and 2.3).

In the early Late Permian Lungtan stage, alluvial plain, littoral and neritic sea were formed successively eastward with progressive transition. The alluvial plain is dominated by rivers and lakes, forming the deposition of sandstone, siltstone, mudstone and coal-bearing series. Whereas the marine-terrestrial transition off-shore environment contained mainly tidal flats and lagoons, accompanied by the small deltas with the coal-bearing deposition of off-shore-plain, dominated by sandstone, siltstone, mudstone, limestone and coal. The neritic sea was a semi-limited carbonates terrace, consisting mainly of micrites and bio-clastic limestone intercalated with few terrestrial clastics.

The Lungtan paleogeographic framework was then inherited during the Changhsingian except that the transgression expanded westward and the continent reduced relatively. However, the area along Xuanwei and Weining (Fig. 2.2) contained deposition from fresh-water lakes, swamps and rivers, composed mainly of sandstone, mudstone and coal. The plants were mainly dominating in the biome. In the late Changhsingian, the continent expanded to some extent and the coast shifted a little eastward.

In the Early Triassic Induan, the transgression again expanded with the coast moving westward. However, the area east to the Kangdian Massif still contained terrestrial sediments, some areas in Xuanwei of Yunnan and Weining of Guizhou (Fig. 2.3) still existed as continental. A series of grayish-green and yellowish-green sandy mudstone interbedded with purple-red ones, containing a few plant fossils with absence of coal were deposited in early Induan. In the Middle-Late Induan, the environment changed into seasonal rivers and lakes, where a series of purple-red sandy mudstone was deposited. The main characteristic in lithology is dominant to purple-red, rare organisms and development of cross-beddings.

In Panxian–Shuicheng area (Fig. 2.3) on the eastern side of the area occur late Induan nearshore tidal clastics deposits bearing limestone with abundant marine fossils dominated by bivalves and ostracodas. This environment then is changed in a gradual transitional style into marine carbonates terrace eastward.

In Olenekian, there was the similar environment as the Induan except that the coast moved further westward, causing further decrease of the continent. The rocks deposited are dominated by tidal facies sandstone, siltstone and mudstone intercalated with argillaceous limestone, oolitic limestone and dolomites. Tidal beddings are well developed in the clastic rocks. Eastwards, the carbonates take up a bigger proportion. In the deposition area predominated by clastics, few fossils are found with the exception of a few bivalves and plant fragments.

Sum up, the Permian–Triassic transition (Late Permian and Early Triassic) in the research area falls into three lithofacies zones, from east to west, respectively marine facies (clastics and carbonates facies), marine-terrestrial alternation facies (clastics facies) and terrestrial facies (clastic facies). Hence, many clear and complete Permian–Triassic boundary sections of these facies are well developed in western Guizhou and eastern Yunnan, which provide ideal area for proposing a precise locality of the Marine and terrestrial Permian–Triassic boundary strata and for correlating between the marine and terrestrial boundaries. Vertically, from Late Permian to latest Induan we can observe:

(1) Late Permian: deposited marine facies strata in Nayong–Liuzhi area, standing for the beginning of the westward transgression;
(2) Early Induan: deposited marine facies strata in Tucheng and Mide areas, standing for the other transgression reached this western area;

(3) Middle-Late Induan: deposited terrestrial strata in Tucheng area;
(4) Latest Induan: occurred marine facies deposition again in Tucheng area.

A typical terrestrial Permian–Triassic boundary section is situated in the area to the west of Xuanwei-Fuyuan (Figs. 2.2 and 2.3), where there are other P–T boundary sections ever reported, such as Laibing section (Fig. 2.1) in Xuanwei, Yunnan (Nanjing Institute of Geology and Paleontology of the Chinese Academy of Science 1980), Zhejue section and Chahe section in Weining, Guizhou (Wang and Yin 2001). To its east, there are sections of marine-terrestrial transition facies, such as Mide section in Xuanwei, Yunnan, and Tucheng section in Panxian, Guizhou and so on (Fig. 2.1).

2.3 Brief Descriptions of Sections Across Permian–Triassic Boundary

To propose an accurate accessory terrestrial Permian–Triassic section and point, three sections had been selected (Fig. 2.1) to be investigated in detail. We have performed more detailed paleontological (especially paleobotanical) and stratigraphical studies on the different facies sections in Guizhou and Yunnan Provinces, corresponding to terrestrial, terrestrial-marine alternative, shallow-water facies and deep-water facies respectively. The focus is hence on the biostratigraphy, the evolution of the paleofloras through time in relation with the sedimentary environments of these sections in this monograph. The following will present a brief introduction of these selected representative sections.

2.3.1 Terrestrial Chahe Section in Weining, Guizhou Province

The Chahe Section in Weining County, western Guizhou Province (Figs. 2.1 and 2.4a) has been studied in numerous previous investigations (Yu et al. 2015; Chu et al. 2016). The bottom of this section is about 2 km away from exposures of the Emeishan basalt. After loose sediments, this section exposes the Xuanwei Formation (Beds 1–70) and overlying Kaiyitou Formation (Beds 71–89) (Fig. 2.4), then conformably overlies the Dongchuan Formation. According to a zircon U–Pb date of 252.30 $\pm$ 0.07 Ma from volcanic ash in Bed 68 of the Xuanwei Formation (Shen et al. 2011), the Permian–Triassic boundary (PTB) may be higher than the lithological boundary between the Xuanwei and Kaiyitou formations in this section.

The Xuanwei Formation at the Chahe section contains numerous plant fossils including the Cathaysian wetland genera *Lepidodendron*, *Pecopteris*, *Fascipteris* and *Gigantopteris*. Plant fossils are not only abundant in each layer, but also this section contains at least 31 plant-bearing layers with their vertical distributions shown in

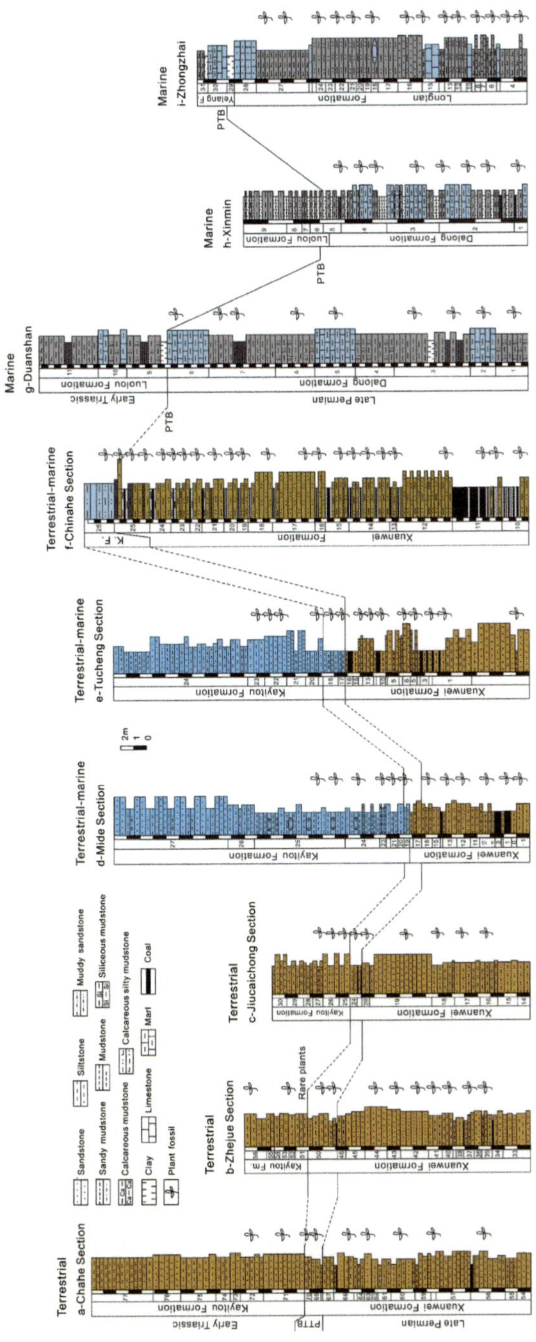

Fig. 2.4 Lithological Column, Fossil plants and palynomorphs sample localities of different facies Permian–Triassic transitional representative sections in South China

Fig. 2.4. In the Chahe section, most of the plants stop at Bed 69 with after the EPPC only *Peltaspermum* sp. occurring in Bed 70. Fossil plants are absent in the Kaiyitou Formation in the Chahe section (Fig. 2.4a).

In the overlying Kayitou Formation, the plant fossils surviving from the Late Permian disappear above Bed 71, whilst plants of the genus *Peltaspermum* occur in association with conchostracans in the middle-upper part of Bed 71, which is a conchostracan-rich horizon. A total of 45 conchostracan specimens were collected in an approximately 8 m-thick interval between Bed 71 and 78, including the genera *Euestheria* and *Palaeolimnadia* (Liao et al. 2020). In the red-purple Dongchuan Formation, except in the basal part, neither conchostracans nor any other fossils have been observed but few plant fragments.

2.3.2 Terrestrial-Marine Transitional Chinahe Section in Xuanwei, Yunnan Province

The Chinahe Section in Haidai town, Xuanwei district of eastern Yunnan Province (Figs. 2.1 and 2.4f) is reported for the first time in this book. This section starts from the Emeishan Basalt (Bed 0) that is unconformably overlain by the upper terrestrial facies of the Xuanwei Formation (Beds 1–25). Marine, grayish green mudstone facies of the Kaiyitou Formation (Bed 26) which in turn overlain by the purple-red Dongchuan Formation conformably overly the Xuanwei Formation (Fig. 2.4f).

The Xuanwei Formation in the Chinahe Section commences with a thick layer of grayish black mudstone (Bed 1), that contains large amounts of well-preserved plant fossils. In Bed 1 *Lobatannularia, Pecopteris* and *Gigantopteris* are common, but the flora in this bed is diverse and also contains a range of sphenopsids (*Lobatannularia cathaysiana, L. heianensis, Paracalamites stenocostatus and Schizoneura amnchuriensis*), *Marattialean ferns* (*Pecopteris* (*Asterotheca*) *guizhouensis, P.* (*A.*) *orientalis, P.* (*A.*) *hemotelioides, P. sahnii, Fascipteris sinensis* and *F. hallei*), gigantopterids (*Gigantonoclea guizhouensis, G. rosulata, Gigantopteris dictyophylloides and Gigantopteris nicotianaefolia*), ferns (*Cladophlebis permica, C. ozakii*) and occasional gymnosperm leaves (*Neuropteridium* sp., *Peltaspermum* sp.,*Taeniopteris multinervis, Rhipidopsis panii*). Above Bed 1, the lithology changes into cycles of grayish yellow or grayish green thin-bedded siltstone to pelitic siltstone and thin coals, as well as pale gray or grayish blue thin-bedded clay or mudstone (Bed 2–25). Usually thin-bedded mudstones on coals [in Beds 3, 12, 16, 25] contain some plants fragments that are not identifiable to species level, including fragments of *Lepidostrobophyllum*, gigantopterids, *Compsopteris, Pecopteris* and *Taeniopteris*. Until Bed 25 there are more grayish black, medium bedded, pelitic siltstones mixed with paper coals as well as grayish yellow or green silty-mudstone. Above Bed 25, there are two gray, thin-bedded mudstones together with a black mudstone layer, which may be comparable with the sandwich-like lithologies in the Chahe Section at the boundary of the Xuanwei and Kaiyitou formations. Above this "sandwich-like" mudstone layer, a layer of grayish yellow, thin-bedded sandstone occurs and upon a layer of black siltstone contains numerous of fragmentary specimens of *Peltaspermum*, *Lepidopteris*, *Pecopteris* and *Gigantopteris*. In the boundary of the Xuanwei and Kaiyitou formations (boundary-Bed 26), there is a monotypic layer of dispersed *Tomiostrobus* in the dark grayish black thin-bedded siltstone (in the bottom of bed 26). After the layer of *Tomiostrobus* comes the grayish blue to greenish blue thick-bedded siltstones of Kaiyitou Formation, which contains no plant fossil but abundant conchostraca and bivalves.

2.3.3 Shallow-Water Zhongzhai Section, Liupanshui, Guizhou Province

The Permian–Triassic interval at Zhongzhai includes the Lungtan Formation and basal part of Yelang Formation. The Zhongzhai section is situated about 1 km northeast of Zhongzhai Village, Liuzhi County, south-western Guizhou Province (Figs. 2.1 and 2.4i). At Zhongzhai section, the lower part (Beds 1–3) of the Lungtan Formation is dominated by fine-grained sandstones and muddy siltstones with coal beds and seams. Lithology of the upper part of this formation (Beds 4–26) is the same as that of the lower part, but there is no coal seam in the upper part, representing a littoral setting (Zhang et al. 2014). The plant fossils that standing for the typical Cathaysian Flora are abundant. The top part (Bed 27) of the Lungtan Formation is dominated by calcareous mudstones, with fine laminations (Zhang et al. 2014). The fauna from Bed 27 at Zhongzhai mainly comprises dense-populated, well-preserved brachiopods (forming shell beds) and lacks radiolarians. These features suggest that this interval represents a low-energy back-barrier shallow-marine setting above the fair-weather wave-base (generally shallower than 50 m deep) (Shen et al. 2011; He et al. 2017, 2019).

The basal Yelang Formation comprises light gray thin-bedded argillaceous limestones, yellowish green calcareous mudstones and silty mudstones, intercalated with pale volcanic ash, with conodonts, bivalves, gastropods and brachiopods (Peng et al. 2007; He et al. 2008; Gao et al. 2009; Zhang et al. 2013; He et al. 2019), no plant fossil.

2.3.4 Deep-Water Xinmin Section in Anshun, Guizhou Province

The Xinmin Section is located at Xinmin valley, Puding County, Guizhou Province (Figs. 2.1 and 2.4h) and is easily accessible, well exposed, and displays a continuous sedimentation from the thin-bedded siliceous mudstone in the Talung Formation and mudstone in the Luolou Formation, indicating a deep water basinal facies. Previously, primary works have been carried on the adjacent area of this section. Yao et al. (1980) have found abundant ammonoids, gastropods, brachiopods and foraminifers at the Jiaozishan Section, which is about 4 km south to Xinmin. Research on Guizhou stratigraphy (Dong 1997) showed that ammonoid *Pseudotirolites* is common in the Talung Formation in this area, associated with brachiopods, bivalves, gastropods, trilobites and fossil plants which mainly come from black, thin-bedded mudstone interlayers in the middle of Bed 2 and the top of Bed 4. These comprise *Lepidostrobophyllum, Paracalamites, Pecopteris, Gigantopteris, Taeniopteris* leaves with good cuticle, *Cordaites, Sphenobaiera* and a large number of fossil conifers with secondary or tertiary branches and well-preserved cuticles have been identified as *Anshuncladus xinminensis, A. contiguous, A. aduncatus, Pseudoullmania frumentarioides* and *Szecladia multinervis.*

Recently, many researches related to the end-Permian event were focused on the Xinmin Section. Feng et al. (2011) first described the strata across the PTB and reported trilobite *Pseudophillip* sp. from the top of Changhsingian at this section. Based on geochemistry works, Shen et al. (2013) discussed the volcanism records at this section. Conodont *Hindeodus parvus* in the upper part of Bed5-3 firstly occurred suggests that the PTB should be placed in Bed 5–3 (Fig. 2.4h) (Zhang et al. 2014).

References

Chu DL, Yu JX, Tong JN, Benton MJ, Song HY, Huang YF, Song T, Tian L (2016) Biostratigraphic correlation and mass extinction during the Permian–Triassic transition in terrestrial-marine siliciclastic settings of South China. Glob Planet Chang 146:67–88

Feng Z, Wang J, Rößler R (2011) A unique gymnosperm from the latest Permian of China, and its ecophysiological implications. Rev Palaeobot Palynol 165(1–2):27–40

He BH, Liu SF, Wu P (2017) LA-ICP-MS U-Pb geochronology and its geological implications of the detrital zircons from the lower strata of upper Permian Xuanwei formation in Zhehai Town, Eastern Yunnan Province. China Acad J Electron Publ House 40(2):126–133 (In Chinese with English abstract)

Li H, Yu JX, McElwain JC, Yiotis C, Chen ZQ (2019) Reconstruction of atmospheric CO_2 concentration during the late Changhsingian based on fossil conifers from the Dalong formation in South China. Palaeogeogr Palaeoclimatol Palaeoecol 519:37–48

Liao ZW, Hu WX, Cao J, Wang XL, Fu XG (2020) Oceanic anoxia through the late Permian Changhsingian stage in the Lower Yangtze region, South China: evidence from sulfur isotopes and trace elements. Chem Geol 532:119371

Nanjing Institute of f Geology and Palaeontology, Academia Sinica (1980) Late Permian coal-bearing strata and palaeontological Fauna in western Guizhou and eastern Yunnan. Science Press, Beijing, pp 1–277

Shen SZ, Crowley JL, Wang Y, Boweing SA, Erwin DH, Sadler PM, Cao CQ, Rothman DH, Henderson CM, Ramezani J, Zhang H, Shen YA, Wang XD, Wang W, Mu L, Li WZ, Tang YG, Liu XL, Liu LJ, Zeng Y, Jiang YF, Jin YG (2011) Calibrating the end-Permian mass extinction. Science 334:1367–1372

Shen J, Algeo TJ, Hu Q, Xu GZ, Zhou L, Feng QL (2013) Volcanism in South China during the Late Permian and its relationship to marine ecosystem and environmental changes. Glob Planet Chang 105:121–134

Sheng JZ (1962) The Permian system in China: 1–95. Science Press, Beijing
Wang SY, Yin HF (2001) Study on terrestrial Permian–Triassic boundary in eastern Yunnan and western Guizhou. China University of Geosciences Press, Wuhan, p 88 (in Chinese with English abstract)
Yao ZQ, Xu JT, Zhen ZG, Mo ZG (1980) Late Permian biostratigraphy and the Permian–Triassic boundary in Western Guizhou and Eastern Yunnan. In: Nanjing Institute of Geology and Palaeontology (ed) Stratigraphy and palaeontology of Late Permian coal-bearing formations in Western Guizhou and Eastern Yunnan. Science Press, Beijing, pp 1–69 (In Chinese)
Yu JX, Broutin J, Chen ZQ, Shi X, Li H, Chu DL, Huang QS (2015) Vegetation changeover across the Permian–Triassic boundary in Southwest China extinction, survival, recovery and palaeoclimate: a critical review. Earth-Sci Rev 149:203–224

Chapter 3
Materials and Methods

Jianxin Yu, Hui Li, Zhen Xu, and Wenchao Shu

A combination of approaches to evaluate plant species turnover from the Late Permian to the Early Triassic in South China to consider the long-term consequences of the profound changes to terrestrial ecosystems across the End Permian Plant Crisis (EPPC) and its recovery is used in this book. The successive floras will be documented in Chap. 4 and for each we review their stratigraphical relationships and depositional sedimentary facies as geological and taphonomic context for the palaebotanical occurrences.

3.1 Sampling Strategy

We have adopted two sampling approaches for this study. For formations representing the End Permian Plant Crisis (EPPC) of the Lungtan, Talung, Xuanwei and the basal part of the Kayitou formations, and the Lowest Triassic lower part of the Kayitou Formation (Fig. 2.1) as a survival period, we selected key localities for

J. Yu (✉)
China University of Geosciences (Wuhan), Wuhan, Hubei, P.R. China
e-mail: yujianxin@cug.edu.cn

H. Li
Jiangxi Key Laboratory for Mass Spectrometry and Instrumentation, East China University of Technology, Nanchang, P.R. China
e-mail: huili_201860188@ecut.edu.cn

Z. Xu · W. Shu
School of Earth Sciences, China University of Geosciences, Wuhan, P.R. China
e-mail: janexu_cug@qq.com

W. Shu
e-mail: shuwenchao@cug.edu.cn

J. Yu et al. (eds.), *Plants and Palynomorphs around the Permian-Triassic Boundary of South China*, New Records of the Great Dying in South China,
https://doi.org/10.1007/978-981-19-1492-8_3

being representative in terms of their fossil plant composition and stratigraphically complete. For each we have undertaken extensive fieldwork over many years to identify and collect fossil plants in situ on a bed-by-bed basis. For these formations all the fossils reported have been identified by the same individuals using reference materials such that the identifications are accurate and internally consistent. Within key formations we study individual sections and then correlate them stratigraphically to determine overall changes in plant composition at the formation level.

3.2 Evaluating Fossil Plant Species Richness and Normalizing Data Occurrences

While ideally our study would seek to identify species diversity, this represents a measure of species richness and species evenness for which often impossible to determine species evenness from the fossil plant record (Cleal et al. 2012, in press). Here we focus on species richness that denotes the number of species present in a particular bed, locality, formation or flora and is measurable from the plant fossil record. To assess plant species richness, it is necessary to evaluate patterns of species change over time as undertaken by numerous past investigations that simply measure the number of plant species present in each stage (Yao et al. 1983; Li et al. 1995; Peng et al. 2009; Yu et al. 2015; Chu et al. 2016; Feng et al. 2020). This is achieved by constructing stratigraphic range diagrams for each species showing their first (origination) and last (extinction) occurrence in each geological section, and then stratigraphically correlating using litho-, bio- or chemostratigrpahic methods between different sections to compile range charts for each period. From this information the total number of species (i.e. alpha diversity) can be determined, and rates of originations and extinction through time calculated.

To meaningfully extract plant species richness patterns from the fossil data, it is important to eliminate duplicates that are an artefact of palaeobotanical nomenclature (e.g., Hilton and Cleal 2007; Cleal et al. 2012). A single reconstructed whole-plant species in the fossil record may include numerous distinct organs (e.g., leaves, stems, roots, cones, seeds), each with their own genera and species name (Bateman and Hilton 2009). The solution we adopt is to normalize the data (Cleal et al. 2012) and evaluate only those organs whose fossil taxonomy is most likely to reflect the original whole organism taxonomy. This represents the first time that this approach has been applied to plant species richness through the PTB interval including the EPPC and its Triassic recovery, with previous studies artificially inflating species richness. Normalized taxa are retained for species richness estimates so that only one organ of each viable whole-plant is included. Examples include omitting accounts of genera that lack species-level identifications (e.g. *Tomiostrobus* sp.) from in assemblage in which one or more identified species of the same genus occurs (e.g. *Tomiostrobus augusta*, *T. brevicystis*), assuming that the specimens identified as "sp." most likely represent poorly preserved or incomplete examples of one of the other or more of the named species of *Tomiostrobus* present. We also omitted from species richness estimates fertile organs including gymnosperm seeds when other organs of the same plants are present in the same assemblage; in all cases where seeds are present they co-occur with one or more species of gymnosperm leaf from which it is assumed that the seeds belonged to one or more of these plants. Lycopsid rootstock (*Stigmaria*), sporophylls (*Lepidostrobophyllum*) and cones (*Lepidostrobus*) are also omitted as in all instances these co-occur with stems (*Lepidodendron*) that are more distinctive and are typically identified to the species level. We accept that conclusions on normalizing fossil plant data are to some extent subjective and cannot readily be tested for their robustness, but we consider these more realistic estimates of species richness and potentially species richness than non-normalized, raw-data based accounts.

3.3 Fossil Plant Abundance

There is no robust method to quantitatively assess plant abundance in terrestrial settings unless fossil floras are preserved in-situ in obrution events such as volcanic ash-fall events (Wang et al. 2012). In a broad sense plant abundance may be indicated by a number of indirect measures including the number of locations that contain fossil plants as well as the palaeobotanical richness in terms of numbers of specimens at each location, but these are subject to a variety of controls including spatial heterogeneity of plant distributions in contemporaneous settings such as different niches, and a variety of physical (biostratinomic, sedimentary) and chemical taphonomic processes controlling what enters the fossil record (Bateman 1991; Allison and Bottjer 2010). Here we focus on relative measure of abundance as well as using specific environmental proxies that provide crude insights into plant abundance in the environment in which they lived.

Firstly, having undertaken extensive fieldwork and collection based on investigations to Permian–Triassic plant bearing sedimentary successions from South China, it is obvious that plant abundance varies from being very abundant and nearly ubiquitous in some sedimentary contexts within individual formations, to being very rare in others in which finding a single fossil plant remain may take several hours in the field. In this we collectively consider all plant material, from fragmented plant debris to intact fossil plant organs as we are interested in assessing the presence of plants in the sedimentary system rather than focusing on details of what the plants are or where they came from. Due to the nature of the outcrops and the distribution of fossils in the sediments, it has not been possible to sample fossil plants using quadrat or other methods that might facilitate more accurate assessments of abundance. Rather, as very crude proxies for plant abundance, we use the relative descriptors absent, very rare, rare, common, abundant and very abundant to describe the amount of plant fragments encountered in each section in terms of (A) plant fossils on individual beds, and (B) the number of beds containing fossil plants.

Secondly, as coals represent accumulated plant matter that built up as peat, we consider coals to indicate high plant abundance for an extended period combined with low levels of decay (e.g., low pH to suppress fungal and bacterial decay). We do not seek to quantify this further, merely indicate that multiple generations of plants needed to grow in waterlogged conditions for the biomass to accumulate and equating this to abundant or very abundant depending on the vertical thickness and lateral extent of the coal. In contrast, the studied interval includes the early Triassic coal-gap (Retallack et al. 1996), a stratigraphic interval in which adverse environmental conditions prevented peat (and hence coal) formation and/or preservation. We consider this to be evidence for very low plant abundance but accept it could also represent widespread adverse preservation conditions for fossil plants (Vajda et al. 2020; DiMichele et al. 2020).

Thirdly, we consider environmental proxies related to soils on the basis that abundant vegetation cover is likely to bind soils together and diffuse water infiltration into the ground, thus having the net result of reducing surface water run-off (Wu et al. 1997; Davies and Gibling 2008). In contrast, bare, un-vegetated ground would be more susceptible to surface water run-off, physical weathering and erosion (Retallack 2005; Algeo et al. 2011; Kaiho et al. 2016). We therefore use the presence of physical weathering in terrestrial settings as a rough proxy for plant abundance in terms of ground cover, highlighting the co-evolution of plants and the environment. For this we use two methods; Average Linear Sedimentation Rates (LSRs) (Algeo et al. 2010) and Chemical Weathering Intensity (Algeo et al. 2011). LSRs (units of m m.y.$^{-1}$), as the thickness of a given time strata divided by the duration of that strata, is a proxy for sediment flux (Algeo et al. 2010). For example, the LSRs increased from 10-18 m m.y.$^{-1}$ in the Lopingian to 105–488 m m.y.$^{-1}$ during the Early Triassic at the Meishan Section in eastern China (Algeo et al. 2010). The enhanced sedimentation rate is common through the world in the Early Triassic and the LSRs of offshore clastic facies increase more substantial than the deeper water sections, revealing the delivery of sediments may be related to the terrestrial soil erosion (Algeo et al. 2010). LSRs in this context indicate input of terrestrial debris into marine systems, and the rate or type of weathering is determined through the elementary data and mineral composition of the sediments. The Chemical Weathering Intensity derives from clay mineral contents of sedimentary rocks on the basis that clays including kaolinite are chemical weathering products and their compositions can be used as proxies for chemical weathering rates (Algeo et al. 2010, 2011). Finally, we very briefly consider Total Organic Carbon (TOC) levels in sediments as a crude proxy for terrestrial biomass and plant abundance. However, this is a proxy from which extreme care needs to be taken when interpreting TOC values that are affected by taphonomic (e.g. transportation, decay, oxidation) and diagenetic controls (e.g., redox conditions).

3.4 Evaluating Plant Ecology

The geological and palaeoecological contexts in which fossil plants occur is important to their interpretation in order to provide links between fossil plants and the environment(s) in which they grew and the processes that affected them. For example, through analyzing the palaeoenvironments in which plants grew it is possible to elucidate the palaeoenvironmental constraints regulating growth of different plant taxa or groups. Through analyzing spatial and temporal distributions in fossil plant occurrences it may be possible to identify migration patterns of individual taxa or floras, and from this identify environmental parameters such as water availability or substrate conditions controlling plant distributions. It also allows greater understanding of the conditions in which extinction survivors lived, identifying features that may have contributed to their resilience as well as identifying the locations and environments of refugia.

Palaeoecology assessments for plant taxa here come primarily from the available literature and is based on plant fossil anatomy and morphology and the sedimentary depositional environments in which they occur (Yao 1983; Meyen 1987; Bateman 1991; Yang 1993, 1994; Taylor et al. 2009). Here, fossil plant occurrences were evaluated for their shape, size and completeness to elucidate their taphonomy in terms of being transported long, medium or short distances from their growth environment. For example, entire or almost entire organs lacking signs of taphonomic fragmentation, abrasion or size-sorting were interpreted as having undergone minimal transportation and being recorded in-situ or very close to their growth environment. This included in-situ plants and fossil plants found in palaeosols, e.g., whole plant of *Annalepis* (*=Lepacyclotes*) in Badong Formation of Hongjiaguan Section and *Pleuromeia* in Dawotang Section (Meng 1995). These kinds of fossils often have exquisite preservation of delicate plant tissues such as leaf tips or sporophylls attached in cones. Greater levels of transport result in higher levels of fragmentation and size sorting and culminate in fine grained well sorted plant debris beds. Depositional environments of the plant fossil were determined by sedimentary features on fieldwork and from the literature (Yao et al. 1983; Bercovici et al. 2015; Yu et al. 2015; Chu et al. 2016), as well as comparison to established palaeogeographic analyses (Yao et al. 1980; Zheng et al. 2010). For instance, an in-situ fossil with attached roots preserved in tidal flat facies would suggests this special plant grew in the facies, for example, specimens of *Annalepis* from the Kayitou Formation in the Mide and Tucheng Sections (Yu et al. 2015).

References

Algeo TJ, Chen ZQ, Fraiser ML, Twitchett RJ (2011) Terrestrial–marine teleconnections in the collapse and rebuilding of early Triassic marine ecosystems. Palaeogeogr Palaeoclimatol Palaeoecol 308:1–11

Algeo TJ, Twitchett RJ (2010) Anomalous Early Triassic sediment fluxes due to elevated weathering rates and their biological consequences. Geology 38:1023–1026

Allison PA, Bottjer DJ (2010) Taphonomy: process and bias through time (Second Edition). Springer

Bateman RM (1991) Palaeoecology. In: Cleal CJ (ed) Plant fossils in geological investigation: the palaeozoic. Ellis Horwood, London, pp 34–116

Bateman RM, Hilton J (2009) Palaeobotanical systematics for the phylogenetic age: applying organ–species, form–species and phylogenetic species concepts in a framework of reconstructed fossil and extant whole–plants. Taxon 58:1254–1280

Bercovici A, Cui Y, Forel M, Yu JX, Vajda V (2015) Terrestrial paleoenvironment characterization across the Permian-Triassic boundary in South China. J Asian Earth Sci 98:225–246

Chu DL, Yu JX, Tong JN, Benton MJ, Song HY, Huang YF, Song T, Tian L (2016) Biostratigraphic correlation and mass extinction during the Permian-Triassic transition in terrestrial-marine siliciclastic settings of South China. Glob Planet Change 146:67–88

Cleal CJ, Uhl D, Cascales-Miñana B, Thomas BA, Bashforth AR, King SC, Zodrow EL (2012) Plant biodiversity changes in Carboniferous wetlands. Earth Sci Rev 114:124–155

Davies NR, Gibling MR (2010) Cambrian to Devonian evolution of alluvial systems: the sedimentological impact of the earliest land plants. Earth Sci Rev 98:171–200

Dimichele WA, Bashforth AR, Falcon-Lang HJ, Lucas SG (2020) Uplands, lowlands, and climate: Taphonomic megabiases and the apparent rise of xeromorphic, drought-tollerant flora during the Pennsylvanian-Permian transition. Palaeogeog Palaeoclimatol Palaeoecol 559:109965

Feng Z, Wei HB, Guo Y, He XY, Sui Q, Zhou Y, Liu HY, Gou XD, Lu Y (2020) From rainforest to herbland: New insights into land plant responses to the End-Permian Mass Extinction. Earth-Sci Rev 204:103153

Hilton J, Cleal CJ (2007) The relationship between Euramerican and Cathaysian tropical floras in the Late Palaeozoic: palaeobiogeographical and palaeogeographical implications. Earth-Sci Rev 85:85–116

Kaiho K, Saito R, Ito K, Miyaji T, Chen ZQ (2016) Effects of soil erosion and anoxic-euxinic ocean in the Permian-Triassic marine crisis. Heliyon 2(8):137

Li XX (ed) (1995) Fossil floras in China through the geological ages (English edition). Guangdong Science and Technology Press, Guangzhou, pp 1–695

Meng FS, Xu AW, Zhang ZL, Lin JM, Yao HZ (1995) Nonmarine biota and sedimentary facies of the Badong Formation in the Yangtze and its neighbouring areas. China University of Geosciences Press, Wuhan, pp 1–76. (In Chinese with English abstract)

Meyen SV (1987) Fundamentals of palaeobotany. Chapman and Hall, London

Peng Y, Shi GR (2009) Life crises on land across the Permian-Triassic boundary in South China. Global Planet Change 65:155–165

Retallack GJ, Veevers JJ, Morante R (1996) Global coal gap between Permian-Triassic extinction and Middle Triassic recovery of peat-forming plants. Geol Soc Amer Bull 108(2):195–207

Retallack GJ (2005) Earliest Triassic claystone breccias and soil-erosion crisis. J Sed Res 75:663–679

Taylor TN, Taylor EL, Krings M (2009) Paleobotany, the biology and evolution of fossil plants, 2nd edn. Academic Press, Amsterdam

Vajda V, McLoughlin S, Mays C, Frank TD, Fielding CR, Tevywa A, Lehsten V, Bocking M, Nicoll RS (2020) End-Permian (252 Mya) deforestation, wildfires and flooding—an ancient biotic crisis with lessons for the present. Earth Planet Sci Lett 529:115875

Wang J, Pfefferkorn HW, Zhang Z, Zhou F (2012) Permian vegetational Pompeii from Inner Mongolia and its implications for landscape palaeoecology and palaeobiogeography of China. Proc Natl Acad Sci 109:4927–4943

Wu Y, Liu SQ, Wang JX (1997) Effect of plant root system on soil anti-erosion. Chinese J Appl Environ Biol 3(2):119–124

Yao ZQ (1983) Ecology and taphonomy of Gigantopterids. J Nanjing Inst Geol Palaeontol 6:62–84 (In Chinese with English abstract)

Yu JX, Broutin J, Chen ZQ, Shi X, Li H, Chu DL, Huang QS (2015) Vegetation changeover across the Permian-Triassic boundary in Southwest China extinction, survival, recovery and palaeoclimate: a critical review. Earth-Sci Rev 149:203–224

Yang SP (1993) Paleoecology: principles and methods. Geological Publishing House, Beijing (In Chinese)

Yao ZQ, Xu JT, Zhen ZG, Mo ZG (1980) Late Permian biostratigraphy and the Permian-Triassic boundary in Western Guizhou and Eastern Yunnan. In: Nanjing Institute of Geology and Palaeontology (Ed.), Stratigraphy and palaeontology of Late Permain coal-bearing formations in Western Guizhou and Eastern Yunnan. Science Press, Beijing, pp 1–69 (in Chinese)

Zheng RH (2010) China pre-Mesozoic structural sequence and lithofacies paleogeography atlas. Geological Publishing House, Beijing (In Chinese)

Chapter 4
Late Permian Changhsingian Floral Assemblage in Varied Palaeogeographic Settings

Jianxin Yu

China is one of the ideal regions for the study of paleofloras owing to the wide distribution of continental or alternately marine continental deposits bearing rich fossil plants with the exception of marine strata in South China in the Permian period. Even though South China was mostly covered by sea in Permian, there still exist marine-continental transitional strata in many localities which bear the famous Permian *Gigantopteris* flora which was part of middle to late stages of the Cathaysian Flora. It was represented by the *Gigantopteris* group and only existed during the Permian.

Fossil plants have not been found in the early Early Permian (Asselian) of South China Province owing to the deposition of only marine strata. From the beginning of the middle Early Permian (Sakmarian) alternately marine-continental strata are sporadically distributed in the Jiangnan Block, the Yangtze Platform and on margins of the Kangdian massif yielding a few fossil plants. From middle-late Early Permian (Artinskian), the palaeogeographic features of the South China Province considerably changed, which not only led to distinctive differentiation of the lithofacies of the Permian marine strata but also to the deposition of alternately marine-continental coal-bearing strata during the late Early Permian to Late Permian from the east or northeast to the west, which yield rich Gigantopterids plants.

In the Latest Permian, the flora was mainly distributed in western Guizhou, eastern Yunnan and southern Sichuan of the Yangtze Platform and sporadically discovered near the Jiangnan Block. To the end of the Permian, the *Gigantopteris* flora stepped down from the historical arena except that several relics persisted in the earliest Triassic in a few places of western Guizhou and eastern Yunnan.

J. Yu (✉)
China University of Geosciences, Wuhan, Hubei, P.R. China
e-mail: yujianxin@cug.edu.cn

J. Yu et al. (eds.), *Plants and Palynomorphs around the Permian-Triassic Boundary of South China*, New Records of the Great Dying in South China,
https://doi.org/10.1007/978-981-19-1492-8_4

4.1 Composition and General Features of the Studied Paleofloras

Fossil plants from the upper part of Xuanwei Formation and Kayitou Formation in western Guizhou and eastern Yunnan not only show a distinctive evolutionary trend through time, and also possess the typical characteristics of the Cathaysian flora. Based on the collected fossil plants of over 20 sections (Fig. 2.1), total of 105 species (including indet. species) assigned to 39 genera were described from the Permian–Triassic transition in western Guizhou and eastern Yunnan, and they were assigned to 11 groups (Tables 4.1, 4.2, 4.3, 4.4 and 4.5).

4.1.1 Lycophyta: Lepidodendrales

The Lycophytes of Late Permian Changhsingian Flora from the Xuanwei and Lungtan formations belong to Lepidodendrales arborescent forms, which consisted of 9 species in 4 genera. Except for their high abundance and diversity in this region, this group seems to have a stage in abundanceand diversity within the entire flora based on the number of genera and species. The vertical distribution of this group indicates that they persist into the lower part of the Kayitou Formation (e. g. *Lepidodendron* and *Stigmaria*). The Lepidodendrales were often adapted to the moist and warm environments such as river mouth bay of abandoned delta, marginal bay or the swamp of tidal flat. This order served as the important coal forming plants in this region during the Late Permian. Genus *Stigmaria* is abundant under the coal bed, occurring in Earliest Triassic. All species of *Lepidodendron* is a typical oriental-type of *Lepidodendron* of the Cathaysian Floristic Region.

4.1.2 Arthrophyta

This group contains 15 species in 6 genera. The fossil plants belonging to the Sphenophyales are very rare in this flora, only 1 indet. species of *Sphenophyllum* occurs. Calamitaceae of Equisetales is also abundant. *Lobatannularia*, a typical genus of the Permian Cathaysian Flora, has 6 species, occurs in 8 layers from these sections. The morphological evolutionary trend was indicated by the following changes: leaf bases from entirely separates to gradually fused to finally with the fusion of about 4/5 of leaf length; the number of leaves increases gradually; and the leaf curvature increases. Most species of *Annularia*, are typical elements of the Cathaysian Flora, has 3 species. Stem pith casts are common. *Calamites* and *Paracalamites* occur in succession with one after the other. In the Xuanwei Formation, one species, *Schizoneura manchuriensis* (Equisetaceae), occurs in the Angara and Gondwana Floristic Regions.

Table 4.1 Vertical distributions of fossil plants across Permian–Triassic boundary in Chahe section

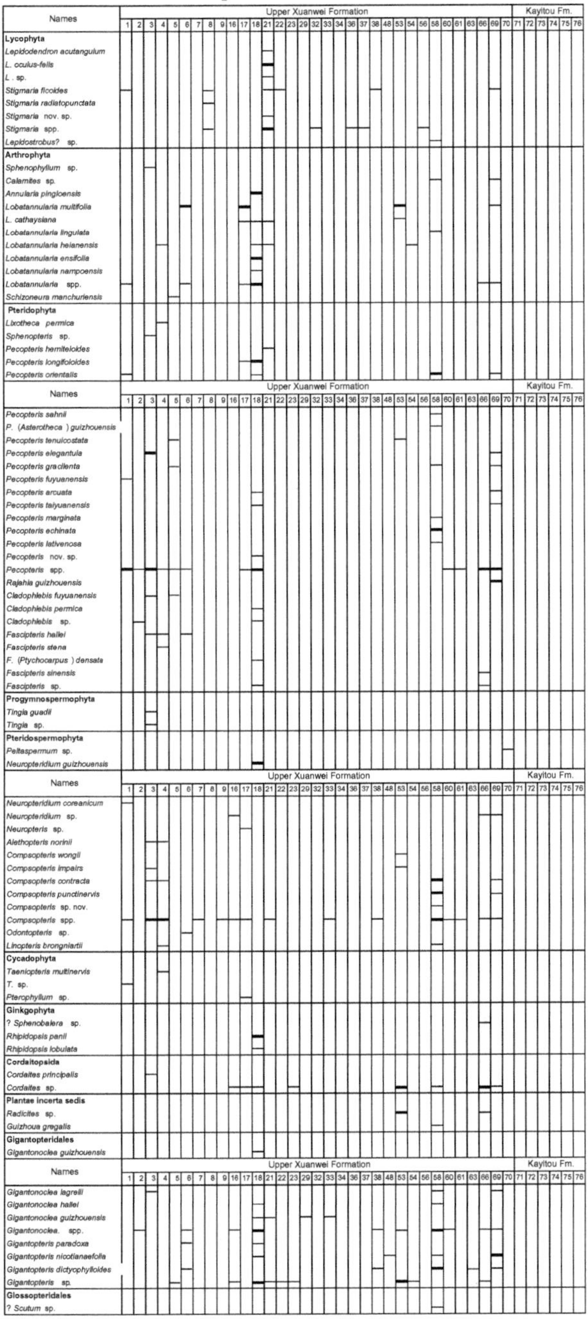

Table 4.2 vertical distributions of fossil plants across Permian–Triassic boundary in Zhejue section.

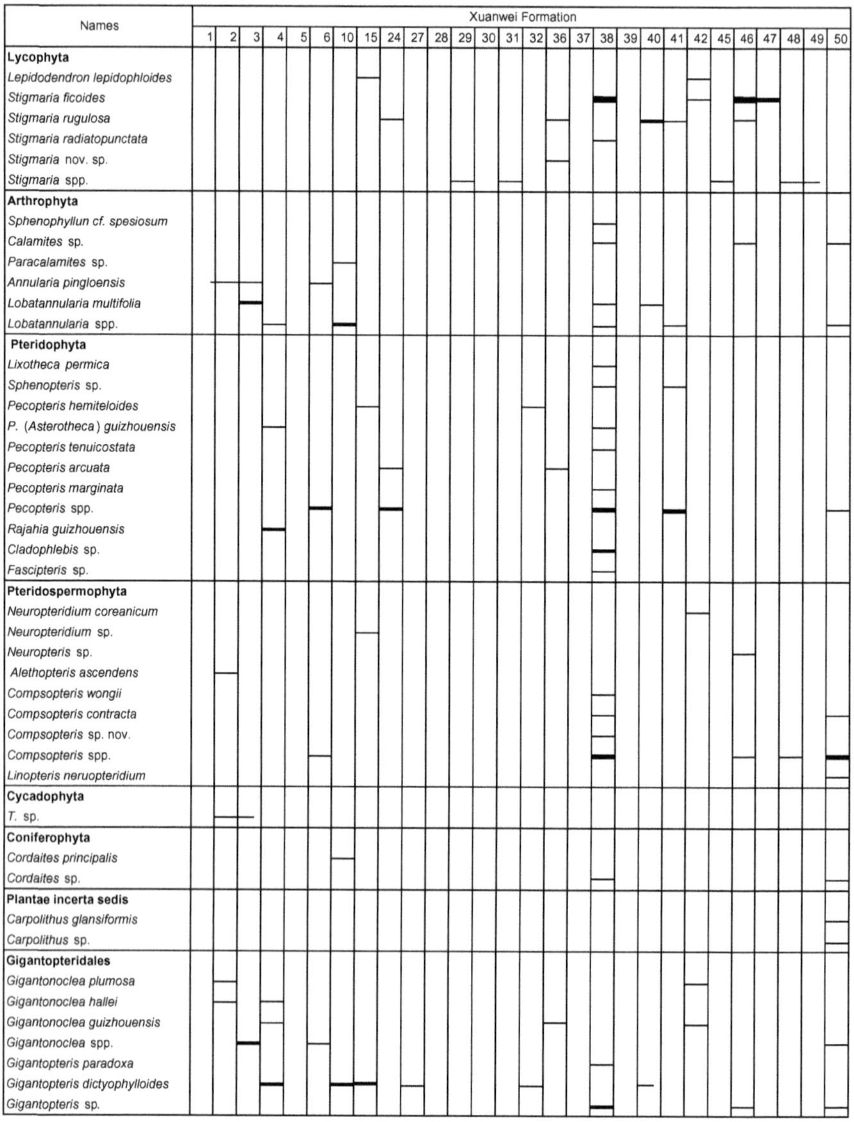

4.1.3 *Pteridophyta*

Pteridophyta represent 29 species in 7 genera. They are the most abundant group in Late Permian Changhsingian Flora in South China. There are some elements of both

Table 4.3 Vertical distributions of fossils Plants across Permian–Triassic boundary in Mide section (A)

Names	Upper Xuanwei Formation																								Kayitou Fm.			
	1	2	3	4	5	6	7	8	9	14	15	16	17	18	19	20	21	22	23	26	27	28	29	30	43	46	47	48	49
Lycophyta																													
Lepidodendron acutangulum																									—				
Stigmaria ficoides								—																					
Stigmaria rugulosa												—		—											—				
Stigmaria spp.					—	—	—									—				—		—	—						
Annalepis zeiller																										—	—		
Annalepis spp.																										—	—		
Arthrophyta																													
Sphenophyllum sp.																							—						
Calamites sp.									—																				
Paracalamites stenocostatus																	—												
Paracalamites sp.				—																			—		—				
Annularia shirakii																													
Annularia pingloensis																													
L. cathaysiana		—											—																
Lobatannularia spp.	—													—			—												
Pteridophyta																													
Sphenopteris sp.														—															
Pecopteris hemiteloides		—																											
P. (*Asterotheca*) *guizhouensis*			—																										
Pecopteris spp.	—	—		—								—	—	—															
Rajahia guizhouensis																													
Cladophlebis fuyuanensis			—																										
Cladophlebis permica																													
Cladophlebis sp.	—																												
Fascipteris sp.														—															
Progymnospermophyta																													
Tingia cf. hamaguchii																	—												
Pteridospermophyta																													
Neuropteridium guizhouensis																	—									—	—		
N. cf. *coreanicum*														—															
Compsopteris wongii																	—												
Compsopteris spp.				—								—	—	—			—						—						
Cordaitopsida																													
Cordaites sp.														—															
Gigantopteridales																													
Gigantonoclea plumosa																	—												
Gigantonoclea guizhouensis		—	—	—																									
Gigantonoclea. spp.	—																		—					—	—				
Gigantopteris paradoxa														—															
G. dictyophylloides										—	—		—	—			—						—						
Gigantopteris sp.									—																				

Euramerican and Cathaysian Floristic provinces. Genus *Fascipteris* is also a typical element of the Cathaysian Flora.

In addition, some pinnules of *Sphenopteris* badly preserved, may belong to some Gleicheniaceae and *Lixotheca permica* is assigned to Hymenophyllaceae (Filicales).

Table 4.4 Vertical distributions of fossil plants across Permian–Triassic boundary in Mide section (B)

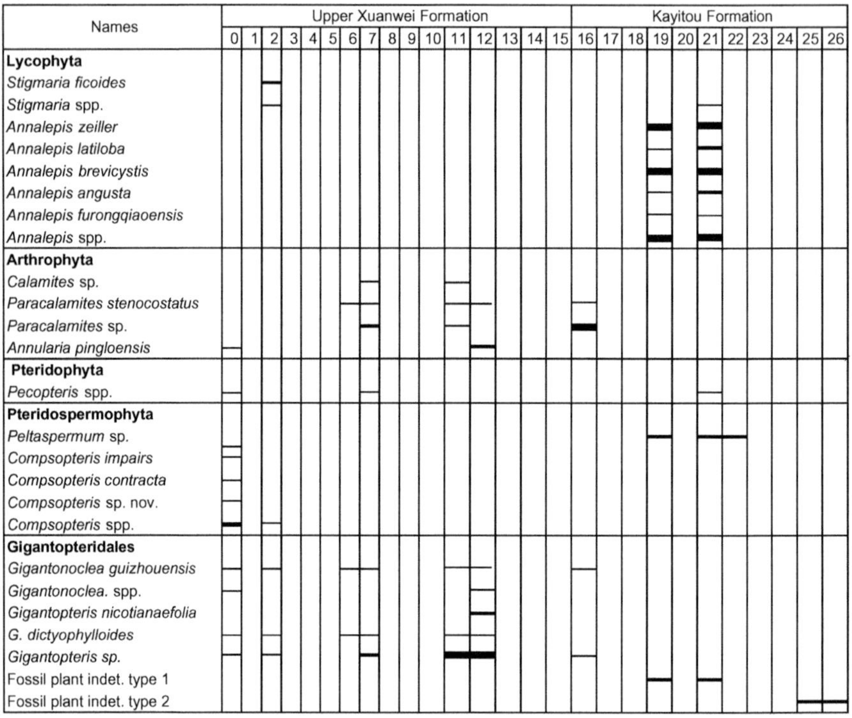

4.1.4 *Progymnospermophyta: Noeggerathipsida*

This group only contains 3 species of 1 genus. Genus *Tingia* is also a typical element of the Cathaysian Floristic Region., and common in the upper part of the Xuanwei Formation. This genus dominates in upper delta plain.

4.1.5 *Pteridospermophyta*

There are 18 Pteridosperm species in 7 genera in Late Permian Changhsingian Flora in South China. They are the secondary in this flora. In general, Pteridosperms are known back to the Devonian and flourished from the Carboniferous of Euramerican Region, were proposed as primitive Pteridosperms (Wang, 1989). In South China, the appearing form groups include Neuropterids, Callipterids, Odontopterids and Alethopterids that are abundant in the Xuanwei and Lungtan fFormations. They are important elements of Euramerican and Cathaysian Regions. Some species of this

Table 4.5 Vertical distributions of fossil Plants across Permian–Triassic boundary in Tucheng section

Names	Upper Xuanwei Formation																Kayitou Formation									
	1	2	3	4	5	6	7	8	9	10	11	12	13	14	15	16	17	18	19	20	21	22	23	24	25	26
Lycophyta																										
Lepidodendron sp.																	—	—								
Stigmaria ficoides										—		—			—											
Stigmaria rugulosa															—											
Stigmaria spp.						━				—		—					—	—								
Annalepis zeiller																	━	━								
Annalepis latiloba																	—	—								
Annalepis brevicystis																	—	—								
Annalepis spp.																	━	━								
Arthrophyta																										
Sphenophyllum sp.														—												
Paracalamites stenocostatus													—	—												
Annularia pingloensis										—		━	━		—											
Lobatannularia multifolia										—																
L. cathaysiana										—																
Pteridophyta																										
Sphenopteris cf. *tenuis*										—																
Sphenopteris sp.																						—	—			
Pecopteris orientalis					—								—													
Pecopteris sahnii				—																						
P. (*Asterotheca*) *guizhouensis*				━																						
Pecopteris tenuicostata										—																
Pecopteris spp.										—		—		━	—		—	—				—	—			
Rajahia guizhouensis															—											
Fascipteris hallei				—																						
Pteridospermophyta																										
Aff. *Peltaspermum martinsii*																	—	—				—	—			
Peltaspermum lobutalum																						—	—			
? *Peltaspermum* sp.																						━	━			
Neuropteridium sp.				—						—																
Alethopteris norinii				—																						
Compsopteris wongii										—																
Compsopteris spp.				━						━		—			━											
Cycadophyta																										
Taeniopteris. sp.															—											
Ginkgophyta																										
? *Sphenobaiera* sp.																	—									
Cordaitopsida																										
Cordaites sp.					—							—	—													
Plantae incerta sedis																										
Rhizomopsis gemmifera															—											
Gigantopteridales																										
Gigantonoclea cf.*guizhouensis*										━																
Gigantonoclea plumosa					—																					
Gigantonoclea cf. *hallei*					—			—																		
Gigantonoclea guizhouensis				━	━					—		—		━	━							—	—			
Gigantonoclea. spp.															━								—			
Gigantopteris dictyophylloides				━	━					━		—			—							—	—			
Gigantopteris sp.												—			—							━	—			
Glossopteridales																										
? *Glossopteris* sp.																										
? *Plumsteadia* sp.																										

group (including *Alethopteris norinii*, *Alethopteris ascendens*, *Compsopteris* sp.) are special elements of the Cathaysian region,

Pteridosperms include Peltaspermales known from the Lower Permian in Europe that flourished during early Mesozoic. This group has representatives during the Permian in every floristic region around the world. In western Guizhou and eastern Yunnan, South China, some elements of Peltaspermaceae appeared in the Lower part of the Kayitou Formation.

4.1.6 Cycadophyta

Cycadophytes consist of 3 species in 2 genera in the total Flora. Fossils leaves of *Taeniopteris* and *Pterophyllum* were found. *Taeniopteris* was often referred to the form group of Filicales and Pteridospermopsida. Mamay (1973, 1976) found some *Taeniopteris*-type Cycadalean megasporophylls from the lower Middle Permian (Leonardian) of America and discussed their relationships with Pteridosperms. In the Permian of the North China Platform, *Taeniopteris* are always found co-occurring with Cycadophyte megasporophyll and microsporophyll, and slender shoots of Cycadophytes, which indicated the existence of Cycadophytes. According to the sudden speciation of the punctuated Equilibrium Theory, the Cycadophytes are supposed to be in the initial appearance and rapidly flourish to enter the abrupt mutational stage. Based on the observations of Taeniopterids and the co-occurring reproductive organs of Cycadophytes, it is suggested that Taeniopterids be more closely related to the form group of Cycadophytes (Yang et al. 2006).

4.1.7 Ginkgophyta

Ginkgophyta or Ginkgophytes consist of 4 species in 2 genera in the total Flora. Only *Rhipidopsis* and *Sphenobaiera* are found. Both *Rhipidopsis panii* and *R. lobulata* are mainly distributed in the Xuanwei and Lungtan formations.

4.1.8 Coniferophyta

There are 7 species in genera 4 (including *Cordaites*, *Anshuncladus*, *Pseudoullmania* and *Szecladia*). Genus *Cordaites* often occurs in non-marine facies Xuanwei and Longtan formations, while genera *Anshuncladus*, *Pseudoullmania* and *Szecladia* only in marine facies Talung Formation. Due to taphonomic parameters, fossil remains of conifers, surprisingly absent in the terrestrial deposits of the Lopingian Xuanwei Formation, often co-occur with marine fossils in the marine coeval deposits of South China.

4.1.9 *Gymnospermarum Seeds*

Few seeds were collected from this area and assigned to 1 species *Carpolithus glansiformis*. They only account for 0.9% in the total Flora. These seeds are mainly reported from the top part of the Xuanwei Formation.

4.1.9.1 Plantae Incertae Sedis

They consist of 4 unidentified species, accounting for 3.8% of the total Flora. The plant organ type of the 2 species of *Radicites* and *Guizhoua* are not known. 2 species in genus *Rhizomopsis* are associated with the leaves of Gigantopterids. Possibly they are the stems or twigs of Gigantopterids.

4.1.9.2 Gigantopteridales and Glossopteridales

During the Latest Permian, the Gigantopteridales played an important role in the Flora. Stratigraphically, they are distributed in over 16 horizons in South China with a total of 11 species in 2 genera. During a period of about 19 million years from early Middle Permian to early Late Permian (Jin et al. 2000) in the middle and southern North China Platform, the Gigantopteridales had rapidly undergone the entire process of evolution from occurrence to flourish to extinction. In South China Platform, the coal-bearing beds preserved Gigantopteridales ranged from late Middle Permian to Latest Permian (Yao 1978; Li et al., 1980). A few elements persisted across the PTB and disappeared in Earliest Triassic. This group of enigmatic plants is characteristic of the Cathaysian floristic region.

The systematic position of the Gigantopterids is still uncertain despite various proposals in several publications, ultimately Yang (2006) proposed to include it into a Preanthophyta group. This is a topic still in controversy. Since the discovery of the Gigantopterids seeds from Fujian (Li et al. 1983b), this group was propsed as "at least partially angiospermized" based on the anatomical structure of the seeds. The order Gigantopteridales was established and was assigned to Pteridosperms. Meyen (1987) established Gigantonomiales based on seeds and assigned this group to Ginkgophytes of the Gymnosperms. Taylor (1994) assigned Gigantopteridales to "other Gymnosperms". Zhang (2000, 2004) Gigantopteridales to Gigantopteridophyte of Proangiospermae and he proposed that it is originated from seed ferns, and it is not the direct ancestor of flowering plants.

Similarly, the Glossopteridales play an important role in Gondwana Flora. Maybe 3 types (glossopterids-like fructification, ?*Scutum* and/or ?*Plumsteadia*) were assigned to the Glossopterids. The species yielded by the Permian coal-bearing formation in Guizhou, South China, which had been named *Glossopteris guizhouensis* Gu et Zhi (Li et al. 1974, p.137), was later renamed *Abrotopteris guizhouensis* (Gu et Zhi) Mo, gen. et comb. nov. (Zhao et al. 1980, p.85). Until now, the real Glossopterids are unknown within Late Permian the Cathaysian flora in South China, but recorded from Arabian Peninsula (e.g. Oman, Saudi Arabia and Turkey) (Berthetin et al. 2003) mixed Cathaysian-Gondwanan Permian floras.

Late Permian Changhsingian Flora from Xuanwei and Lungtan Formations is a typical Permian Cathaysian Flora. This is well developed in the southernmost part of the South China Platform with its distinctive features. The characteristics of the Changhsingian flora are reflected in the following aspects. (1) It is distributed in the margin of the Kangdian massif. The vertical distributions obviously are influenced by marine regression and transgression. (2) In composition, about 2/3 of the genera and species in the flora are endemic to the Cathaysian Flora. Many genera consist of enigmatic plants that had undergone a long evolution and a prolonged extinction (see Chap. 5). (3) It is the last evolutionary stage of the Cathaysian Flora. After the PT mass extinction, some elements of the Gigantopteridales survived into the Earliest Triassic and died out.

Based on the developing and declining history of different plant groups and their combinations in different horizons, especially on the evolutionary history of the Gigantopterids, Lycopsids and Sphenopsids, the appearance and flourishing of Pteridosperms and Conifers, the paleoflora is divided into two assemblages in contemporaneous heterotopic facies, as the following.

4.2 Floral Assemblage of Late Permian Changhsingian Xuanwei and Lungtan Formations

During the Late Permian Changhsingian, the paleogeographic position of western Guizhou and eastern Yunnan was in the upper Yangtze Plateform. Similar to North China at that time, it was located on the Tethys coastal belt under tropical humid warm climate for a long time. Many coal swamps were formed at that time. These swamps were suited for the preservation of fossils. Coal-bearing strata of Late Permian Xuanwei and Lungtan formations in this region are continuous, yield abundant and well-preserved fossil plants.

The Assemblage is named *Gigantonoclea guizhouensis—Annularia pingloensis* Assemblage, standing for the last stage of the Cathaysian Flora. It is distributed in Xuanwei and Lungtan formations, consists of 68 known species in 39 genera.

The main elements of this assemblage are *Gigantopteris dictyophylloides, Gigantonoclea guizhouensis, Lobatannulariamultifolia* and *Stigmaria ficoides.* The common occurrences are *Annularia shirakii, Lobatannularia cathaysiana, Paracalamites stenocostatus, Fascipteris stena, Pecopteris marginata, Compsopteris contracta, Rhipidopsis pani* and, while a few species of *Lepidodendron.*

The features of this assemblage include: (1) Few genera or species of Lepidodendrales are found but abundant individual taxa; (2) Of the Sphenophyta, except for *Schizoneura* of the Equisetaceae, there were many elements of *Annularia* and

Lobatannularia and their stem (e.g. *Calamites* and *Paracalamites*); (3) In Pteridophyta, *Pecopteris* and *Fascipteris* is main in the abundance and the diversity. The number of Mesozoic-type ferns such as *Cladophlebis* increased from bottom to top. (4) Elements of Noeggerathiales were exiguous, only occurred *Tingia* in the basal part of the upper Xuanwei Formation. (5) *Compsopteris* of the Pteridospermophyta flourished with 6 species. (6) Gigantopteridales flourished, with diverse leaf architecture and venation patterns, the main is a few species of *Gigantopteris* and *Gigantonoclea*. (7) The elements of Glossopteridales are rare and distributed in the upper part of upper Xuanwei Formation. (8) *Rhipidopsis* of the Ginkgophyta is common in the middle part of upper Xuanwei Formation.

In a word, this assemblage is still dominated by Filices, Pteridospermopsida, especially, the genus *Pecopteris* is well developed with obvious specific diversity. The Gigantopterids are different from the Lungtan flora (early Late Permian) in composition of genera and species. The species number of Gigantopterids markedly decreased and the famous species *Gigantopterisnicotianaefolia* became quite rare and was sporadically discovered in a few strata. The genus *Protoblechnum* that was common in the Permian floras of China disappeared in the present assemblage. From the Changhsingian time, the Gigantopteids flora was gradually poverty stricken and declined. To the end of the Permian, the Cathaysian flora stepped down from the historical arena except that several relics persisted in the earliest Triassic in western Guizhou and eastern Yunnan.

4.3 Floral Assemblage of Late Permian Changhsingian Marine Deep-Water Talung Formation

Although there is no coal-bearing strata of Late Permian deep-water facies Talung Formation in South China, a diverse plant fossil.assemblage (including Lycopsida, Sphenopsida, Pteridopsida, Pteridospermopsida, Cordaitopsida and Pinopsida) was obtained, accompanied by marine biota including Ammonites, brachiopods, bivalves, conodonts, and foraminifera. Fossil conifers and *Taeniopteris* are the dominant components of the assemblage, and are distributed in the majority of beds, while usually common components of the Cathaysian flora such as *Gigantopteris* and Lepidodendron are only sporadically present. In general, the gymnosperm fossils are well-preserved in the depositional facies while the typical Cathaysian floral elements are more fragmentary and abraded. These appear to be transported from further distance into the depositional setting. These plants arrived in the marine basin probably after rafting on surface currents from the basin margins and sinking to the basin floor. Anoxia likely protected the plant remains from decomposition. The notably thick leaves with robust, waxy cuticles contributed to the excellent preservation of fossil conifers and *Taeniopteris* after transportation. A similar taphonomic explanation is provided for peltasperm leaves in deep-water, basinal sediments (Baumgardner et al. 2016). In addition, the conifers which are commonly categorized as an upland group may shift downslope under the influence of climate change. We discuss this more fully later in the book.

In spite of the large number of specimens, the diversity of fossil conifers is considerably low. Only four species in three genera are identified, namely *Pseudoullmannia frumentarioides* He, Liang and Shen (He et al. 1996), *Szecladia multinervia* Yao, Liu, Rothwell and Mapes (Yao et al. 2000), *Anshuncladus sp. 1* and *Anshuncladus sp. 2* (tentatively denominated after Song et al. 2013). *Anshuncladus* is a new genus established after extensive studies of fossil conifers from the Permian deposits in South China. Up to now, the materials of this new genus have all been collected from the Talung Formation in South China.

These fossils conifers include a secondary or tertiary branch that can reach up to 50 cm long with well-preserved cuticles and *Taeniopteris* leaves with good cuticle. In addition, smaller, fragmentary fossils of *Lepidostrobophyllum*, *Paracalamites*, *Pecopteris*, *Gigantopteris*, *Cordaites* and *Sphenobaiera* (Li et al. 2019).

4.4 Characters of Representative Taxonomic Groups of the Cathaysian Flora in South China

Why does the Late Permian floras belong to late Cathaysia one? It is that there are some special groups in present paleofloral assemblage which different from the other paleofloras. They are:

4.4.1 Lobatannularia *Kawasaki, 1927*

One of the most prosperous and typical sphenopsids of the middle and late Cathaysian floras is represented only by one genus *Lobatannularia*, but nonetheless numerous in species. It first appeared in the Shansi Formation, reaching its evolutionary peak in early Late Permian, with only a single form (*L. nampoensis*) survived in the Early Triassic of Korea. This group seems to have relatively strong adaptability for various ecological environments and a very widely geographical distribution; in addition to the widespread occurrence in China, they extend southward to Kashmir, westward to Anatolia of Turkey and in Asia Minor, northward to Heilongjiang Province of Northeast China, etc., and can be traced to almost anywhere of the Permian Cathaysian-Gondwana or the Permian Cathaysian-Angara mixed floras (Oman in Arabia, northwestern China).

Before the name of *Lobatannularia* was proposed by S. Kawasaki in 1927, similar specimensfrom Eastern Asia had been reported under such names as, Genera *Annularia* Sternberg (e.g. *Annularia maxima* Schenk 1883, p. 231, pl. 31, Fig. 3-6), *Annulariopsis* Zeiller (e.g. *Annulariopsis inopinata* Yabe 1922, p. 2, pl. 4, Figs. 2-3) and *Schizoneura* Shimper (e.g. *Shizonerura heianensis* Kodaira 1924, p. 163, pl. 23,

Figs. 1-6) and so on. Kawasaki in detail discussed the relationships of the above mentioned three genera, considered genus *Lobatannularia* as the transitional plant between *Annularia* Sterberg and *Schizoneura* Shimper. The characters of this genus are distinct (Fig. 4.1), and Genus *Lobatannularia* was established.

According to the statistics, this genus has 21 species that official published until now. There are 4 known species (namely *L. multifolia*, *L. cathaysiana*, *L. ensifolia* and *Lobatannularia heianensis*) and some incertae species, which distributed in over 8 horizons in thesis researching area, *L. multifolia* is the main elements. They mainly appeared in Late Permian Xuanwei and Lungtan formations, few relic species entered into Earliest Triassic.

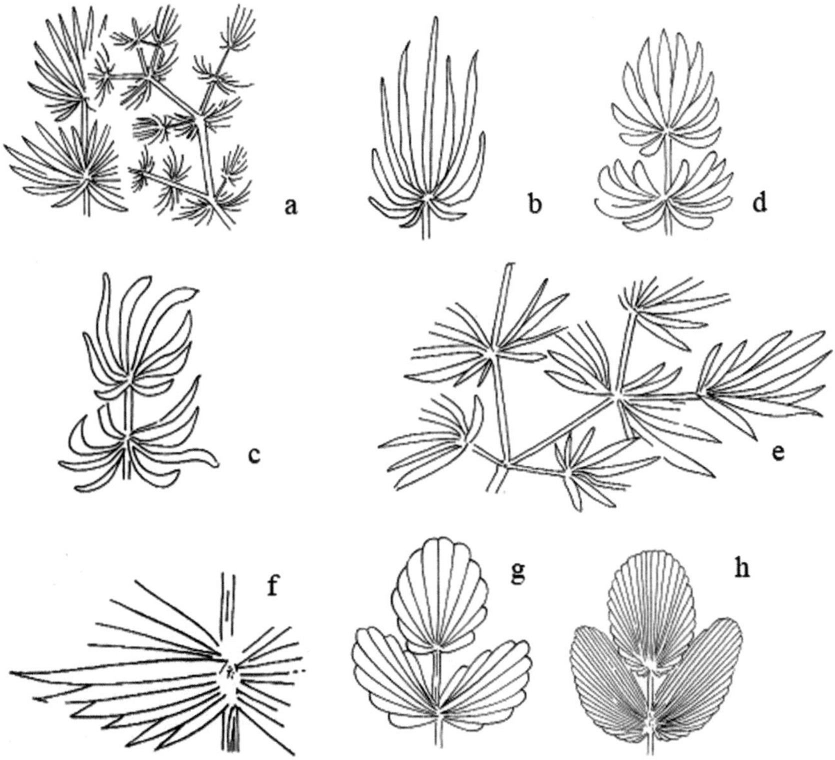

Fig. 4.1 Schematic drawing showing diversity of Genus *Lobatannularia*. **a** *L. sinensis;* **b** *L. inequifolia;* **c** *L. lingulata;* **d** *L. sinensis* var.*curvifolia;* **e** *L. ensifolia;* **f** *L. nampoensis;* **g** *L. heianensis;* **h** *L. multifolia* (according to Kon'no and Asama, 1950)

4.4.2 **Fascipteris** *Gu Et Zhi 1974*

A particularly diagnostic group of the early late Cathaysian flora, represented by a single genus *Fascipteris*, includes 6–7 species. One of the species (*F. densata*) belongs to Filices owing to its fertile pinnae being covered with *Ptychacarpus synangia*. This taxon is widespread in North China and South China, extends as far as to west New Guinea of Southeast Asia. Nearly all the species are confined to early Late Permian, except one form *F. hallei* which is recorded from the Early Permian in the Korea Peninsula.

Fascipteris was established by Gu and Zhi in 1974 (see [Paleozoic plants of China], p.99). Its holotype is *Fascipteris hallei* (Kawasaki) Gu et Zhi. Four species (*F. sinensis*, *F. stena*, *F. hallei* and *F. densata*) of this genus are occurred in Late Permian Xuanwei and Lungtan formations.

4.4.3 Gigantopterids

There have been 100 s years since the *Gigantopteris nicotianaefolia* firstly was established by Schenk in 1883. This species formally cited by Potonie in 1902. Subsequently, the Gigantopterides was continuously discovered in South, North and Northeastern China. In addition, this group is reported abroad, such as Korea (Kodaira et Kon'no, 1932); Japan (Asama 1967, 1976); Thailand (Asama et al. 1968), Malaysia (Kon'no et al. 1971), Sumatra (Jongmans et Gothan 1935) in Southeastern Asia; Turkey (Wagner 1962); Oman in Arabia (Broutin et al. 1995); North America (White 1912; Mamay 1986; Mamay et al. 1986, 1988).

It is the roost important characteristic and representative group of Cathaysian flora. With the exception of *Gigantonoclea fukienensis* with its male and female fructifications named *Gigantomomia* and *Gigantotheca* respectively, nearly all of this group is described as form genera or species based chiefly on the difference of their leaf shape, type of netted venation and developing degree of pinnae. They firstly appeared in late Early Permian in a large quantity, reaching the climax in early Late Permian and persisting to the end of Permian or locally to the beginning of Triassic. The gigantopterids are widely distributed in East Asia; some, however, appeared with different frequency and in wider or restricted areas. For example, *Gigantopteris* existed chiefly in South China, rarely found in North China; many species discovered in central Henan have not yet been recorded from other regions; the two genera known from the Djambi flora of Sumatra occur also in a restricted area. There are about 10 genera and nearly 40 species of gigantopterids described in East Asia. The relationships among the forms of the group are still obscure, although some authors have proposed that the genus *Gigantonoclea*, characterized by the simple netted venation, might have originated from a mariopterid plant in the transition from Carboniferous to Permian and evolved through plants of *Emplectopteris* type.

Due to the similar venation between *Gigantonoclea* and *Emplectopteris*, the latter genus being identified to have a *Callipteridium* framework as that of *Mariopteris*.

Some gigantopterids found in the Early Permian of southwestern America (Texas, Oklahoma and New Mexico) are considered likely to be closely related with those of East Asia. Recent studies, however, provide the evidence that such a close relation seems to be much exaggerated. In the new discoveries made by Mamay (1988, 1989, etc.) from southwestern America, there are approximately 6 genera, all limited to the Early Permian, which can be ascribed to the gigantopterids, some of them identical with those of East Asia. Yet the typical genus *Gigantopteris* has not hitherto been found in North America. The gigantopterid relationship between North America and East Asia remains unclear. Most recently, some authors (Li Hongqi et al. 1994) have demonstrated that based on a study of foliage anatomy of *Gigantonoclea guizhouensis* and a comparison with the American genera *Delnortea* and *Gigantopteridium*, there is no evidence for the affinity of gigantopterids between North America and East Asia. The southwestern American region yielding the Permian gigantopterids is recently named "West American Subprovinee" (Cleal et al., 1991).

Both *Gigantopteris* and *Gigantonoclea* of Gigantopterids are the most typical elements in the present Floral assemblage, with the abundant specimens and widespread distributions, which indicated that this group had developed a new phase. Genus *Gigantonoclea* is abundant in Carboniferous and Permian in North China. Really the specimens of *Gigantopteris* occasionally appeared in Late Permian Upper Shihhotse Formation in the south margin of North China Platform. The specimens of both *Gigantonoclea guizhouensis* and *Gigantopteris dictyophylloides* largely occurredin western Guizhou and eastern Yunnan not only were finely preserved, but also their complicated net veins were very clear. This group of gigantopterids consists of 9 known species, which vertically distributed in over 16 horizons, and a few appeared in Earliest Triassic Kayitou Formation through Permian–Triassic boundary.

4.4.4 *Tingia* Halle 1925

Another very significant group of the Cathaysian flora, *Tingia* established by Halle in 1927 according to the samples from Shanxi Province, was most prosperous in late Early Permian in both North China and South China. Holotype species is *Tingia carbonica* (Schenk) Halle (1925). There are two types of foliar shoot fossils; one is represented by the type species of *Tingia carbonica*, generally held as typical Tingialeans, of which the shoot axis has leaves arranged in four rows, two larger on the upper and two smaller on the lower side of the axis; another type is represented generally by *Tingia hamaguchii*, in which whether the two rows of smaller leaves as seen in *Tingia carbonica* are also present remains a subject of much dispute; they are commonly found in close association with a peculiar strobiloid fossil named *Tingiostachya*. This type of strobiloid fossil has never been found in accociation with the first type of Tingiaieans. Accordingly, the second type of Tingialeans (*T.*

hamaguchii) might represent a new taxon. The structure of *Tingiostachya* has been intensively studied recently (Gao Zhifeng et al. 1987) and proved to be quite different from that formerly described by Japanese authors. Typical Tingialeans have never been seen in any other phytogeoprovinces. Recently this group (*Tingia* sp., cf. *Tingia hamaguchi* and fertile cones *Tingiostachya* sp.) also has been recorded from Late Permian Gharif Formation in Oman (Broutin et al. 1995; Berthelin et al. 2003).

In North China, Genus *Tingia* is the most abundant in Permian strata. By contrarily, the specimens of *Tingia* are sporadically appeared in the thesis researching area, including fragments of *T. multinervis* and *T. crassinervis.*

4.4.5 Taeniopterides

One of the common taxa of Mesozoic plants, including numerous, jumbled forms, is tentatively ascribed to the Cycadophytes. There are more than twenty species of *Taeniopteris* (nearly all endemic forms), appearing first in the upper part of the Taiyuan Formation and most prosperous in early Late Permian. A few Cathaysian forms of *Taeniopteris* are also found in the Shihchienfeng flora of North China and in the Permian Angora flora of Turpan, Xinjiang. This indicates that they might be more adaptable for various environments. In other countries, there are also *Taeniopteris* in the Permian, but the species number and abundance are far less as compared with those in the Cathaysian flora.

4.4.6 Oriental Lepidophytes

The appearance of this type of Lycopsida in the Cathaysian flora can be regarded as a special character for the flora (Li Xingxue 1980). It is noteworthy that in the early Early Permian Taiyuan Formation (North China) and Late Permian Xuanwei Formation (South China), there are still abundant lepidophytic plants), in contrast, in approximately coeval strata in the Euramerican Phytogeoprovince, the lepidophytes are almost completely extinct. In addition, the larger leaf-scars, most commonly rhomboid-quadrate in shape, and the absence of lower parichnos are the characteristics for the oriental lepidophytes, distinguished from those of Euramerican forms. Furthermore, some of the oriental species, such as *Lepidodendronoculus-felis*, grew well since Late Carboniferous through Early Permian and persisted prosperously to Late Permian in South China. This unusual lepidophytic record seems to have never occurred in any other Permian phytogeoprovinces (Li et al. 1980). It is noticed that a single leaf cushion of *Lepidodendron* occurred in Earliest Triassic Kayitou Formation in associated with *Annalepis* and marine bivalves and so on.

4.5 Correlation with Domestic and Foreign Coeval Flora

4.5.1 Correlation of the Late LatePermian Floras in China

Combined with the geostructural features of China and differences in composition of floras in different places, the geographical distribution of the late Late Permian flora in China may be divided into 6 Provinces (Fig. 4.2).

1. **Comparison with late Late Permian Cathaysian, Euramerian and Angara mixed flora in North China Province (Fig. 4.2, II)**

Fossil plants from the late Late Permian Shihchienfeng Formation in North China were little known in the past because it is mainly compound of red beds. The assemblage was once tentatively named by Li Xingxue (1963) the *Rhipidopsis lobata-Ullmannia* aff. *bronnii* Assemblage on the basis of materials from North China. Before 1970s, few fossil plants had been studied from the formation of North China, and only one kind of wood fossil *Dadoxylon leei* recorded by H. C. Sze (1952) from Yaoxian County, Shaanxi Province, possibly came from the formation. The so-called *Yuania striata* and a few other plants described by H. C. Sze (1953) from the Shihchienfeng Formation of Linyou County, northern Shaanxi may be from the Upper Shihhotse Formation. Wang Ziqiang (1983, 1986) discovered quite rich fossil plants from the Sunjiagou Formation (namely, the Shihchienfeng Formation in a

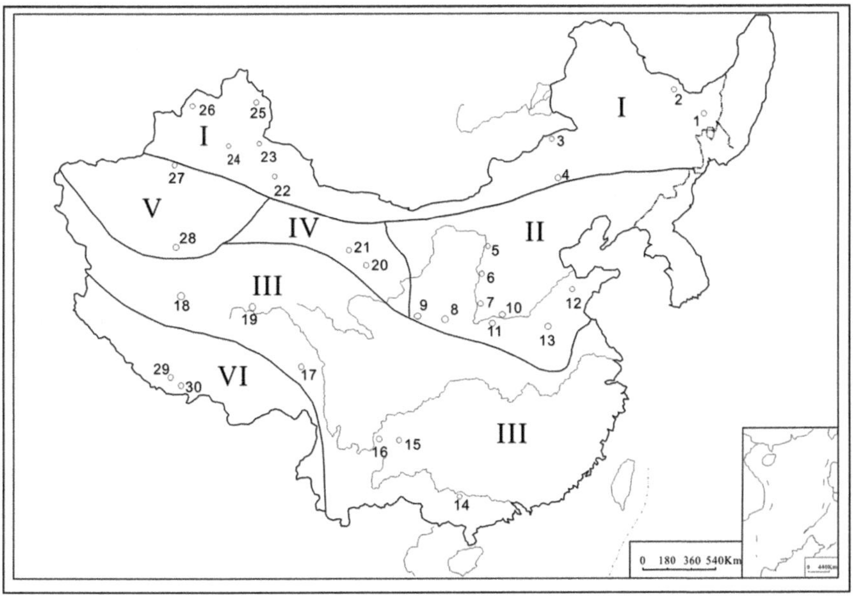

Fig. 4.2 Phytogeoprovinces and important localities of the late Late Permian floras in China (According to Li Xingxue 1995)

narrow sense) in Shanxi, Shaanxi, Yiyang and Jiyuan of Henan (Fig. 4.2, Nos.5–11). The assemblage named by Wang Ziqiang and others is used to summarize the late Late Permian fossil plants in the North China Province in this paper. Based on data by Wang Ziqiang et al. (1983, 1986, 1989), the fossil plants from the Shihchienfeng Formation in North China include *Algites junduensis*, *Sphenophyllum* spp., *Calamites* sp., *Phyllotheca* sp., *Asterophylliteseqnisetiformis*, *Yuania magnifolia*, *Discinites*? *fimbriata*, D. sunjiagouensis, *Sphenopteris* sp., *Pecopteris* cf. *arcuata*, *Pecopteris* spp., *Cladophlebis* sp., *Callipteris lobulota*, *C. martinsii*, *C. papillosa*, *Tatarina* cf. *sinuosa*, *T.*? *mirabilis*, *Phylladoderma* (*Aeguistomia*) cf. *aequalis*, *Gaussia*? *shanxiensis*, *Sphenobaiera micronervis*, *Walchia*? sp., *Pseudovoltzialiebeana*, *Ullmannia bronnii*, *Ull. frumentaria*, *Quadrocladus solmsii*, *Q. heterodermus*, *Quadrocladus* sp., *Piryospermum dafengshanense*, *Esterella* sp., *Lesleya anastomosis*, *Scytopyllumsunjiagouense*, *Peltaspermum dafungshanense*, *Taeniopteris liulinensis*, *T. taiyuanensis*, *T. longifolia*, *T. nystroemii*, *T.* cf. *serrulata*, *Norinia* sp., *Cardiocarpus* cf. *triangularis*, *Carpolithus* sp., *Squarmacarpus cuneiformus*.

I. late Late Permian Angara flora in Junggar-Hingan Provinces;
II. late Late Permian Cathaysian, Euramerian and Angara mixed flora in North China Province;
III. late Late Permian Cathaysian flora in South China Province;
IV. late Late Permian Cathaysian and Angara mixed flora in Northwest China Province;
V. late Permian Angara flora in Tarim Province;
VI. late Late Permian Gondwana flora in southern Tibet and western Yunnan

1. Baoqing, Heilongjiang; 2. Yichun, Heilongjiang; 3. Jarud Qi, Heilongjiang; 4. Linxi, Inner Mongolia; 5. Hequ, Shaaxi; 6. Liulin, Shanxi; 7. Xiangning, Shanxi; 8. YaoXian, Shaanxi; 9. Linyou, Shaanxi; 10. Jiyuan, Henant; 11. Yiyang, Henan; 12. Zibo, Shandong; 13. Yongcheng, Henan; 14. Laibin, Guangxi; 15. Anshun, Guizhou; 16. Huize, Yunnan; 17. Toba, Xizang; 18. Shuanghu, Xizang; 19. Wuli, Qinghai; 20. Qilian, Gansu; 21. Sunan, Gansu; 22. Turpan, Xingjiang; 23. Karamay, Xinjiang; 24. Urumq, Xinjiang; 25. Fuyun, Xinjiang; 26. Tuoli, Xinjiang; 27. Kuche, Xinjiang; 28. Qiemo, Xinjiang; 29. Tingri, Xizang; 30. Tinggye, Xizang.

The above list shows that the Shihchienfeng Formation contains a typical Late Permian flora, of which most genera and species were confined to the Permian except species of a few genera appearing in the Carboniferous and a few could extend to the Mesozoic. The Shihchienfeng flora in the North China Province includes algae, Sphenopsida, Filicopsida, Pteridospermopsida, Coniferopsida and Ginkgopsida and a few components of Lycopsida. Filicopsida which were quite rich in the Carboniferous and Permian and even in the Mesozoic is only represented by a few fossil remains. Compared with other earlier Permian floras, the Shihchienfeng flora is obviously devoid of coal forming plants, which can be regarded as another character of the flora and indicates its peculiar ecological background. In respect to the generic and specific composition, except a few species of *Sphenopteris*, almost all important forms of the Zechstein flora in West Europe have been discovered in the assemblage.

It is also noticeable that a few Peltaspermae commonly seen in the early Mesophyte are discovered in the assemblage, which implies that ancient Pteridosperms were already extinct prior to the late Late Permian.

Based on the above mention, the features of the Shihchienfeng flora are: (1) the genera and species representing the Late Permian upland or arid surroundings in West Europe are not dominant but rather extensively distributed; (2) a few genera and species of the Upper Shihhotse Formation flora in the North China Province still continue to exist, which makes the flora with an aspect of the Cathaysian flora; (3) the late stage representatives of the Cathaysian Flora such as *Gigantonoclea, Lobatannularia* and *Fascipteris* disappeared; (4) Compared with other earlier Permian floras, the Shihchienfeng flora is obviously devoid of coal forming plants, which can be regarded as another character of the flora and indicates its peculiar ecological background.

The *Gigantonoclea guizhouensis—Annaularia pingloensis* Assemblage of the upper part of Xuanwei Formation in western Guizhou and eastern Yunnan is obviously different from the Shihctfienfeng flora in the North China Province in general features. Of the Shihctfienfeng flora, the xerophytes of Zechstein type and Peltaspermae with a Mesophytic aspect all well developed besides that a few common forms of the Shihhotse Group are occasionally discovered, which forms a unique feature of the flora in the North China Province. However, of the present assemblage of the upper part of Xuanwei Formation, the representatives of the Cathaysian Flora such as *Gigantopterids, Lobatannularia, Fascipteris* and Lepidodendrales are abundant, while a few of Gigantopterids survived into the Earliest Triassic. They formed the upperest strata of coal –bearing plants. The existence of some conifers such as *Ullmannia bronnii* in both the western Guizhou and eastern Yunnan [Zhao et al. (1980) reported that this species was discovered in the upper part of Xuanwei Formation.] and North China Provinces indicates that the two provinces can be correlated to a certain degree.

2. Comparison with late Late Permian Cathaysia flora in northern Xizang (**Fig. 4.2, III**)

The plant fossils were collected from the Upper Rejuechaka Formation in the Shuanghu district of Northern Xizang (Fig. 4.2, No.18) and came from the same horizon and locality. They were nearly all preserved as impressions on grey to dull, brownish black, fine-grained sandstones and claystones as well as a workable coal-seam intercalated in the lower part of the Upper Rej- uechaka Formation.

The flora or plant-assemblage from this locality consists of 17 species: Selaginellites tibeticus, Sphenophyllum aff. speciosum, Annularia pingloensis, Lobatannularia sp., Ptychocarpus (Pecopteris) tibeticus, Rajahia (Pecopteris) calceiformis, Pecopteris shuanghuensis, Cladophlebis cf. permica, Gigantonoclea guizhouensis, G. meridionais, Gigantonoclea sp., Rhizomopsis gemmifera, Alethopteris sinensis, Neuropteridium nervosum, Compsopteris contracta var. punctinervis, ? Pterophyllum sp.

At the first glance the flora has rather the appearance of that commonly known from the Lungtan Formation or the so-called *Gigantopteris* coal series of South China, but it is distinguished by the presence of many new forms. The flora is definitely of a slightly younger age. It seems that this flora is the latest stage of the Cathaysia flora in Northern Xizang (Li et al. 1982). The Shuanghu flora might be considered to be the uppermost plant-assemblage of the Cathaysia flora and probably represents really a latest phase of the Late Permian flora in Eastern Asia (Li et al. 1982).

By the floral composition, the small flora established by Li et al. (1982) suggests a more close relationship with the Late Permian flora or the Lungtan-Changhsingian flora of South China rather than that of the Upper Shihhotse Formation of North China. It is also to be noted that none of the described plants give any indication of affinity of the Angara flora, nor do they show a close relationship with the *Glossopteris* flora.

The present assemblage are compared with the paleoflora assemblage from the northern Xizang, both have some similarities, for instance, the floral composition, existing the coal-bearing plants, which explained they belong to the same phytogeographical province.

3. Comparison with late Late Permian Cathaysian and Angara mixed flora in Northwest China Province (**Fig. 4.2, IV**)

The plant assemblage of this floristic region is named *Iniopteris sibirica-Lobatannularia lingulata* assemblage. The strata bearing the assemblage are represented by the Sunan Formation in the North Qilian Mountain (Fig. 4.2, No.21). The Late Permian Angara flora discovered by the Joint Chinese Swedish Scientific Expedition was actually found from this stratigraphic unit. The common and important fossil plants of the assemblage are *Paracalamites tenuicostatus, Annularia shirakii, A. gracilescens, A. longissima, Lobatannularia lingulata, Pecopteris lativenosa, P. tenuicostata, Rajahia minor, Cladophlebis nystroemii, Callipteris bexellii, Protoblechnum wongii, Compsopteris wongii, Pursongia lanceolata, Zamiopteris glossopteroides, Noeggerathiopsis derzavinii, Peltaspermum buevichae, P. multicostatum, Rhipidopsis panii, Psygmophyllum multipartitum.*

In the assemblage, the features are: (1) some important and common elements of the Cathaysian and Angara Floras were compounded, they consititute the late Late Permian Cathaysian-Subangara mixed flora; (2) Lycopsida might have disappeared, while Filices and Pteridospermopsida, especially *Callipteris* with large-type pinnules and other fossil leaves such as *Zamiopteris* and *Pursongia* possibly belonging to Peltaspermae were rather developed; (3) there are abundant gymnosperms, especially Cordaitopsida and Ginkgopsida being very common and associated with Coniferopsida and rather advanced Peltaspermae, which indicates that the assemblage is dominated by Gymnospermae with a strong Mesophytic aspect and is unique compared with other Permian floras in China.

The flora in the Northwest China Province contains both Angara forms *Iniopterts, Comia, Zamiopteris, Nephropsis* and *Callipteris* with various kinds of large pinnules and the important Cathaysian representative *Lobatannularia lingulata.* We compare the assemblage of the upper part of Xuanwei Formation with the *Iniopteris sibirica-Lobatannularia lingulata* assemblage of Sunnan Formation, other coexisting elements are little besides *Annularia shirakii, Lobatannularia lingulata, Compsopteris wongii,* moreoverthere are a great amount of *Lepidodendron, Gigantonoclea guizhouensis, Gigantopteris dictyophylloides, Lobatannularia multifolia, Compsopteris contracta* in the assemblage of the upper part of Xuanwei Formation. This floral assemblage of this age in western Guizhou and eastern Yunnan is obviously different from that of the Northwest China Province, so they are hard to be directly correlated. But the coexistence of some Cathaysian elements and Subangaran Peltaspermae of the late Late Permian flora both in the Northwest China and North China Provinces make it possible for floral correlation.

4. Comparison with the late Late Permian Angara flora in Junggar-Hingan Provinces (**Fig. 4.2, I**)

The late Late Permian flora in the Junggar Hinggan Province (Fig. 4.2, No.22, etc.) still retains the phase of the rather typical Subangara flora and is hard to be compared with the coeval floras in the South China and North China Provinces. But it should be pointed out that there was once a record of *Lobatannularia multifolia* in the Upper Permian in the Da Hinggan Mountains and Xiao Hinggan Mountains, which indicates that there occurred mixture of some forms of both floras.

5. Comparison with the late Late Permian Angara flora in Tarim Province (**Fig. 4.2, V**)

The Upper Permian has not been reliably discovered in the Tarim Basin Province (Fig. 4.2, Nos.27, 28) and is only outcropped of continental facies in the southern Tianshan Mountains, e, g, Kuqa closely next to the margin of north Tarim yielding Subangara plants. The Late Permian flora of the province is tentatively named by "*Callipteris*"*zeiller-Comia kuqaensis* assemblage. Its representative strata is the Biyoulebaoguzi Group which yields fossil plants including *Paracalamites bilinearis, Schizomenura manchuriensis, Koretrophyllites deliquescens, K. linearis, Callipteris acutifolia, C. heilongjiangensis, C. tianshanensis, C. altaica, C. zeiller, Prynadaeopteris anthriscifolia, Compsopteris* sp., *Zamiopteris glossopterides, Pursongia zamiopteroides, Comia kuqaensis, C. tarimuensis, Iniopteris sibirica, Noeggerathiopsis tieliensis, N. derzavinii, N. latifolia, N. subangusta, Pterophyllum eratum, Yavorskyia mungatica, Rhipidopsis* sp..

Due to the dominance of Pteridospermopsida and absence of components of Lycopsida, the assemblage is quite peculiar. Of Sphenopsida, *Sphenophyllum, Calamites* and *Annularia* extremely common in Carboniferous and Permian are not present and are replaced by such genera as *Paracalamites, Schizoneura* and *Koretrophyllites* commonly present in the Angara Realm. Of Pteridospermopsida, ancient pteridosperms are extinct and the fossil leaves of Peltasparmae with Mesophytic aspect are quite common, including *Callipteris, Zamiopteris* and *Pursongia*, of which *Callipteris* with large pinnules most development, composed of 5 species. Of Gymnospermae, Cordaitopsida is dominated by Noeggerathiopsis. Cordaitales with this kind of leaf form is not present in the Cathaysian flora in North and South China and is one of the common elements of the Angara Realm. Cycadopsida includes 2 species, of which *Yavorskyia mungatica* is also an important member of the Angara flora.

On the northern margin of the Tarim Basin, the late Late Permian flora is nearly the same as that in the Junggar Hinggan Province in compositions, the two floras can be directly correlated. The reliable Cathaysian members have not yet been discovered in the late Late Permian flora in the Tarim Basin, which indicates that the flora is different from that of the same age in its adjacent Northwest China Province to a certain extent. It is more difficult correlations between the late Late Permian flora of Tarim Province and the floral assemblage of the upper part of Xuanwei Formation.

4.5.2 Correlation with Late Permian Euramerican Flora

(1) AlthoughtheEuramerican Flora in Permian continued the late Carboniferous flora, it showed the great change in macroflora compositions: the dry and upland flora substituting for the forming coal plants in warm environment. In South China, the climate was warm and humid during the whole Permian. The features of flora in study area, so much as South China, are very different from the Euramercian Flora in this stage, the important differences between the two realms are as follow:

(2) Lepidodendrales are exceeding declining in Euramercian Realm but abundant in South China Cathaysian Flora, which composed mainly oriental type of *Lepidodendron.*

(3) Of Sphenopsida, leaves of a whorl unequal in size and leaf mosaic are quite remarkable in China but not in the Euramerican Realm. In addition, *Lobatannularia* with strong endemic color quite thriving in North China, Northwest China and South China has not been reliably discovered in the Eurameriean Realm.

(4) Of Noeggerathiopsida, *Tingia*, special in China, is also not present in the Euramerican Realm, while the common genus Noeggerathia in the Euramerican Realm has not yet been reliably reported in South China.

(5) Of Filices, *Pecopteris* rather common in the Euramerican Realm also quite developed in South China and there are many common species between the two regions. But plants of this kind in Euramerica are still different in the type of fertile organs from those in China to a certain degree. *Asterotheca, Acitheca, Ptychocarpus, Senftenbergia* and *Nemejcotheca* were common in Europe and America while *Daneites* and *Rajahaia* developed in China, especially in the South China Province.

(6) *Emplectopteris, Emplectopteridium, Gigantopteris, Gigamonoclea and Cathaysiopteris*, being of great significance in phytogeography in the Permian flora of China were not present in West Europe and only gigantopterids similar to those mentioned above are discovered in western North America.

(7) *Taeniopteris* is quite developed in the Early Permian flora of China but is quite rare in Europe and America.

From the above-mentioned facts, it can be seen that the present paleoflora assemblage from Latest Permian Xuanwei Formation exhibits more distinct regional characterstics. On the one hand, it possess of the typical characters of the Cathaysian flora, Gigantopterids and *Lobatannularia* are the main elements of this assemblage. On the other hand, it shows its own feature, the wet and swamp plants are flourish, forming the latest coal-bearing plant assemblage in South China. All of these show that humid and warm climate continued to exist in Latest Permian, the aspect of the paleoflora keeps with the characters of the Cathaysian flora. But the upland and dry Euramerican plants invaded into North China, standing for the beginning of dry climate during the Late Permian.

References

Asama K, Iwai J, Veeraburas M, Hongnusonthi A (1968) Permian plants from Loei, Thailand. J Geol Paleont Southeast Asia 4:82–99

Asama K (1967) Permian plants from Maiya, Japan. I. Cathaysiopteris and Psygmophyllum. Bull Nat Sci Mus Tokyo 10(2):139–153

Berthelin M, Broutin J, Kerp H, Crasquin-Soleau S, Platel JP, Roger J (2003) The Oman Gharif mixed paleoflora: a useful tool for testing Permian Pangea reconstructions. Palaeogeogr Palaeoclimatol Palaeoecol 196(1):85–98

Cleal CJ, Thomas A (1991) Carboniferous and Permian palaeogeography. In: Cleal CJ (ed) Plant fossils in geological investigation, the Palaeozoic. Ellis Horwovod Ltd., New York, pp 151–181

Gao ZF, Thomas BA (1987) A re-evaluation of the *Tingia* and *Tingiostachya* from the Permian of Taiyuan, China. Palaeontology 30(4):815–825

Halle TG (1925) *Tingia*, a new genus of fossil plant from the Permian of China (preliminary note). Bull Geol Soc China 7:1–12

He XL, Liang DS, Sheng SZ (1996) The Permian Floras from Jiangxi, China. China University of Minings Press, Xuzhou, pp 1–201

Jin YG, Wang Y, Wang W, Shang QH, Cao CQ, Erwin DH (2000) Pattern of marine extinction near the Permian-Triassic boundary in South China. Science 289:432–436

Jongmans WJ, Goth W (1935) Die Palaeobotanischen Ergebnisse der Djambi - Expedition 1925. Jahrbook von het Mijnwenzen in Nederlandsch-Indie, Verhandelingen Tweede Gedelte, Batavia, pp 71–201

Kodaira R (1924) Note on a new species of *Sohizonewra* from Chosen (Korea). Jap Jowr Geol Geogr 3(3–4):163–165

Kon'no E, Asama K (1950) On the genus *Lobatannularia* Kawasaki 1927 from Permian Beds in South Manchuria and Shansi, China. Short Pap IGPC, 1:18–31

Li XX (1963) Fossil plants of the Yuehmenkou Series, North China. Palaeont Sinica, Whole No 148. N S A 6:1–185 (in Chinese and English)

Li XX (ed) (1995) Fossil floras in China through the geological ages (English edition). Guangdong Science and Technology Press, Guangzhou, pp 1–695

Li XX, Yao ZQ (1980) Permian coal-bearing formations in South China. J Stratigr 4:241–255 (In Chinese)

Li XX, Yao ZQ (1983) Current studies of Gigantopterids. Palaeontologia Cathayana 1:319–326

Li HQ, Tian BL, Taylor EL, Taylor YN (1994) Foliar anatomy of *Gigantonoclea guizhouensis* (Gigantopteridales) from the upper Permian of Guizhou province, China. Am J Botany 81(6):678–689

Li H, Yu JX, McElwain JC, Yiotis C, Chen ZQ (2019) Reconstruction of atmospheric CO_2 concentration during the late Changhsingian based on fossil conifers from the Dalong Formation in South China. Palaeogeogr Palaeoclimatol Palaeoecol 519:37–48

Li XX, Yao ZQ, Deng LH (1982) An early late permian flora from Toba, Qamdo district, eastern Xizang. In: Comprehensive Scientific Expedition to the Qinghai-Xizang Plateau, Academia Sinica, ed. Palaeontology of Xizang, No. 5. 17–44. Beijing: Science Press. (in Chinese with English abstract)

Li XX (1980) New advances of studies in the Cathaysia flora. In. Scientific paper on geology for international exchange. Prepared for the 26th International Geological Congress (4), Stratigraphy and Palaeontology. Beijing: Geological Publishing House, pp 73–81 (in Chinese with English summary)

Mamay SH (1973) *Archaeocycas* and *Phasmatocycas* - new genera of Permian cycads. US Geol Surv J Res 1:687–689

Mamay SH (1986) New species of Gigantopteridaceae from the lower Permian of Texas. Phytologia 61(5):311–315

Mamay SH (1988) *Gigantonoclea* in the lower Permian of Texas. Phytologia 64(5):330–332

Mamay SH (1989) *Evolsonia*, A new genus of Gigantopteridaceae from the lower Permian vale formation, North-central Texas. Am J Botany 76(9):1299–1311

Mamay SH, Miller JM, Darid MR, William EJ (1988) Foliar morphology and anatomy of gigantopterid plant *Delnortea abbottiae*, from the lower Permian of West Texas. Am J Bot 75(9):1409–1433

Mamay SH, SH M (1976) Paleozoic origin of the cycads. US Geol Survey Profess Paper 934:1–48

Meyen SV (1987) Fundamentals of palaeobotany. Chapman and Hall, London

Potonie H (1902) Ueber die fossilen Filicales im Allgemeinen etc. Emgler & Frantl: Die naturlichen Pflanzenfamilien. Teil 1, Abteil 4, 513

Schenk A (1883) Pflanzen aus der Steinkohlen Formation. In: Richthofen F von. China, 4. Berlin, pp 238–239

Song QQ, Feng JP, Yu JX, Huang QS (2013) Study on palaeophytoecology of the Dalong Formation (Late Permian) in south Guizhou. Guizhou Geol 30(4):255–261 (In Chinese with English abstract)

Sze HC (1952) A petrified wood from northern Shensi, with special reference on the age of the Shihchienfeng Series in North China. Sci Record 5(1–4):171–182

Sze HC (1953a) Atlas of palaeozoic plants from China. Beijing. Chinese Academy of Sciences, pp 1–148 (in Chinese)

Taylor TN, Taylor EL (1994) The biology and evolution of fossil plants. Prentice hall, New Jersey, p 981

Wagner RH (1962) On a mixed Cathaysia and Gondwana flora from SE Anatolia (Turkey). In: C R 4th Congessr International Stratigraphy Carbon, Heerlen, 1958, 3:745–752

Wang ZQ (1983) New materials of fossil plants from Shiqianfeng Group in N. China. Bull Geol Soc Tianjing 1(2):27–80 (in Chinese)

Wang ZQ, Wang LX (1986) Late permian fossil plants from the lower part of the Shiqianfeng (Shihchienfeng) Group in North China. Bulletin Tianjin Institute Geology and Mineral Resource 15:1–80 (in Chinese with English abstract)

White D., 1912. The character of the fossil plant *Gigantopteris* Schenk and its occurrence in North America. Proc US Nat Mus, 41

Yabe H (1922) Note on some mesozoic plants from Japan, Korea and China, in the collection of the Institute of Geology and paleontology of the Tohoku Imperial University. Sci Rep Tohoku Imp Univer (ser. 2), 7(25):1–28

Yang GX (2006) Botanical characters of the Yuzhou flora in western Henan. In: Yang et al. (eds) The Permian Cathaysian flora in western Henan Province, China–Yuzhou Flora. Beijing: Geological Publishing House, pp 26–50

Yao ZQ (1978) On the age of "Gigantopteris coal series" and Gigantopteris-flora in South China. Acta Palaeontol Sin 17:81–89 (in Chinese with English abstract)

Yao ZQ, Liu LJ, Rothwell WG et al (2000) Szecladia new genus, a Late Permian conifer with multiveined leaves from South China. J Paleontol 74:524–531
Zhang HD (2000) New systems of spermatophyte. Chinese Bull Botany 17:152–160
Zhang HD (2004) Subdivision v. preanthophytinae.In Zhang et al. eds., Systems of seed plants. Beijing: Science Press, pp 66–71 (in Chinese with English abstract)
Zhao XG, Mo ZG, Zhang SZ, Yao ZQ (1980) Late Permian fossil floras of western Guizhou and eastern Yunnan. In: Nanjing Institute of Geology and Palaeontology, Academica Sinica, (Ed.), Late Permian coal-bearing strata and palaeontological fauna in western Guizhou and eastern Yunnan. Science Press, Beijing, pp 70–122 (in Chinese)

Chapter 5
Earliest Triassic (Induan, Kayitou Formation) Floral Assemblage in South China

Jianxin Yu, Jean Broutin, Xiao Shi, and Zhen Xu

The Early-Middle Triassic flora was some widely distributed both in the northern and southern China. Phytogeographically, the flora of China is attributed to the American–European Region during the time.

The earlier reports on the Early-Middle Triassic plants from China were made in the late 1970s. Wang et al. (1978) originatively reported *Pleuromeia* from the Heshanggou Formation of Shanxi Province, North China. Zhou et al. (1979) first discovered the Early Triassic flora from the Jiuqujiang of Hainan Island, South China. Since 1980s, there are several new contributions made to the study of Early Triassic flora from North China, particularly from the Liujiagou and Heshanggou Formations (the middle and upper part of Shichienfeng Group) Wang et al. (1982), Wang (1989), Wang and Wang (1990a). These studies not only make clear, in a certain degree, the close affinity of the Chinese Early Triassic flora with the Bunter Sandstone flora of Europe, but also provide convincing evidence for a correct division on the Late Permian–Early Triassic Shichienfeng Group. Besides, the discovery of the Lowest Triassic bearing plants *Annularia shirakii* and *Gigantopteris dictyophylloides* (Yao et al. 1980), associated with the Early Triassic bivalves from the border area between Yunnan and Guizhou, is also most significant for understanding the Earliest Triassic

J. Yu (✉) · Z. Xu
School of Earth Sciences, China University of Geosciences, Wuhan, Hubei, P. R. China
e-mail: yujianxin@cug.edu.cn

Z. Xu
e-mail: janexu_cug@qq.com

J. Broutin
Faculty of Sciences and Engineering (CR2P, MNHN, CNRS), Sorbonne Université, Paris, France
e-mail: jean.broutin@upmc.fr

X. Shi
College of Earth Sciences, Jilin University, Changchun, Jilin, P. R. China
e-mail: xiaoshi@jlu.edu.cn

J. Yu et al. (eds.), *Plants and Palynomorphs around the Permian-Triassic Boundary of South China*, New Records of the Great Dying in South China,
https://doi.org/10.1007/978-981-19-1492-8_5

assemblage in Southwest China and for studying the transition strata spanning the Permian–Triassic.

In recent years, a lot of fossil plants from the Early Triassic Kayitou Formation in western Guizhou and eastern Yunnan are collected (Yu et al. 2010, 2015; Feng et al. 2020). The discovery of many new fossils is of great significance in studying the properties of the Early Triassic flora in South China and the plant succession at the turn of Permian and Triassic.

5.1 Composition and General Features of Floral Assemblage of Induan Kayitou Formation

Compared with the Xuanwei Formation, the Kayitou Formation is very clear in paleofloral assemblage character although it has fewer species variety with broken plant fossils with smaller leaves. However, its plant components are significantly different from those of the upper member of Xuanwei Formation.

This paleofloral assemblage is distributed in Kayitou Formation, which contains 18 species from 10 genera (Tables 4.1–4.5) (Yu et al. 2015). Of these, 6 species from 4 genera were discovered in terrestrial facies sections, while 16 species from 12 genera were found in terrestrial–marine transitional facies sections. Dispersed and in situ megaspores of *Lepacyclotes* (=*Annalepis*), and unidentified types (Fig. 5.1) were also extracted from the same strata. They are *Lepacyclotes* (=*Annalepis*) *zeilleri*, *Lepacyclotes* (=*Annalepis*) *brevicystis*, *Lepacyclotes* (=*Annalepis*) *latiloba*, *Lepacyclotes* (=*Annalepis*) *furongqiaoensis* and *Lepacyclotes* (=*Annalepis*) spp. in Isoetaceae genus *Lepacyclotes* (=*Annalepis*), *Lepidodendron* sp., *Stigmaria ficoides* and *Stigmaris* spp. in Lepidodendrales, *Paracalamites stenocostatus* in Arthrophytes, *Pecopteris* sp. and *Sphenopteris* sp. in Pteridophyta, aff. *Peltaspermum martinsii, P.*

Incertae type 1 Incertae type 2

Fig. 5.1 Incertae plant fossils from the Kayitou Formation in marine–terrestrial transitional sections

lobulatum spp. in Peltaspermaceae, *Sphenobaiera* sp. in Ginkgopsida, Gigantopteridales of *Giantonoclea* sp. and *Gigantopteris* sp., and a few unidentified fossil plants. From the view of the class level, the major components of this assemblage are Lycopsida *Lepacyclotes* (=*Annalepis*) and Pteridospermopsida *Peltaspermum*, respectively appearing in stability from layers 2–4 in the lower part of Kayitou Formation.

The essential characteristic of this assemblage is the abundance of the new types. Large number of sporophylls and megaspore in situ of the *Lepacyclotes* (=*Annalepis*) and suspicious root or stem fossils, appears at the bottom of the Kayitou Formation. All species of *Lepacyclotes* (=*Annalepis*) that had appeared in the Middle Triassic strata of the Lower Yangtze Platform are also found in western Guizhou–eastern Yunnan region. Also seen are single cushion of Lepidodendrales and *Stigmaria* yielded from the same bed of *Lepacyclotes* (=*Annalepis*). Upwards are discovered many broken sterile pinna and fertile organs of Peltaspermaceae as well as the fragments of the Gigantopterids. Thus, the paleoflora from the Kayitou Formation is named as *Lepacyclotes* (=*Annalepis*)–*Peltaspermum* and the relicts of gigantopterids assemblage.

This earliest Triassic floral assemblage is dominated overall by both *Lepacyclotes* (=*Annalepis*) and *Peltaspermum*, which occur in the lower part of the Kayitou Formation in South China (Fig. 5.2). Rare, small fragments of *Gigantopteris* and *Gigantoclea* were found in 2–3 horizons in the upper part of the terrestrial–marine transitional facies of the Kayitou Formation (Fig. 5.2).

5.2 Biostratigraphic and Paleogeographic Implications of the First Discovery of Induan *Lepacyclotes* (=*Annalepis*) in Terrestrial–Marine Transitional Facies

5.2.1 *History of Lepacyclotes (=Annalepis) Research*

Lepacyclotes (=*Annalepis*) is a rare proliferous organ of Triassic lycopods in the North Hemisphere and plays an important role in the Early and Middle Triassic. It is first discovered in the Middle-Late Triassic clay coal layer in eastern France. The term was given by Fliche (1910) based on its scattered sporophyll. However its classification position was not apparent and nor was its ligule scar when it was established.

Originally the genus *Lepacyclotes* (=*Annalepis*) was thought to have gymnospermous (*Araucarites*-like) affinities (Fliche 1910). In contrast, Schuster (1931) suggested that this genus belongs to the Cycadaceae but finally abolished it. According to Ye (1979), who described some *A. zeilleri* specimens within a Middle Triassic flora in Hebei and Sichuan Provinces, South China, this taxon belongs to the Lepidodendrales rather than to the Gymnosperms. Grauvogel-Stamm and Duringer (1983) also described *A. zeilleri* from the Ladinian of eastern France and showed that they are sporophylls of lycopsids containing microspores of the *Aratrisporites*

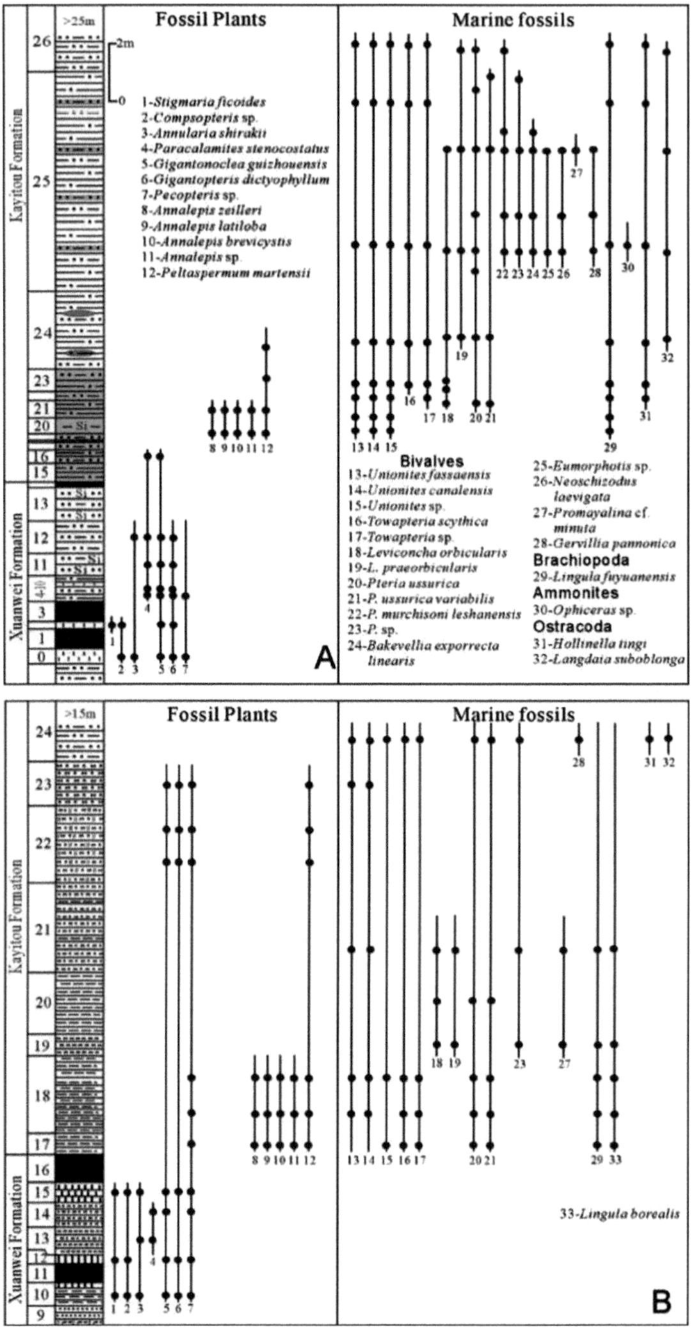

Fig. 5.2 Stratigraphical distribution of plant and marine fossils in the Permian–Triassic boundary interval and the Lower Triassic Kayitou Formation of the selected Mide and Tucheng Sections, Western Guizhou and Eastern Yunnan, South China. (**A**) Mide Section, (**B**) Tucheng Section

type. These authors similarly concluded that *A. zeilleri* belongs to the Lepidodendrales. Meng (1998) definitely indicated that *Lepacyclotes* (=*Annalepis*) resembles in many aspects the living plant *Isoetes*, such as its herbaceous habit with a corm, its more or less clustering sporophylls, the presence of a ligule, its monolete microspores of the *Aratrisporites*-type and its triletes megaspores, etc. Ultrastructural studies of its microspores have shown that they have characteristic isoetalean features (Grauvogel-Stamm and Lugardon 2001).

Lepacyclotes (=*Annalepis*) is firstly found in the Middle Triassic strata in Hubei and Sichuan of South China (Ye 1979), with other identifiable materials also reported in Huaining of Anhui, Nanjing of Jiangsu, and Luodian and Bianyang of Guizhou. While the North China *Lepacyclotes* (=*Annalepis*) was firstly reported by Wang Fuxi (1984a), with the sample collected from the upper part of Early Triassic Heshanggou Formation in Chengde, Hebei, where there is also incomplete strobile. At the meantime, similar sporophylls of *Lepacyclotes* (=*Annalepis*) and megaspora in situ are found in the upper part of Heshanggou Formation in Shanxi (Wang 1984a, b; Wang and Wang 1990b), at the bottom of Middle Triassic Ermaying Formation in Wubu, Shaanxi Wang and Wang (1990c). *Lepacyclotes* (=*Annalepis*) was also seen in the late Middle Triassic in Jingyuan, Gansu of North China (Sun 1992). From 1995 to 2000, Meng and Meng et al. (Meng 1994, 1996, 1998, 1999; Meng et al. 1995; Meng and Li 2002) made a study on the Middle Triassic *Lepacyclotes* (=*Annalepis*) from Huaining of Anhui, Xianfeng of western Hubei, Fengjie of eastern Chongqing and Sangzhi of northwestern Hunan (Fig. 5.3) of South China.

The progressive development of the *Lepacyclotes* (=*Annalepis*) fossils has basically clarified its origin and evolution, spatial distribution and ecologic environment. However, the entire characteristics of its mother plant are still far from identified and the size and shape of the strobile are also difficult to judge. Lately, we firstly have collected numerous samples of *Lepacyclotes* (=*Annalepis*) in the Early Triassic

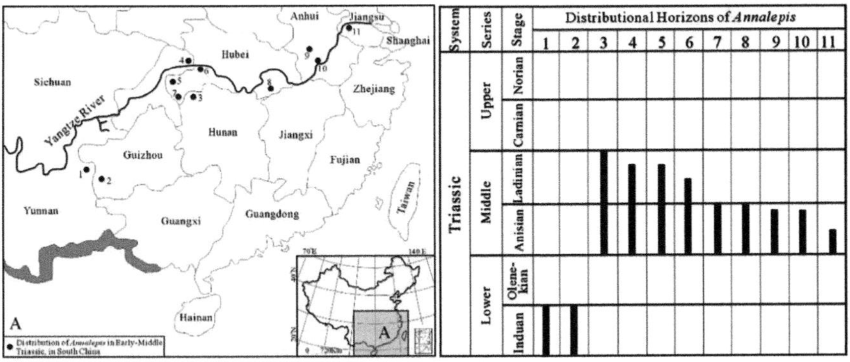

Fig. 5.3 Time and space distribution pattern of *Lepacyclotes* (=*Annalepis*) in South China. 1-Xuanwei, Eastern Yunnan 2-Panxin, Western Guizhou 3-Sangzhi, Northwestern Hunan 4-Fengjie, Eastern Sichuan 5-Lichuan, Western Hubei 6-Badong, Northwestern Hunan 7-Xianfeng, Western Hubei 8-Puqi, Eastern Hubei 9-Huaining, Southern Anhui 10-Anqing, Southern Anhui 11-Nanjing, Southern Jiangsu

Kayitou Formation in western Guizhou and eastern Yunnan. These rare specimens will provide new evidence for the Permian–Triassic evolution of lycopods and the geological time of Kayitou Formation.

5.2.2 *Distribution and Migration of Lepacyclotes (=Annalepis)*

In Induan stage, *Tomiostrobus*, a subgenus of *Lepacyclotes* (=*Annalepis*) (Doburskina and Zapfe 1994) or a synonymous genus as *Lepacyclotes* (=*Annalepis*) (Meng et al. 2000) is widely spread in Siberia, *Cylostrobus* and *Skilliostrobus* in southeastern Australia (Retallack 1975; Ash 1979). While *Lepacyclotes* (=*Annalepis*) found in the Kayitou Formation in western Guizhou and eastern Yunnan coexists with early Early Triassic marine fauna fossils (Figs. 5.2 and 5.4). This may suggest that true *Lepacyclotes* (=*Annalepis*) occurred in early Early Triassic (Induan) at a global scale.

In late Early Triassic (Olenekian), *Lepacyclotes* (=*Annalepis*) developed to some extent, which is common in the Early Triassic Heshanggou Formation in North China, for example in Hebei and Shanxi. However, there is no such report in South China.

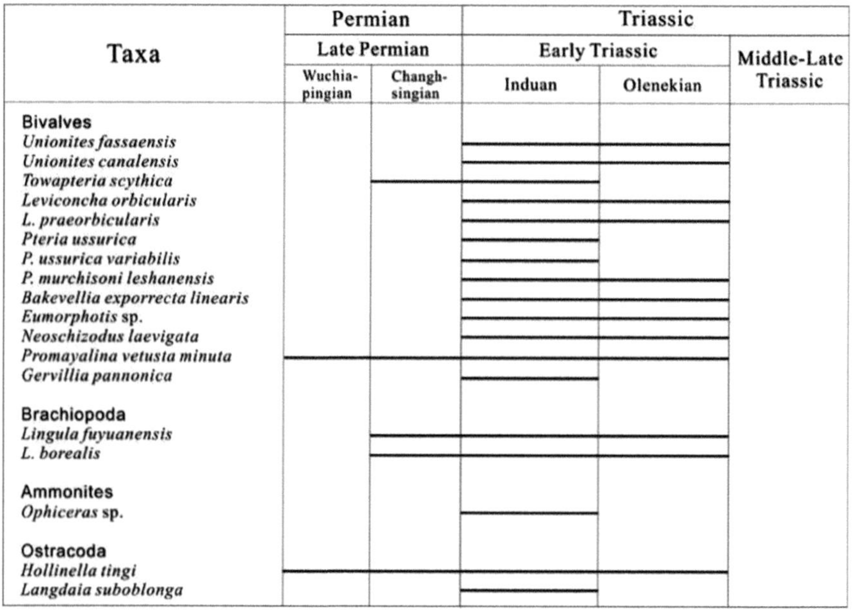

Fig. 5.4 Global stratigraphical ranges of marine fossils from Kayitou Formation of the Mide and Tucheng sections, western Guizhou and eastern Yunnan, South China

Middle Triassic should be the prime age for *Lepacyclotes* (=*Annalepis*), which developed widely. *Lepacyclotes* (=*Annalepis*) is a typical element in the plant assemblage at the bottom of Ermaying Formation in North China. In contrast, *Lepacyclotes* (=*Annalepis*) is commonly found in the Middle Triassic in the Yangtze region, especially distributed in 1–5 members of Badong Formation in Northwestern Hunan. The co-existence of this genus with the Middle Triassic marine bivalves reflects that it was rather prosperous from Anisian to Ladinian.

It should be noted that Wang (1984b) reported two types of *Lepacyclotes* (=*Annalepis*) of Yanchang Group in Shanxi: *Lepacyclotes* (=*Annalepis*) *zeiller* Fliche and *A. shanxiensis* Wang, which show clear lingular pit in the sporophyll and a longitudinal groove in the sporangia. Additionally, *Tomiostrobus convexus* (Brick) found in Middle-Upper Triassic in Kazakhstan (Dobruskina 1985) is also hard to distinguish from the holotype of *Lepacyclotes* (=*Annalepis*). Evidently, this genus occurred in Late Triassic.

To sum up, it is certain to attribute the geologic time of *Lepacyclotes* (=*Annalepis*) to Early Triassic and early Late Triassic. Or rather, it started in Early Triassic, flourished in Middle Triassic and disappeared in early Late Triassic.

From the distribution pattern in time and space of the *Lepacyclotes* (=*Annalepis*)-type fossils (Fig. 5.2) and the related genera (Grauvogel-stamm and Lugardon 2001), coupled with the cladogram of extant *Isoetes* (Hoot and Taylor 2001), Liu et al. (2004) inferred that the *Lepacyclotes* (=*Annalepis*)-type sporophylls originated almost simultaneously in the Lower Triassic from East-South Siberia of Russia (*Tomiostrobus* Neuburg 1936) and Australia (*Cylostrobus* Helby and Martin 1965 and *Skilliostrobus* Ash 1979). Moreover Liu et al. (2004) suggested that they have migrated along the northern and southern coasts of the Tethys, and spread to China (*Lepacyclotes* (=*Annalepis*)), Germany (*Lepacyclotes* (=*Annalepis*)) and France (*Lepacyclotes* (=*Annalepis*)) on the northern coast of the Tethys during the Middle Triassic. Now a problem with the South China is that *Lepacyclotes* (=*Annalepis*) is the first occurrence in Induan (e.g. Mide and Tucheng sections in western Guizhou and Eastern Yunnan). Similarly, Liu et al. (2004) suggested that the Middle Triassic *Lepacyclotes* (=*Annalepis*) of China might have derived from the *Lepacyclotes* (=*Annalepis*)-type of the Lower Triassic of Australia. However, our data indicates that it could derive from Southwestern China. Moreover, according to the distribution characteristics of the horizon-bearing *Lepacyclotes* (=*Annalepis*) in the Yangtze River valley (Meng et al. 2000), it seems that the taxon *Lepacyclotes* (=*Annalepis*) appeared almost simultaneous in this region during the early Anisian (early Middle Triassic) and that its distribution gradually shrunk during the middle Anisian–Ladinian (middle-late Middle Triassic), in response to the marine regression that occurred from east to west on the Yangtze Platform during that period (Meng 1998; Meng et al. 2000).

5.2.3 Ecologic Environment of *Lepacyclotes* (=*Annalepis*)

Lepacyclotes (=*Annalepis*) is likely to be the ancestor of modern *Isoetes* (Meng 1998). It is that many of its characters are similar to those of the modern ones. For instance, both are herbaceous with ambipolarcorm, sporophylls are somewhat clustered with a leaf-tip. Also sporangia be closed to the ventral side of the sporophyll, on top of which there is a ligule and a special labellum. Megaspore has a three ray with exine of verrucas or short ridge-type, while microspore produced by *Lepacyclotes* (=*Annalepis*) is genus *Arartisporites*.

In 1980s, Gauvogel-Stamm and Duringer (1983) studied the specimens from eastern France and collected by herself in a limestone quarry. She believed that this fossil found in Coquinaian and clay coal bed bottom, should be a kind of terrestrial swamp plant on the shore, living in an environment with decreasing sea level variation and sea water influence. Wang induced that *Isoetes* (*Lepacyclotes* (=*Annalepis*)) *ermayingensis* Wang should be the terrestrial plant growing in the temporary shallow water of upland or mountains after his careful discussion of the ecology of *Lepacyclotes* (=*Annalepis*) from Ermaying Formation in North China.

While Meng (1998) and Meng et al. (2000) assumed that *Lepacyclotes* (=*Annalepis*) should have lived in coastal area with more or less wave or tide impact based on his study of the *Lepacyclotes* (=*Annalepis*) structure. Its sporophyll is usually lanceolate with an acute apex and a truncate base, and smooth subsurface. While the sporangia are stick-typed, closed to the middle-lower part of the sporophyll back, whose weight caused the apex of the sporophyll a little upwardly bended. All these characteristics demonstrate the sporangia of *Lepacyclotes* (=*Annalepis*) are easy to float on water. Apart from this, its ligule and labellum take the advantage of *Lepacyclotes* (=*Annalepis*) to breathe or give out water when affected by the splashes (Meng et al. 2000). While in Early and Middle Triassic *Lepacyclotes* (=*Annalepis*) became widely developed owing to its physiological texture. These characteristics of its organs enable it to spread over long distances with the movement of sea water.

5.2.4 Implication of the First Occurrance of *Lepacyclotes* (=*Annalepis*) in Early Triassic Induan Kayitou Formation, Western Guizhou–Eastern Yunnan

It is very significant that in eastern Yunnan and western Guizhou border *Lepacyclotes* (=*Annalepis*) occurred in marine sediments together with typical Induan bivalves (Yin 1985, 1991), ostracods, brachiopods and ammonites (Wang 1984a, b; Yang 1991) (Fig. 5.4). These observations indicate that the first occurrence of *Lepacyclotes* (=*Annalepis*) is as early as Induan in Southwestern China, thus much earlier than in the Yangtze River area (Fig. 5.3). During late Early and Middle Triassic, they might have migrated to the north and widely spread to Middle-Lower Yangtze Platform, in relation to the marine regressions and transgressions.

The fossil plants of Olenekin Lingwen Formation from the Qionghai District, Hainan Island was firstly reported by Zhou and Li (1979). Although the fossil plants preserved in this formation are fragmentary, they are rich in genera and species. This flora is made up of more than 20 species belonging to 18 genera. Later, Meng (1992) added 38 species in 28 genera, among which there are *Equisetites, Phyllotheca, Neocalamites, Neuropteridium, Albertia, Voltzia, Glyptolepis* etc. The fossil plants of the Lingwen Formation are not only very similar to those of the Buntsandstein paleofloras of Western Europe, but also the lithologic characteristics of the plant-bearing bed can be compared. So namely the *Neuropteridium–Albertia–Voltzia* assemblage present in the Lingwen Formation undoubtedly belongs to the late Early Triassic, i.e. the Olenekian in China.

Yao (1978) indicated that the fossil fauna (including the bivalves: *Towapteria* aff. *scythica, Leptochondris bittneci, Neoschizodus ovata, Unionites fassaensis, Pteria ussurica variabilis*; the brachiopods: *Lingula borealis*; the insect: *Tomia*) and these fossils plants (including *Paracalamites stenocostatus, Annularia shirakii, Lobatannularia multifolia, Pecopteris* sp., *Gigantopteris* sp., and present *Lepacyclotes* (=*Annalepis*) which was not yet reported) appeared simultaneously in the lowest part of the Kayitou Formation along the Yunnan and Guizhou border. This stratum corresponds to the *Claraia wangii* zone of the lowest part of the Feixianguan Formation, which is Induan in age. The important elements of the *Gigantopteris*-flora may have survived in the early Lower Triassic as relics (Yao 1978; Yu et al. 2007). Our study indicates that the fossils showed in Fig. 5.3 from the Mide and the Tucheng sections are consistent with the above-mentioned fossils, except that *Lepacyclotes* (=*Annalepis*) was not recorded in Yao's paper (Yao 1978). Then we propose that the residual components of the relics of the former Late Permian *Gigantopteris*-flora and the Early Triassic *Lepacyclotes* (=*Annalepis*) co-formed the floral assemblage of the early Lower Triassic in South China. This assemblage is undoubtedly older than the *Neuropteridium–Albertia–Voltzia* assemblage recorded from the Lingwen Formation. The discovery of this flora therefore, not only fills the blank of Induan plants between the Changhsingian *Gigantonoclea guizhouensis–Ullmannia* cf. *bronnii* assemblage (Li 1995) and the Olenekian Triassic *Neuropteridium–Albertia–Voltzia* assemblage, but it is also significant in the evolution of the Triassic flora, the palaeobotanical and geographical regionalization and in the correlation of the global Early Triassic floras.

References

Ash SR (1979) Skilliostrobus gen. nov., a new lycopsid cone from the Early Triassic of Australia. Alcheringa 3:73–89

Dobruskina IA, Zapfe H (1994) Triassic floras of Eurasia (No. 56: 581 (4+5) DOB)

Dobruskina IA (1985) The Madygen flora (USSR, Middle Asia) as a typical representative of Keuper floras. Symp Sobr Floras Trias. Tard Fitogeogr, Mexico 11–19

Feng Z, Wei HB, Guo Y, He XY, Sui Q, Zhou Y, Liu HY, Gou XD, Lu Y (2020) From rainforest to herbland: new insights into land plant responses to the End-Permian Mass Extinction. Earth Sci Rev 204:103153
Fliche P (1910) Flore fossile du Trias en Lorraine et Franche-Comté (avec des considérations finales par M. R. Zeiller. Bulletin De La Société Des Sciences De Nancy III, XI 1:222–286
Grauvogel-Stamm L, Duringer P (1983) Annalepis zeilleri Fliche 1910 emend., un organe reproducteur de Lycophyte de la Lettenkohle de l'Est de la France. Morphologie, spores in situ et paléoécologie. Geologische Rundschau 72(1):23–51
Grauvogel-Stamm L, Lugardon B (2001) The Triassic lycopsids Pleuromeia and Annalepis: relationship, evolution and origin. Am Fern J 91:115–149
Helby R, Martin ARH (1965) *Cylostrobus* gen. nov., cones of Lycopsidean plants from the Narrabeen group (Triassic) of New South Wales. Aust J Bot 13:389–404
Hoot SB, Taylor WC (2001) The utility of nuclear ITS, a Leafy homolog intron and chloroplast atpB-rbcL spacer region data in phylogenetic analyses and species delimitation in Isoetes. Am Fern J 91(3):166–167
Li XX (ed) (1995) Fossil floras in China through the geological ages (English edition). Guangdong Science and Technology Press, Guangzhou, pp 1–695
Liu X, Gituru WR, Wang QF (2004) Distribution of basic diploid and polyploid species of Isoetes in East Asia. J Biogeogr 31:1239–1250
Meng FS (1994) Discovery of *Pleuromeia—Annalepis* Flora in South China and its significance. Chin Sci Bull 39(2):130–134
Meng FS (1998) Studies on *Annalepis* from Middle Triassic along the Yangtze River and its bearing on the origin of Isoetes. Acta Bot Sin 40(8):768–774
Meng FS (1999) Studies on *Pleuromeia* from Middle Triassic along the Yangtze River Valley and systematics of Isoetales. Acta Geosci Sin 20(2):215–222 (in Chinese with English abstract)
Meng FS, Xu AW, Zhang ZL, Lin JM, Yao HZ (1995) Nonmarine biota and sedimentary facies of the Badong Formation in the Yangtze and its neighbouring areas. China University of Geosciences Press, Wuhan, pp 1–76. (In Chinese with English abstract)
Meng FS (1992) Fossil plants. In: Yichang Institute of Geology and Mineral Resources, et al (eds) Geology of Hainan Island, 1. Stratigraphy and palaeology. Geological Publishing House, Beijing, pp 166–170, 175–183. (in Chinese)
Meng FS (1996) Floral palaeoecological environment of the Badong Formation in the Yangtze gorges area. Geol Miner Resour South China 1–13. (In Chinese with English abstract)
Meng FS, Li XB (2002) Declension and survival strategy of the middle Triassic Pleuromeia in the Yangtze valley. Geol Min Resour South China 2:60-65 (in Chinese with English abstract)
Meng FS, Zhang ZL, Niu ZJ, Chen DY (2000) Primitive lycopsid flora in the Yangtze valley of China and systematics and evolution of Isoetales. Hunan Science and Technology Press, Hunan (in Chinese with English abstract)
Neuburg MF (1936) On stratigraphy of the coal-bearing deposits in Kuznesk Basin. Bull Acad Sci USSR Ser Geol 4:469–510 (in Russian)
Retallack GJ (1975) The life and time of a Triassic lycopod. Alcheringa 1:3–29
Schuster J (1931) Ueber das Verhaltnis der systematischen Gliederung, der geographischen Verbreitung und der palaontologischen Entwieklung der Cycadaceen. Englers Bot Jahrb 64(1/2):165–260
Sun BN (1992) The assemblage succession of the Early Mesozoic fossil plants from Jingyuan coal field in Gansu. J Lanzhou Univ (Nat Sci) 28(3):163–167
Wang ZQ, Wang LX (1982) A new species of the lycopsid Pleuromeia from the Early Triassic of Shanxi, China and its ecology. Palaeontology 25(1):215–225
Wang ZQ, Wang LX (1990a) Late Early Triassic fossil plants from upper part of the Shiqianfeng Group in North China. Shanxi Geol 14(1):23–40 (in Chinese with English abstract)
Wang ZQ, Wang LX (1990b) Late Early Triassic fossil plants from upper part of the Shiqianfeng Group in North China. Shanxi Geol 5(2):97–154 (in Chinese with English abstract)

Wang ZQ, Wang LX (1990c) A new plant assemblage from the bottom of the Mid-Triassic Ermaying Formation. Shanxi Geol 5(4):303–318 (in Chinese with English abstract)
Wang LX, Xie ZM, Wang ZQ (1978) On the occurrence of *Pleuromeia* from the Qinshui Basin in Shanxi Province. Acta Palaeont Sin 17(2):195–212 (in Chinese with English abstract)
Wang XF (1984a) Supplement of Mesozoic plants from Hebei. In: Tianjin Institute of Geology and Mineral Resources (ed) Palaeontological atlas of North China (II). Geological Publishing House, Beijing, pp 297–302 (in Chinese)
Wang ZQ (1984b) Plant kingdom. In: Tianjin Institute of Geology and Mineral Resources (ed) Palaeontological atlas of North China (II). Geological Publishing House, Beijing, pp 223–295 (in Chinese)
Wang ZQ (1989) Permian gigantic palaeobotanical events in North China. Acta Palaeont Sinica 28(3):314–337 (in Chinese with English abstract)
Yang FQ (1991) Ammonoids. In: Yang Z, Wu S, Yin H, Xu G, Zhang K, et al (eds) Permo-Triassic Events of South China. Geological Publishing House, Beijing, pp 111–116
Yao ZQ (1978) On the age of "Gigantopteris coal series" and Gigantopteris-flora in South China. Acta Palaeontol Sin 17:81–89 (in Chinese with English abstract)
Yao ZQ, Xu JT, Zhen ZG, Mo ZG (1980) Late Permian biostratigraphy and the Permian–Triassic boundary in Western Guizhou and Eastern Yunnan. In: Nanjing Institute of Geology and Palaeontology (ed) Stratigraphy and palaeontology of Late Permian coal-bearing formations in Western Guizhou and Eastern Yunnan. Science Press, Beijing, pp 1–69. (In Chinese)
Ye MN (1979) On some Middle Triassic plants from Hupeh and Szechuan. Acta Palaeotol Sin 18(1):73–81 (In Chinese with English abstract)
Yin HF (1985) Bivalves near the Permian–Triassic boundary in South China. J Paleontol 59(3):572–600
Yin HF (1991) Bivalves. In: Yang Z, Wu S, Yin H, Xu G, Zhang K, et al (eds) Permo-Triassic Events of South China. Geological Publishing House, Beijing, pp 107–111
Yu JX, Peng YQ, Zhang SX, Yang FQ, Zhao QM, Huang QS (2007) Terrestrial events across the Permian–Triassic boundary along the Yunnan–Guizhou border, SW China. Glob Planet Chang 55:193–208
Yu JX, Broutin J, Huang QS, Grauvogel-Stamm L (2010) *Annalepis*, a pioneering lycopsid genus in the recovery of the Triassic land flora in South China. Comptes Rendus Palevol 9:479–486
Yu JX, Broutin J, Chen ZQ, Shi X, Li H, Chu DL, Huang QS (2015) Vegetation changeover across the Permian–Triassic boundary in Southwest China extinction, survival, recovery and palaeoclimate: a critical review. Earth Sci Rev 149:203–224
Zhou ZY, Li BX (1979) A preliminary study of the Early Triassic plants from the Qionghai District, Hainan Island. Acta Palaeotol Sin 18(5):444–466 (In Chinese with English abstract)

Chapter 6
Palynology Across the PTB in South China

Jianxin Yu, Wenchao Shu, and Jean Broutin

6.1 Palynofloral Assemblages Across the PTB in South China

The materials on which the present study is based were collected from four non-marine PTB sections (Fig. 1.1) across Permian–Triassic boundary, as fresh as possible. The samples were processed using conventional palynological acid maceration techniques, as described by Wood et al. (1996) 0.185 samples in all (66 from the Chahe Section, 41 from the Zhejue Section, 40 from the Mide Section A, 38 from the Mide Section B) were processed for this study. Until now, only 20 samples produced enough palynomorphs to enable statistical analysis near the Permian–Triassic boundary (Figs. 6.1, 6.2 and 6.3). However, the taxa recognized are still considered to reflect the basic palynofloral variations across the PTB associations of similar characteristics, three palynological assemblages across the PTBST have been recognized Fig. 6.2.

The Upper Permian to Lower Triassic strata in the terrestrial exposure sections of the WGEY region yields very few spores and pollen grains (Peng et al. 2006; Bercovici et al. 2015). In contrast, they are considerably abundant in three terrestrial–marine transition drilling sections of WGEY (Ouyang 1982; Fig. 1). Previously, Ouyang (1982) recognized three palynofloral assemblages from the Late Permian to Early Triassic succession in WGEY based on qualitative and quantitative analysis

J. Yu (✉) · W. Shu
School of Earth Sciences, China University of Geosciences, Wuhan, Hubei, P. R. China
e-mail: yujianxin@cug.edu.cn

W. Shu
e-mail: shuwenchao@cug.edu.cn

J. Broutin
Faculty of Sciences and Engineering (CR2P, MNHN, CNRS), Sorbonne Université, Paris, France
e-mail: jean.broutin@upmc.fr

J. Yu et al. (eds.), *Plants and Palynomorphs around the Permian-Triassic Boundary of South China*, New Records of the Great Dying in South China,
https://doi.org/10.1007/978-981-19-1492-8_6

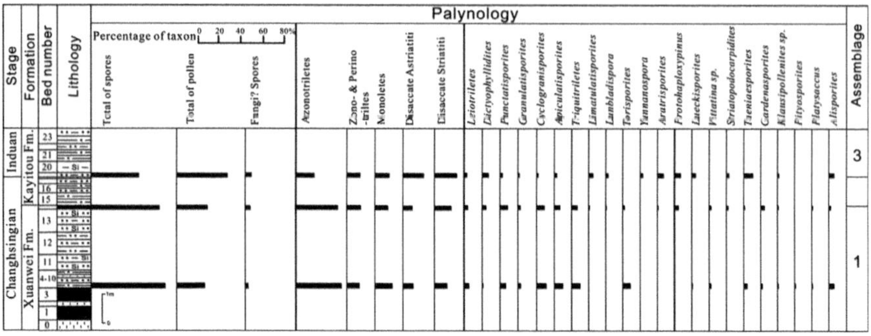

Fig. 6.1 Percentage of selected genera and suprageneric groups near Permian–Triassic Boundary from the Mide Section B

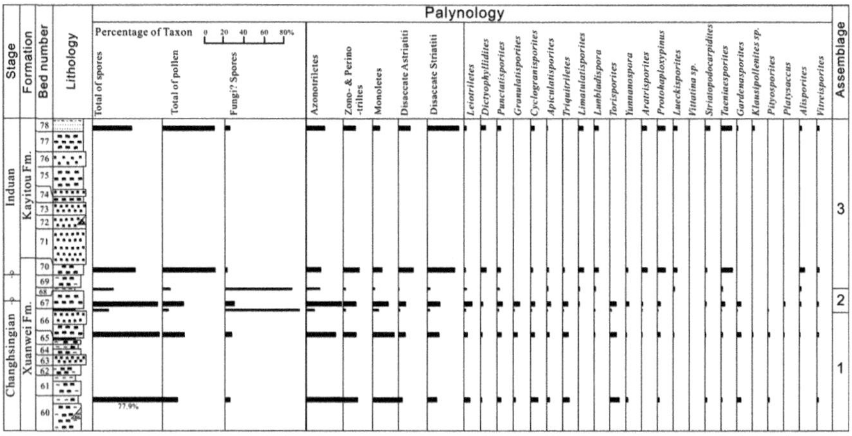

Fig. 6.2 Percentage of selected genera and suprageneric groups near Permian–Triassic Boundary from the Chahe Section

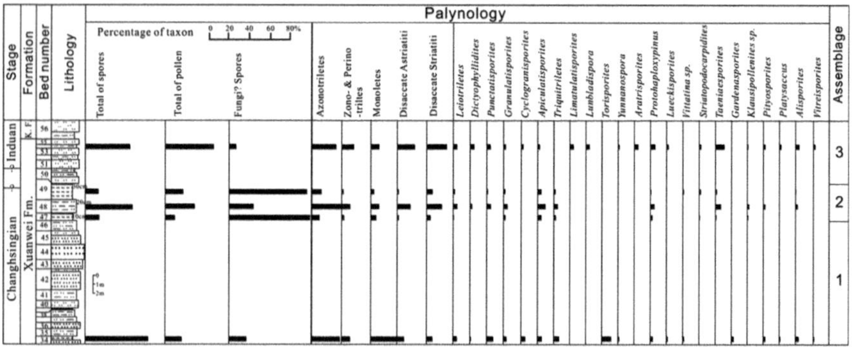

Fig. 6.3 Percentage of selected genera and suprageneric groups near Permian–Triassic Boundary from the Zhejue Section

of palynomorph's distributions. Of these, the *Torispora gigantean–Patellisporites meishanensis* Assemblage was established from the lower Xuanwei Formation of Wuchiapingian age and contains abundant pteridophyte and pteridosperm spores, and few gymnosperm pollen grains. The *Yunnanospora radiate–Gardenasporites* Assemblage from the upper Xuanwei Formation of Changhsingian age is dominated by the Palaeozoic elements, but also characterized by the presence of typical Mesozoic genera and species such as *Dictyophyllidites mortoni* and *Aratrisporites yunnanensis*. The *Lundbladispora–Aratrisporites–Pteruchipollenites* Assemblage from the lower Kayitou Formation of Induan age, includes numerous pteridophyte and pteridosperm spores and a few gymnospermous pollen grains, appearing a truly transitional or "mixed microflora" (term of Ouyang 1982). It should be noted that most of the Palaeozoic-type miospores persisted into the lowest Triassic (Ouyang 1982).

Peng et al. (2006) have also established three distinct palynological assemblages across the PTB stratigraphic set (PTBST, term of Peng et al. 2006) in two terrestrial PTB sections, namely Chahe and Zhejue, in WGEY. Assemblage 1 of Peng et al. (2006) was established from the upper Xuanwei Formation of Changhsingian age is dominated by ferns and pteridosperms, with few gymnosperms. Most taxa are typical long-ranging Palaeozoic forms, but the appearance of *Lueckisporites* in Bed 24 of the Zhejue section and Bed 62 of the Chahe section confirms a latest Permian age for this assemblage (Peng et al. 2006). Assemblage 2 was recognized from the PTBST and is marked by an abrupt decrease in palynomorph abundance and diversity, and thriving fungal/algal(?) spores. Although palynological Assemblage 2 is still dominated by ferns and pteridosperms, with a few gymnosperms, it is characterized by a mixed palynoflora containing both the Late Permian and Early Triassic elements. Most taxa are typical Late Permian elements, which also occurred in Assemblage 1. However, some taxa of the Early Triassic aspect such as *Dictyophyllidites* appeared for the first time in Assemblage 2. Assemblage 3, ranging from the upper most Xuanwei Formation to lowest Kayitou Formation, saw a rapid increase in the proportion of gymnospermpollens, exceeding that of ferns and pteridosperms. Nevertheless, abundance of palynomorphs still remains rather low. Typical Early Triassic taxa such as *Lundbladispora* and *Aratrisporites* are present in a greater abundance and confirm an Early Triassic age for this assemblage.

In this paper, the microspores were also collected from three non-marine PTB sections (Mide, Zhejue and Chahe sections) and considered to reflect the basic palynofloral variations across the PTB. Similarly, three palynological assemblages across the PTBST have been recognized (Figs. 6.1, 6.2 and 6.3). Assemblage 1 may be called the *Leiotriletes–Triquitrites–Laevigatosporites–Gardenasporites* assemblage. This encompasses the major part of the Xuanwei Formation to the lower boundary of the PTBST. The palynoflora is dominated by an association of pteridophytic spores of the Late Permian aspect, with minor gymnosperms. The spores are dominant, comprising more than 65.0% of the palynological assemblage, with gymnosperms pollen usually less than 30.0%. Appearance of some endemic elements such as genera *Patellisporites* represents the characteristic Paleozoic Cathaysian vegetation in South China. Most of the palynomorphs are Paleozoic taxa that are also common in West Europe (Visscher 1971) and North America (Utting 1989), such as

Leiotriletes, Waltzispora, Cyclogranulatisporites, Thymospora and *Torispora*. The appearance of some typical Permian, especially the Late Permian, palynomorphs such as *Leiotriletes exiguus, Granulatisporites adnatoides, Gardenasporites* spp. and *Vittatina* spp., indicates its Late Permian affinity (Ouyang and Utting 1990). Among these Late Permian taxa, the genera *Lueckisporites* and *Vittatina* are highly characteristic of the Late Permian (Mangerud 1994; Ouyang and Norris 1999).

Assemblage 2 is confined to the PTBST in the upper Xuanwei Formation, is tentatively named the fungi "spike" assemblage. It is marked by an abrupt drop in the abundance of palynomorphs and the appearance of fungal/algal(?) spores. The terrestrial palynofloral assemblage is still dominated by spores, with a few pollens. Most palynomorphs are still typical of Late Permian taxa also found in Assemblage 1, but some palynomorphs of Early Triassic aspect (e.g. *Lundbladispora communis, Aratrisporites yunnanensis*) appear in low abundances. Thus, this assemblage contains a mixed flora featuring both Late Permian and Early Triassic elements. In the meantime, both in the Chahe and Zhejue sections, the palynofloras in the clay beds of the PTBST record the highest abundance of fungal/algal(?) remains. High abundance of these fungal/algal(?) spores commonly appears in many PTB strata, though not at the GSSP Meishan Section (Ouyang and Utting 1990).

Assemblage 3 from the top part of Xuanwei Formation and the succeeding Kayitou Formation is called the *Aratrisporites–Lundbladispora–Lunatisporites* assemblage. In assemblage 3, the pollens increase to 55.0% of the assemblage, exceeding the spores (40.0%) for the first time in the study area, as is found in the Fuyuan district of South China (Ouyang 1986). Although still containing many Paleozoic palynomorph elements, just above the PTBST in the terrestrial PTB sections (stratigraphically above Assemblage 2), palynomorphs of Early Triassic aspect (e.g. *Aratrisporites yunnanensis*) appear in higher abundances (>5.0%) than in Assemblage 2 of the PTBST. *Lunatisporites* is particularly abundant in the Zhejue section where it reaches more than 12.0% of the assemblage. Ouyang (1986) found that the abundance of *Aratrisporites* was already 18.0% of the assemblage from the lower part of the original Kayitou Formation (i.e. the upper part of the Xuanwei Formation according to the definition of the boundary between the Xuanwei and Kayitou formations used herein) in Fuyuan District, eastern Yunnan.

The above palynological studies also verified that some Paleozoic spores and pollen survived the main end-Permian mass extinction event as well. The palynoflora recovered by the palynological assemblage 2 strongly shows transitional characteristics across the Paleozoic–Mesozoic transition. Some newcomers (disaster elements) appeared in Assemblage 2, together with typical Paleozoic survivors (mainly spores). Similar transitional palynomorph assemblages are found all over the world, such as at the Kap Stosch and Jameson Land in East Greenland (Balme 1980), Meishan section of Changxing, China (Ouyang and Utting 1990).

In addition, we have also found an abundant and diverse megaspore assemblage from the Early Triassic successions in WGEY (Li et al. 2014; Yu et al. 2010). In particular, the dispersed and in situ megaspores of *Annalepis* and *Plumeria* are very common in association with the Early Triassic fossil plants, although the fossil sporophyllum of genus *Plumeria* was not discovered until now in WGEY.

They include the lycopsids *Maexisporites pyramidalis*, *M. rotundus*, *M. parvus*, *Triangulatisporites* sp., *T. dalongkonensis*, *Trileites levis*, *T. vulgaris*, *Trileites* sp., *Horstisporites sulcatus*, *H.* sp., *Aneuletes rotundus*, *A.* sp., *Otynisporites eotriassicus*, *Hughesisporites simplex*, *Banksisporites pinguis*, *Zerndtisporites* sp. Of these, the genus *Trileites* is most abundant and contains 8 species (35.5%). The genus *Maexisporites* ranks the second most diverse group, including 3 species (12.9%) in Lycopsida order.

The palynomorphs obtained from the different sedimentary facies sections of WGEY show an abrupt change of palynofloras across the PTB, marked by a distinct drop of palynomorphs in both abundance and diversity. However, it also indicate a stepwise extinction pattern as marine fauna did, for mixing palynofloral elements from both Late Permian and Early Triassic found in the PTBST following the end-Permian mass extinction, and most Late Permian survivors disappeared shortly there after in the Earliest Triassic.

6.2 Comparison the Coeval Palynoflora with Other Region

6.2.1 Comparison with Fuyuan District, Eastern Yunnan, China

The features of Assemblage 1 and 2 described in the present paper are largely the same as the *Yunnanospora radiate–Gardenasporites* spp. assemblage (the upper member the Xuanwei Formation) in Fuyuan district, eastern Yunnan (Fig. 1.1) (Ouyang and Li 1980; Ouyang 1982, 1986). They are characterized by the dominance of Pteridophytic and many Pteridospermous spores. Gymnospermous, chiefly saccate, pollen grains occupy a subordinate position. Some species are restricted to the upper member of the Xuanwei Formation though they are often represented by very rare individuals. The Mesozoic elements are more abundant than in the assemblage from the lower part of the Xuanwei Formation (Ouyang 1986) (Fig. 6.4). However, the species with a Paleozoic aspect still dominated the assemblages. The difference between them is that no fungi spike event was reported in the *Yunnanospora radiate–Gardenasporites* spp. assemblage (the upper member the Xuanwei Formation) (Ouyang 1986). Considering other stratigraphical (Peng et al 2005; Zhang et al. 2006; Yin et al. 2007) and paleontological evidences, the Assemblages 1 and 2 from the upper part of the Xuanwei Formation are Late Permian Changhsingian.

Assemblage 3 is comparable with the *Aratrisporites–Lundbladispora* assemblage (the Kayitou Formation) reported from the Lower Triassic strata of Fuyuan District, eastern Yunnan (Ouyang 1986). They showed: (1) about a half of these species continue from the Permian assemblages. The association of fossil plants in the lower part of the Kayitou Formation, as well as other considerations, indicate that the miospores with a Paleozoic aspect present in these assemblage are not reworked (Ouyang 1982). (2) The lower part of the Kayitou Formation, is characterized by

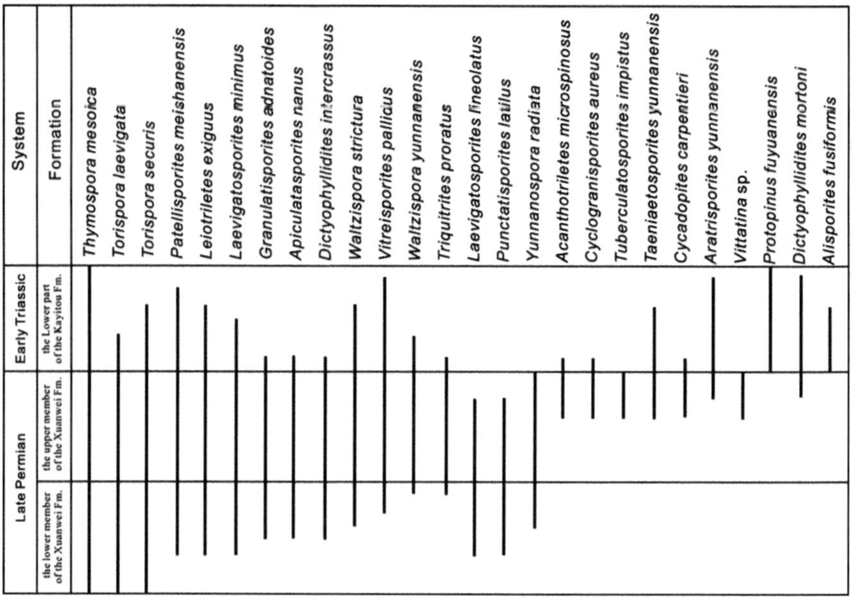

Fig. 6.4 Vertical ranges of the same taxa of this paper and Ouyang (1986) from in the assemblage from the Xuanwei Formation and lower part of Kayitou Formation

the dominance of the spores (including *Aratrisporites* produced by *Annalepis* strobili which totals from 10.0 to 18.0%), but the middle-upper part is dominated by gymnosperm pollen. (3) Another striking feature of Assemblages 3 is that many chiefly Mesozoic types are also present. So Assemblage 3 is most likely to be younger than Permian and older than late Early Triassic and may be regarded as Earliest Triassic (Scythian), a conclusion supported by the faunal evidences from bivalves, ostracods, brachiopods and ammonites etc. as previously stated.

6.2.2 Comparison with the Meishan Section D, Changxing County, Zhejiang Province, China

The Meishan D section in Changxing County of Zhejiang Province, South China, has been officially accepted as the Global Stratotype Section and Point (GSSP) of the PTB by the International Union of Geological Sciences (IUGS) Executive Committee (Yin et al. 2001).

Ouyang and Utting (1990) reported three assemblages zones recognized at Meishan Sections. In ascending order they are: (1) *Leiosphaeridia changxingensis–Micrihystridium stellatum* Assemblage from the Changhsing Formation (Changhsingian); (2) *Vittatina–Protohaploxypinus* Assemblage from the basal part of the Lower Yinkeng Formation (Earliest Griesbachian); and (3) *Lunatisporites–Ephedripites*

Assemblage Zone from the upper part of the Yinkeng Formation (Griesbachian). According to the characters of the lithology and "mixed" fauna described by Ouyang and Utting (1990), the strata in which this assemblage occurs correspond to the PTBST. We analyzed the palynology of the Permian–Triassic transition from Meishan Section D in 2004 and 2006 (Table 6.1). Two assemblages have been recognized, they are: Assemblage 1 is named by the *Leiotriletes–Vittatina–Protohaploxypinus* assemblage (Bed 26, the basal part of the Lower Yinkeng Formation), Assemblage 2 is called the *Lundbladispora–Lunatisporites–Equisetosporites* assemblage (Beds 33–53, the Yinkeng Formation).

The most prominent features of the *Leiotriletes–Vittatina–Protohaploxypinus* assemblage established by Ouyang and Utting (1990) from the upper part of the Changxing Formation and the basal part of the Lower Yinkeng Formation is its transitional or "mixed" nature; on the one hand, it includes many taxa which are more typical of the Permian, such as *Lueckisporites virkkiae*, but on the other hand, it includes elements more characteristic of the Triassic, such as *Aratrisporites* sp. *A. yunnanensis* and *Lundbladisporanejburgii* (Fig. 6.5). This characteristic is similar to Assemblage 2 (PTBST) named by the writer in Western Guizhou and Eastern Yunnan. But both assemblages have some differences. Within the *Leiotriletes–Vittatina–Protohaploxypinus* assemblage, the gymnosperm pollen is dominant, whereas fungal spores are rare, but the terrestrial palynofloral is still dominated by ferns and pteridosperms, with a few gymnosperms in the Assemblage 2 from the PTBST in the upper Xuanwei Formation. The main reason making this otherness is the depositional environment. In the beginning of the Changhsingian, coal deposits and swamp had not been found in the lower Yangtze Region, but coal bed is continuously found until the end Changhsingian in Western Guizhou and Eastern Yunnan. These explained that both grew different flora.

A palynological assemblage from the Yinkeng Formation (Beds 33–53, Meishan Section D), the *Lundbladispora–Lunatisporites–Equisetosporites* assemblage (Zhang et al. 2007), is herein proposed. This assemblage is dominated by pollen (59.0–79.6%) and spores (20.4–41.0%) (Table 6.1). Among the gymnosperm pollens, the non-striate bisaccate pollen types are the dominant forms (19.7–54.4%) in which the bisaccate pollen is the most common, and the striate bisaccate pollen types followed the former in abundance. The polyclicate pollen is represented only by one genus *Equisetosporites* but this genus has a rather high abundance (up to 24.3%). Among the spores, *Leiotriletes* is quite common (14.3–24.4%). Echinate, granular and verrucose trilete spores are also present but less frequent in the samples analyzed (1.9–21.2%). Fastigiate trilete spores, mainly *Lundbladispora*, are relatively low in proportion but were found in most studied horizons. Zonotrilete spores are rare in abundance and diversity. This assemblage is the same as *Lunatisporites–Ephedripites* Assemblage Zone (Ouyang and Utting 1990).

The *Aratrisporites–Lundbladispora–Lunatisporites* assemblage from the Kayitou Formation of western Guizhou and eastern Yunnan can be precisely correlated with the assemblage from the Meishan Section (Ouyang and Utting 1990) (Table 6.2): (1) the Paleozoic and Mesozoic miospores coexisted in both assemblages; (2) spores of Pteridophyta and some Pteridospermae are common, but gymnospermous pollen is

Table 6.1 Palynological categories and content statistic of Early Triassic in Meishan D Section, Changxing County, Zhejiang Province, China

Name	26–2	33–1	38–1	40–1	45–1	43–1	44–1	47–2	51–3	52–3	53–1	53–5	53–10
Leiotriletes sp.	3.8	4.8	1.9	1 g.	6.4	4.8	4.0	4.6	6.1	3.0		3.6	2.7
L. adrienniforumis	2.6		0.9										
L. concavus	3.8				1.3	1.6							1.4
L. directus								0.9					
L. exiguous	2.6	3.2	0.9		1.3		1.0			1.5			
Dictyophyllidites sp.	1.3		0.9										
D. mortoni								0.9	1.5				1.4
Calamospora sp.		1.6			1.3				1.5				
Punctatisporites sp.	6.4	8.1	6.5		6.4	6.4	7.0	6.5	4.5	6.1	1 g.	8.9	6.8
P. microtumulosus	1.3		0.9		1.3		1.0	0.9	3.0				
P. triassicus	1.3		1.9		1.3		3.0		1.5			1.8	1.4
P. minutus	1.3	1.6			1.3	3.2		1.9	3.0	1.5			4.1
Cyclogranisporites sp.	2.6	3.2			3.8	1.6	3.0	2.8	3.0	4.5			
C. crassirimosus	1.3								1.5			3.6	
C. major										1.5			
C. pseudoxonotriletes										1.5			
C. aureus	3.8				1.3								
Apiculatisporites sp.	1.3	3.2			2.6	1.6	1.0			4.5			
A. perirugosus								1.9		1.5			
A. decorus								0.9					
A. belliensis		1.6			1.3		1.0	0.9		1.5			

(continued)

Table 6.1 (continued)

Name	26–2	33–1	38–1	40–1	45–1	43–1	44–1	47–2	51–3	52–3	53–1	53–5	53–10
Osmundacidites sp.			0.9				1.0				1 g.		2.7
O. fissus								0.9			1 g.	1.8	
Lunzisporites sp.		1.6	0.9		1.3		2.0		1.5			1.8	
L. pallidus							1.0			1.5			
Convolutisporites sp.	1.3							0.9	3.0	1.5		1.8	1.4
Neoraistrickia irregularis										3.0			
Acanthriletes sp.	2.6							2.8					
Gulisporites cochlearius			0.9										
Lophozontriletes sp.	1.3	1.6			1.3	1.6						1.8	
Annulisporites sp.			0.9									1.8	1.4
A. folliculosa	1.3					3.2	1.0						
Lundbladispora sp.	1.3	3.2	0.9		2.6	1.6	2.0	1.9		3.0		1.8	2.7
L. nejburgii		1.6			1.3				1.5				1.4
L. watangensis						1.6				1.5			
Limatulasporites sp.					1.3								
Muerrigerisporis sp.							1.0						
Endososporites sp.	2.6				2.6				1.5				
Retusotriletes sp.					1.3		1.0						
Aratrisporites sp.	1.3	1.6	0.9						1.5				
A. yunnanensis			0.9			1.6				1.5			1.4
Netusotriletes rigidus	2.6											3.6	1.4

(continued)

Table 6.1 (continued)

Name	26–2	33–1	38–1	40–1	45–1	43–1	44–1	47–2	51–3	52–3	53–1	53–5	53–10
Cycadopites sp.	1.3	4.8	0.9		2.6	1.6		2.8	1.5	3.0			
C. complanatus			0.9										
C. nitidus	1.3							0.9				1.8	1.4
Chasmatosporites sp.	3.8	1.6	0.9					2.8					
C. hians										0.5			
Equestosporites sp.	2.6	8.1	0.9		6.4	3.2	5.0	4.6	3.0	15.2	2 g.s	7.1	8.1
E. chacheutensis		1.6			3.8	3.2	3.0	1.9	1.5	9.1		3.6	4.1
Alisporites sp.	3.8	4.8	0.9		3.8	6.4	5.0	7.4		3.0		1.8	2.7
A. australis	1.3		1.9		1.3			0.9					
A. toralis		1.6	1.9			1.6	4.0	3.7	3.0			1.8	1.4
A. thomasii	1.3					3.2							
A. parvus	1.3				1.3		2.0						
Pinuspollenites sp.	1.3	9.7	11.1		2.6	1.6	2.0	6.5	4.5	1.5		16.1	16.2
P. divulgatus	2.6		2.8	1 g.				1.9	3.0				
Piceites sp.	1.3	1.6	3.7		3.8	4.8	3.0	3.7	6.1			3.6	6.8
P. notialis		3.2											
Podocarpidites sp.	3.8	1.6	6.5					0.9	3.0		1 g.		1.4
Pityosporites sp.		1.6	4.6		6.4	3.2	6.0	1.9	1.5		2 gs	1.8	
P. devolvens	6.4		1.9						1.5				
Platysaccus sp.	1.3		4.6		1.3	1.6		1.9		1.5			1.4
P. alatus	1.3	1.6	1.9										

(continued)

Table 6.1 (continued)

Name	26–2	33–1	38–1	40–1	45–1	43–1	44–1	47–2	51–3	52–3	53–1	53–5	53–10
Caytonipollenites sp.					1.3		4.0	0.9		1.5		3.6	1.4
Caytonipollenites pallidus	1.3		0.9					0.9					
C. subtilis								0.9					
Klausipollenites schauberger	2.6					1.6							
Chordasporites pallidus		1.6	3.7										
Chordasporites sp.	2.6						2.0	2.8					1.4
Lunatisporites pellucides			0.9										
T. labdcus		1.6	2.8										
T. kraeuselis			0.9		2.6		1.0						1.4
T. sp.	1.3	3.2	10.2	1 g	3.8	4.8	5.0	4.6	4.5	4.5	1 g	5.4	4.1
Protohaploxypinus sp.	5.2												
Luckisporites vikkiae	2.6		1.9				1.0						
L. sp.	1.3	1.6			1.3	3.2	1.0	0.9		3.0			
Striatopodocarpidites sp.		1.6				1.6	2.0	1.9	1.5			1.8	4.1
Psophosphaera sp.			2.8		1.0					1.5		3.6	
Gardenasporites minor	1.3			1 g									
Vittatina sp.	3.8				1.3		1.0	2.8		1.5		1.8	4.1
Total content of spores	48.4	37.1	20.4	1 g	41.0	30.6	30.0	29.6	33.3	40.9	3 gs	32.2	32.4
Total content of pollens Gymnospermes	51.6	62.9	79.6	5 gs	59.0	69.4	70.0	70.4	66.7	59.1	5 gs	67.8	67.6
Baltisporites sp.		1 g									1 g		
Polygonium sp.						2 gs							1 g
Multicellaesporites sp.	2 gs												

g.—grain gs.—grains

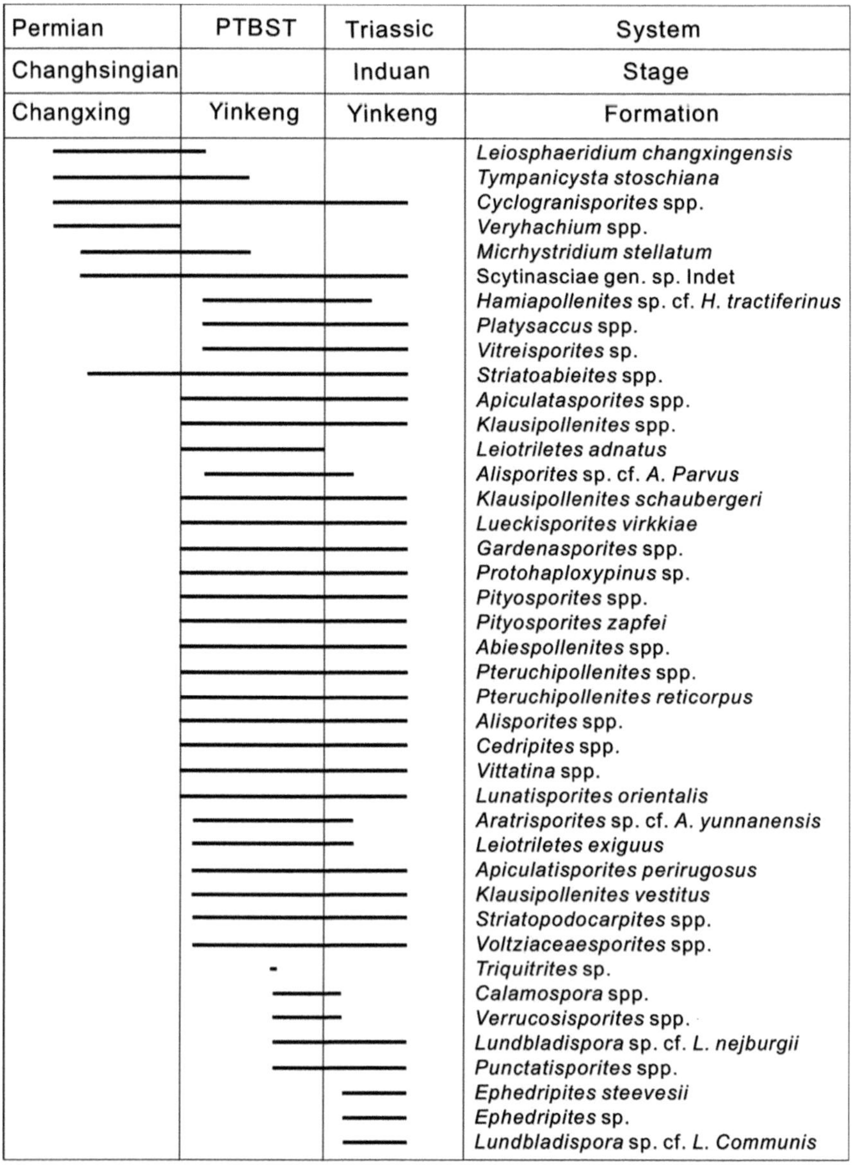

Fig. 6.5 Vertical ranges of selected taxa in the assemblage from the Changhsing and Yinkeng Formation, Meishan, Changxing County (modified after Ouyang and Utting 1990)

Table 6.2 Palynological correlations of late Permian and early Triassic from different sedimentary facies in South China

Stage	Meishan Section (According to Ouyang and Utting 1990)	Series	Western Guizhou and Eastern Yunnan (in the paper)	Series	Fuyuan district (according to Ouyang 1986)	Bed No.	Meishan Section D (the author)
Griesbachian	*Lunatisporites–Ephedripites* Assemblage	Lower Triassic	*Aratrisporites–Lundbladispora–Lunatisporites* Assemblage	Lower Triassic	*Aratrisporites–Lundbladispora* Assemblage	33–53	*Lundbladispora–Lunatisporites–Equisetosporites* Assemblage
PTBST	*Vittatina–Protohaploxypinus* Assemblage (Earliest Griesbachian)	PTBST	Few palynomorph but fungi spike Assemblage			26	*Leiotriletes–Vittatina–Protohaploxypinus* Assemblage
Changhsingian	*Leiosphaeridites Changxingensis–Micrhystridium stellatum* Assemblage	Upper Permian	*Leiotriletes–Triquitrites–Laevigatosporites–Gardenasporites Assemblage*	Upper Permian	*Yunnanospora radiate–Gardenasporites* Assemblage		

dominant in the proportion of individuals; (3) the index fossil for the Early Triassic, such as *Lundbladisporanejburgii* and *Aratrisporitesyunnanensis* has been discovered in both assemblages.

References

Balme BE (1980) Palynology and the Carboniferous–Permian boundary in Australia and other Gondwana continents. Palynology 4(1):43–55

Bercovici A, Cui Y, Forel M, Yu JX, Vajda V (2015) Terrestrial paleoenvironment characterization across the Permian–Triassic boundary in South China. J Asian Earth Sci 98:225–246

Li H, Yu JX, Huang QS, Shi X, Huang C (2014) Geological significances of the lycopods megaspores from the early Triassic Kayitou formation of the Western Guizhou and Eastern Yunnan Region. J Lanzhou Univ (Natural Sciences) 50(2):1–11

Mangerud G (1994) Palynostratigraphy of the Permian and lowermost Triassic succession, Finnmark Platform, Barents Sea. Rev Paleobot Palynol 82:317–349

Ouyang S, Li ZP (1980) The micro-floral in the Kayitou Bed of Fuyuan, Yunnan and its stratigraphic and paleobotanic significance. In: Nanjing Institute of Geology and Palaeontology, Academica Sinica (ed) Late Permian coal-bearing strata and palaeontological fauna in western Guizhou and eastern Yunnan. Science Press, Beijing, pp 123–194 (in Chinese)

Ouyang S (1986) Palynology of upper Permian and lower Triassic strata of Fuyuan district, eastern Yunnan. Science Press, Beijing, p 122 (in Chinese with English Summary)

Ouyang S (1982) Upper Permian and lower Triassic palynomophs from eastern Yunnan. Can J Earth Sci 19:68–80

Ouyang S, Norris G (1999) Earliest Triassic (Induan) spores and pollen from the Junggar Basin, Xinjiang, northwestern China. Rev Palaeobot Palynol 106:1–56

Ouyang S, Utting J (1990) Palynology of upper Permian and lower Triassic rocks, Meishan, Changxing county, Zhejiang province, China. Rev Palaeobot Palynol 66:65–103

Peng YQ, Zhang SX, Yu JX, Yang FQ, Gao YQ, Shi GR (2005) High-resolution terrestrial Permian–Triassic eventostratigraphic boundary in western Guizhou and eastern Yunnan, southwestern China. Palaeogeogr Palaeoclimatol Palaeoecol 215:285–295

Peng YQ, Yu JX, Gao YQ, Yang FQ (2006) Palynological assemblages of non-marine rocks at the Permian–Triassic boundary, western Guizhou and eastern Yunnan, South China. J Asian Earth Sci 28:291–305

Utting J (1989) Preliminary palynological zonation of surface and subsurface sections of Carboniferous, Permian and lowest Triassic rocks, Sverdrup Basin, Canadian Arctic Archipelago. In: Current research, Part G. Geological survey of Canada, Paper 89-1G, pp 233–240

Visscher H (1971) The Permian and Triassic of the Kingscourt outlier, Ireland. Geology Survey of Ireland, Special Paper, No. 1

Wood GD, Gabriel AM, Lawson JC (1996) Palynological techniques—processing and microscopy. In: Jansonius J, Mcgregor DC (eds) Palynology: principles and applications. Publishers Press, Salt Lake, pp 29–50

Yin HF, Feng QL, Lai XL, Baud A, Tong JN (2007) The protracted Permo-Triassic crisis and multi-episode extinction around the Permian–Triassic boundary. Glob Planet Chang 55:1–20

Yin HF, Zhang KX, Tong JN, Yang ZY, Wu SB (2001) The Global Stratotype Section and Point (GSSP) of the Permian-Triassic boundary. Episodes 24:102–114

Yu JX, Broutin J, Huang QS, Grauvogel-Stamm L (2010) *Annalepis*, a pioneering lycopsid genus in the recovery of the Triassic land flora in South China. C R Palevol 9:479–486

Zhang KX, Tong JN, Shi GR, Lai Xl, Yu JX, He WH, Peng YQ, Jin YL (2007) Early Triassic conodont—palynological biostratigraphy of the Meishan D Section in Changxing, Zhejiang Province, South China. Palaeogeogr Palaeoclimatol Palaeoecol 252(1–2):4–23

Zhang SX, Peng YQ, Yu JX, Lei XR, Gao YQ (2006) Characteristics of claystones across the terrestrial Permian-Triassic boundary: evidence from the Chahe section, western Guizhou, South China. J Asian Earth Sci 27:358–370

Chapter 7
Extinction Pattern and Recovery of the Permian–Triassic Flora in South China

Jianxin Yu, Zhen Xu, Jean Broutin, Xiao Shi, and Hui Li

7.1 On Researching the Extinction Pattern and Recovery of Plants During the Permian–Triassic Transition

The end-Permian mass extinction and its subsequent survival and recovery patterns have been studied by many geologists and paleontologists in the last two decades. A number of papers involving extinction patterns and controlling factors of the mass extinctions, have been published (Peng and Shi 2009; Fielding et al. 2019; Vajda et al. 2020). To this extinction pattern, geologists made suggestions, such as catastrophe, step-wise and gradualism (Kauffman 1987); pulse pattern and press pattern (Erwin 1996); two episodes (Rong and Fang 2004; Xie et al. 2005) and so on. This great extinction event affected not only marine realm, but also the terrestrial ecosystem. The Cathaysian Flora, for example, which thrived in the Permian, became extinct as a whole. In contrast, little is known about the Early Triassic flora of South China, suggesting a desolated and impoverish environment on the ancient land. This displays the disruption of land ecosystem that led to the degradation and the delay of recovery after the mass extinction.

J. Yu (✉) · Z. Xu
School of Earth Sciences, China University of Geosciences, Wuhan, P.R. China
e-mail: yujianxin@cug.edu.cn

J. Broutin
Faculty of Sciences and Engineering (CR2P, MNHN, CNRS), Sorbonne Université, Paris, France

X. Shi
College of Earth Sciences, Jilin University, Changchun, Jilin, P.R. China
e-mail: xiaoshi@jlu.edu.cn

H. Li
Jiangxi Key Laboratory for Mass Spectrometry and Instrumentation, East China University of Technology, Nanchang, P.R. China
e-mail: huili_201860188@ecut.edu.cn

J. Yu et al. (eds.), *Plants and Palynomorphs around the Permian-Triassic Boundary of South China*, New Records of the Great Dying in South China,
https://doi.org/10.1007/978-981-19-1492-8_7

In discussing terrestrial P–T vascular plants extinction, some scholars tend to put them together with the transformation from paleophyte into mesophyte. As a matter offact, the two problems are not similar though they may be related to each other (Rong and Fang 2004). It is known that the transformation of paleophyte into mesophyte should have suffered a long evolution over 40 Ma, related to the change of climate, characterized by diachronic developments in different areas (Meyen 1973; Knoll 1984; Traverse 1988). For instance, the wetland or swamps well developed in Carboniferous disappeared rapidly; the coal swamp forest dominated by woody lycopods began to vanish at the beginning of Early Permian in this Euramerican domain. This started the transformation from paleophyte into mesophyte marked by widespread development of highland flora and the large-scale occurrence of Peltaspermales represented by *Autunia* in this period. In contrast, South China only witnessed this transformation in Early Triassic though the area also lies in the ocean-island block of the equatorial zone, where the coal-forming flora maintained the basic paleophytic outlook until Changhsing stage (Ouyang 1982).

As described above, the vegetations in Euro-America and in South China all changed greatly at the end of Permian although the former had entered earlier mesophytic conditions and the latter, the Cathaysian area, still remained in paleophytic paleoenvironments, both of which are characterized in common by the occurrence of new type of lycopods and the substitution of the original plants (Fang 2004). It has been assumed that the vegetation transformation should be the result of the global sudden catastophe and should have no relation with whether they entered mesophytic time, for the characteristics of this event is different from the progressive change from paleophyte into mesophyte. Meanwhile, the palynofloras also prove that the vegetations in the world experienced sudden changes to a great extent at the turn from Permian to Triassic. It is found that the Late Permian palynomorphic assemblages in the most areas of the northern hemisphere are dominated by gymnospermae pollens, with the typical examples found in middle Euramerican flora and Angara flora (Utting and Piasecki 1995). At the traditional marine Permian–Triassic boundary, gymnospermae pollen grains from the upland floras are suddenly replaced by the coastal low diversity assemblages characterized by lycopods (Balme 1970; Utting and Piasecki 1995; Eshet et al. 1995; Visscher et al. 1996; Looy et al. 1999, 2001; Twitchett et al. 2001). However, in South China are dominated by Pteridophytic and Pteridospermous spores in Late Permian, dominated by gymnospermae pollen grains in Early Triassic.

The previous geological workers also researched the paleofloral changeovers across the Permian–Triassic boundary in non-marine depositional areas of South China (Yao et al. 1980; He et al. 1996; Li 1995; Yu et al. 2007, 2015). They considered that the paleofloral evolution is a gradual change in this period. Ouyang (1986) indicated that the evolution of plants during Late Permian and Early Triassic had a sharp continuity by analyzing the Late Permian and Early Triassic palynomorph assemblages in Fuyuan District, Yunnan Province. Wang and Yin (2001) proposed that the plants showed an obvious break, while having a shade during the break by researching the fossil plants from non-marine Permian–Triassic transitional strata in western

Guizhou and eastern Yunnan. Fang (2004) pointed out that terrestrial Permian–Triassic vascular plants presented two-episodes extinctions during the Permian–Triassic transition. Now based on the data from both macroflora and microflora studied from the non-marine PTB sections, including four non-marine PTB sections in present dissertation and several other non-marine in western Guizhou and eastern Yunnanstudied by previous workers. (Yao et al. 1980; Zhao et al. 1980; Ouyang and Li 1980; Ouyang 1982, 1986), the changing patterns of land vegetation across the non-marine PTB in South China will be discussed in the following.

7.2 Changes of Mega-Flora Fossils from the Late Permian to Early Triassic

Fossil Plants are the most common megafossils found in the non-marine PTB sections in western Guizhou and eastern Yunnan,and are usually highly concentrated in the Upper Permian associated with abundant coal beds/seams. They are composed mainly of Filicopsida, Pteridospermopsida, Sphenopsida, Lycopsida, Gigantopteridales and some members of Cycadopsida, Ginkgopsida and Cordaitopsida. In comparison, the Lower Triassic in the study area is poorly fossiliferous; until now only a few macrofloral fossils have been found from the Lower Triassic Kayitou Formation. This disparity in the diversity of early Triassic plant species clearly contrasts with the diverse flora in the Late Permian.

7.2.1 Short-Term Mega-Flora Changes Across the PTB

In western Guizhou and eastern Yunnan, the abundance and diversity of plants demonstrate profound changes across the PTB interval. Plant fossils are abundant in the Late Permian strata in the whole area, with up to 59 genera and 138 known species recorded (data after Li 1995). The most common plant genera found are those typical of the Cathaysian *Gigantopteris* flora; notable examples are *Gigantopteris, Gigantonoclea, Fascipteris, Taeniopteris, Lobatannularia* and *Tingia*, etc. In contrast, the Earliest Triassic vegetation was low in abundance and diversity, with only 15 genera and 19 known species represented (data after Zhao et al. 1980), and locally, such as in the paralic Mide and Tucheng sections, they are associated with typical Early Triassic marine bivalve species such as *Pteria ussurica variabilis* and *Pteria ussurica*. In general, the sparse and monotonous Early Triassic vegetation is mainly composed (found up to date) of *Annalepiszeilleri* Fliche, *A. latiloba* Meng, *A. brevicystis* Meng, *Peltaspermum lobulatum* Wang and Wang, cf. *Peltaspermum martinsii* sterile foliage (Germar) Poort and Kerp., *Annularia shirakii* Kawasaki, *Compsopteriscontracta* Gu and Zhi, *Cordaites principalis* Geinitz, *Gigantonoclea* sp., *Gigantopteris* sp., *Lobatannularia* sp., *Paracalamites stenocostatus* Gu and Zhi, *Pecopteris* sp.,

Rajahia guizhouensis Zhang. Most of the listed Early Triassic species are relicts of the Permian Cathaysian *Gigantopteris* flora. They extended to the early Early Triassic as the survivals, lastly gradually disappeared along with the occurrences of *Annalepis* and *Peltaspermum.* Only two genera *Annalepis and Peltaspermum,* considered to be Triassic newcomers, were found in the Lower Triassic of the Mide and Tucheng sections.Thus, the regional disappearance of plants across the PTB in the study area appears to be gradual.

7.2.2 Long-Term Changes of Megaplants from Late Permian to Early Triassic

As any temporal patterns of diversity changes are very much a function of time scale used, it is therefore considered important to also consider the patterns of megaplant changes across a longer timescale, which in this case refers to the interval from Wuchiapingian to the Early Triassic Griesbachian. To do this, we compiled a stratigraphic database of megaplant distributions in western Guizhou and eastern Yunnan, (mainly based on data from Zhao et al. 1980; Li 1995; collecting by the author etc.). As statistically illustrated in Table 7.1, plant genera and species gradually decreased

Table 7.1 Genus and species number of the groups of the paleofloras and percentage of species in western Guizhou and eastern Yunnan, South China

Type	Genus		Species		Percentage of species (%)	
	Changhsingian	Induan	Changhsingian	Induan	Changhsingian	Induan
Lycophyta, Lepidodendrales and Isoetales	4	3	13	6	12.4	33.3
Sphenophytes	6	2	15	2	14.3	11.1
Pteridophytes	7	1	29	1	27.6	5.6
Progymnospermophyta, Noeggerathiales	1		3		2.9	
Pteridospermophyta	7	3	18	4	17.1	22.2
Cycadophyta	2		3		2.9	
Ginkgophyta	2	1	4	1	3.8	5.6
Cordaitopsida	1		3		2.9	
Gymnospermum seeds	1		1		1.0	
Gigantopteridales	2	2	11	2	10.5	11.1
Glossopteridales	2		2		1.9	
Plantae incertae sedis	4	2	4	2	3.8	11.1
Other forms	2		2		1.9	
Total	**39**	**14**	**105**	**18**	100.0	100.0

from 47 genera and 114known species in the Wuchiapingian, to 45 genera and 87 known species in the Changhsingian, and only 15 genera and 19 known speciesin the Induan. Regionally, up to date the PTBST (Beds 66f-68 in Chahe section; Beds 47–49 in Zhejue section), has only recorded two genera and three species. In other words, about 24% of megaplant species were eliminated at the end-Wuchiapingian, and those species that survived into the Changhsingian mostly also decreased in abundance. About 90% species of the Changhsingian flora were the descendents of Wuchiapingian taxa, and only about 15% of species originated in the Changhsingian. At the end of the Changhsingian, about 80% of species vanished (Table 7.1). The surviving 12 known species in the earliest Triassic also decreased in abundance and most of them were descendents of the Late Permian Cathaysian types.

Stratigraphic distribution shows that both the macro- and microfloras from the WGEY region have suffered a dramatic change in composition, abundance and diversity (Fig. 7.1) across the PTB. The Wuchiapingian flora is characterized by abundant gigantopterids, which were rather abundant and diverse and commonly present in most of the Upper Permian successions in South China (Fig. 7.2). The coeval palynoflora was dominated by spores of the Pteridophyta and Pteridospermae. Most genera are characteristics of the Permian Cathaysian elements (i.e., *Nixispora* and *Patellisporites*). During the Changhsingian stage, the *Gigantopteris* flora markedly

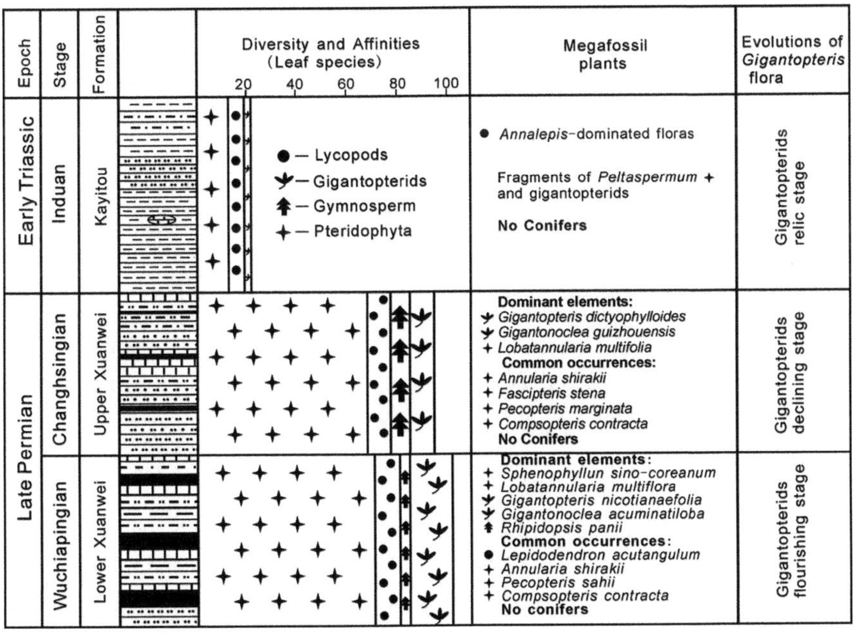

Fig. 7.1 Changes in fossil floras of non-marine facies deposits across the Permian–Triassic boundary within the Southwestern China, showing abrupt decline of the coal-forming *Gigantopteris* flora and replacement by a low-diversity lycopod flora

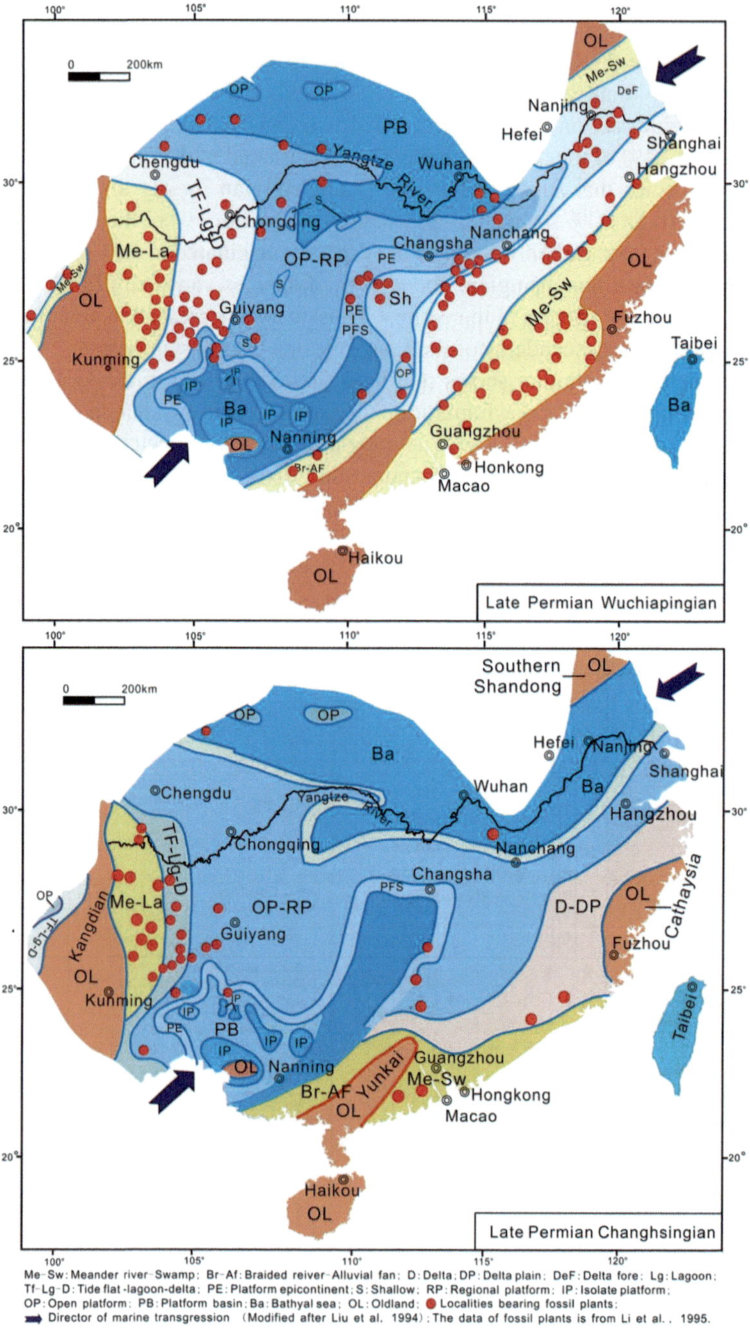

Fig. 7.2 Distribution of the Late Permian *Gigantopteris* flora localities related with marine transgressions in South China

decreased in species richness and localities (Fig. 7.2). In contrast, increasing representatives of gymnosperms occurred in the Changhsingian palaeoflora. It should be noted that the WGEY flora contains not only many taxa typical of the Late Permian *Gigantopteris* flora (such as *Lobatannularia, Fascipteris, Gigantoclea,* and *Gigantopteris*), but also many Triassic-type elements, and thus, appears a mixture nature of the Permian-type and Triassic-type floras, like the well-known mixed faunas near the PTB in marine regime in South China (Sheng et al. 1984; Chen et al. 2005).

7.3 Changes of Palynomorph Assemblages Across the PTB

Like the megaplants, the palynomorph assemblages in the study area also demonstrate significant changes across the PTB (Peng et al. 2006). In general, the palynomorphs changed from fern-dominated spores to pteridosperm and gymnosperm-dominated pollen across the PTBST, which itself contains a mixed flora with Late Permian relicts and Early Triassic elements, such as *Aratrisporites yunnanensis, Lundbladispora communis,* as well as abundant spores of lower organisms (e.g. fungi and/or algae). The latter is comparable with the so-called fungal event (Eshet et al. 1995; Visscher et al. 1996) or algae event.

However, the long-term, Wuchiapingian to Early Triassic Induan, temporal changes in the palynomorph assemblages in south China show a different pattern with regards to the megaplants over the same interval in that it first recorded a small increase in diversity from the Late Permian Wuchiapingian to the Changhsingian, followed by a decrease from the Changhsingian to early Early Triassic Induan (Fig. 7.3). However, from Wuchiapngian to Induan the palynomorphs decreased in abundance in a stepwise pattern, as do the mega-flora records, with most disappearing in the Induan. Therefore, again the temporal changing patterns of palynomorphs from the Late Permian to the Early Triassic do not suggest a synchronous mass extinction at the end-Permian.

Data on terrestrial floral changes across the Permian–Triassic boundary are presented separately from megaplant fossils and palynomorphs across several non-marine PTB sections in western Guizhou and eastern Yunnan. The changing trend in higher plant vegetation across the Paleozoic–Mesozoic transition is characterized by a change from fern-dominated vegetation in the Upper Permian to pteridosperm and gymnosperm-dominated vegetation in the Lower Triassic. Temporarily, the decline of land vegetation in southwest China began in the early Late Permian (Wuchiapingian) and persisted through to the Early Triassic, and the rate of decline in both species and generic diversity was accelerated at the Permian–Triassic transitional interval.

Overall, the temporal trend in the change of plants in the study area appears to be indicative of a long, protracted extinction process that may have commenced well before the Permian–Triassic boundary. Such a "drawn out" extinction process spanning the entire Late Permian and across the PTB therefore does not support a globally synchronous end-Permian catastrophic event such as meteor impacts;

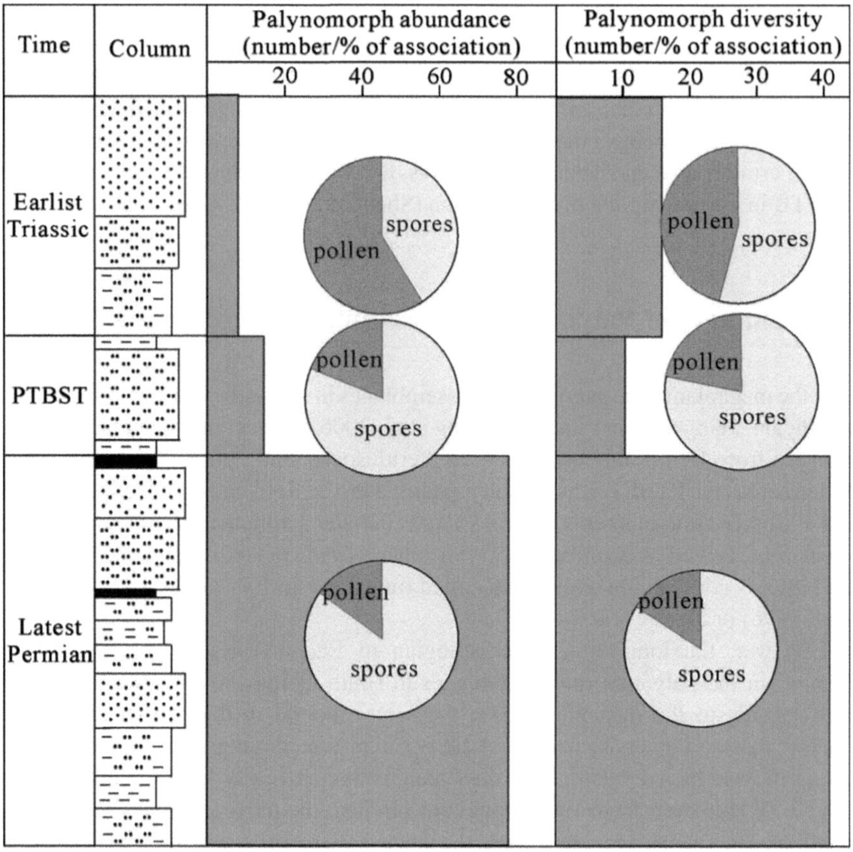

Fig. 7.3 Palynomorph changing patterns across the PTB (versus the lithostratigraphic base of the Chahe section) in western Guizhou and eastern Yunnan, South China (data from the Zhejue and Chahe sections)

rather it is more consistent with a protracted global climate change that may have been initiated by Pangea formation, and then exacerbated bythe Siberian Trap (and also possibly the Emeishan Basalt) volcanism at the end-Permian.

7.4 Recovery of Flora from South China After the End-Permian Biotic Crisis

The geologic past holds many examples of extinction and recovery of both terrestrial and marine ecosystems. At the beginning of the 1990s, new research on the problems of biotic survival and recovery during (after) great crisis has become a subject of increasing scientific interest and has resulted in two international projects: the Global

Changes Program (Ricklefs et al. 1990) and the IGCP Project 335 Biotic recoveries from mass extinctions (Kauffman and Erwin 1994). However, because interest in recoveries has been less than in the crisis events themselves, the recuperation intervals have been only superficially examined and therefore are poorly understood (Erwin 2001). This is particularly true of the Triassic survival and recovery intervals. Furthermore, most studies concern only marine invertebrates (Erwin 2001). Although land plants are known to be particularly sensitive to environments and paleoclimate and therefore very likely to be instructive for evaluating the patterns and processes of biotic recovery, they have been rarely analyzed in detail from this perspective. This part focuses on the post-crisis recovery of vegetationdestroyed by physical events at the Permian–Triassic boundary.

Wang Ziqiang (1996) is the first paleobotanist who has thoroughly studied the land-plant recovery and noted its decimation at the Permian–Triassic transition and its subsequent recovery. Looy et al. (1999), who analyzed the Triassic land-plant recovery on the basis of palynological data, also noted that the end-Permian flora, particularly the conifers, has been decimated, and showed that the most suitable fossils for such studies are pollen and spores, because of their unique fossilization potential. Grauvogel-Stamm and Ash (2005) summarized Early and Middle Triassic land-flora, and pointed out thatthe recovery of the Triassic floras of Europe began, as in North China, with the proliferation of the lycopsid *Pleuromeia* during the Early Triassic and that it proceeded with the resurgence of the conifers in the early Middle Triassic (Early Anisian), the return of the cycadophytes and the pteridosperms in the Late Anisian and the progressive evolutionary modernization of the subsequent. This lycopsid was an opportunistic pioneer plant, as *Annalepis* did in South China which took advantage of the post-crisis vacuity of the environment to proliferate, preparing sites for re-colonization with the plants having survived in refugia. The Triassic land-plant recovery is a long stepwise process.

The above analysis clarifies that Permian–Triassic Xuanwei flora in western Guizhou-eastern Yunnan is consistent in its transformation with quality of the global flora. Part of the Permian prosperous elements survived the extinction as the residual elements and continued to Early Triassic, for instance, gigantopterids of Cathaysian Flora and Glossopterids of Gondwana Flora (Pant and Pant 1987; McManus et al. 2002). While the new type of lycopods appeared in a sudden above the P–T boundary, became predominant at the beginning of Triassic (Balme 1970; Dobruskina 1987), which is marked by the appearance of representatives of the genus *Annalepis* in early Early Triassic in western Guizhou and eastern Yunnan of South China, *Pleuromeia* became the advantageous elements of North China florain the same period, and the rapid spread of *Peltaspermum* and *Lepidopteris* to the whole globe in Early Triassic (Dobruskina 1987; Shah 1987; Mogutcheva 1996; Paland Ghosh 1997; Mcloughlin et al. 1997). However, the Xuanwei flora has its own characteristics in the evolution.

7.4.1 Triassic Land-Plant Recovery in North China: A Model

Recovery process after mass extinction is the prelude of a new radiation. With the exception of a few groups, recovery for biotic groups appeared based on the survival stage. It is characterized by an increase in origination rate, endemic taxa.

In North China, the end-Permian life crisis was caused by increasing desertification, which wiped out up to 75% of the land flora at the genera level and 98% at the species level, including most of the typical Palaeozoic species Wang (1989, 1993, 1996). The total species number passed from 146 taxa in the lower Upper Permian (Tianlongsi Formation) to 54 taxa in the late Upper Permian (Sunjiagou F.), 16 taxa in the early Lower Triassic (Liujiagou F. = Induan) and 58 taxa in the late Lower Triassic (Heshangou F. = Olenekian) where it began again to increase (Wang 1989), showing that the end- Permian life crisis was a long process that began well before the end of the Permian, and that the subsequent Triassic recovery was also a long process lasting until the end of the Middle Triassic.

Wang (1996) recognized two stages in the Triassic land-plant recovery: the first was the Early Triassic *Pleuromeia* stage with three sub-zones, which was dominated by the lycopsid *Pleuromeia* [Fig. 7.4 (1).II–III] and the second was the Middle

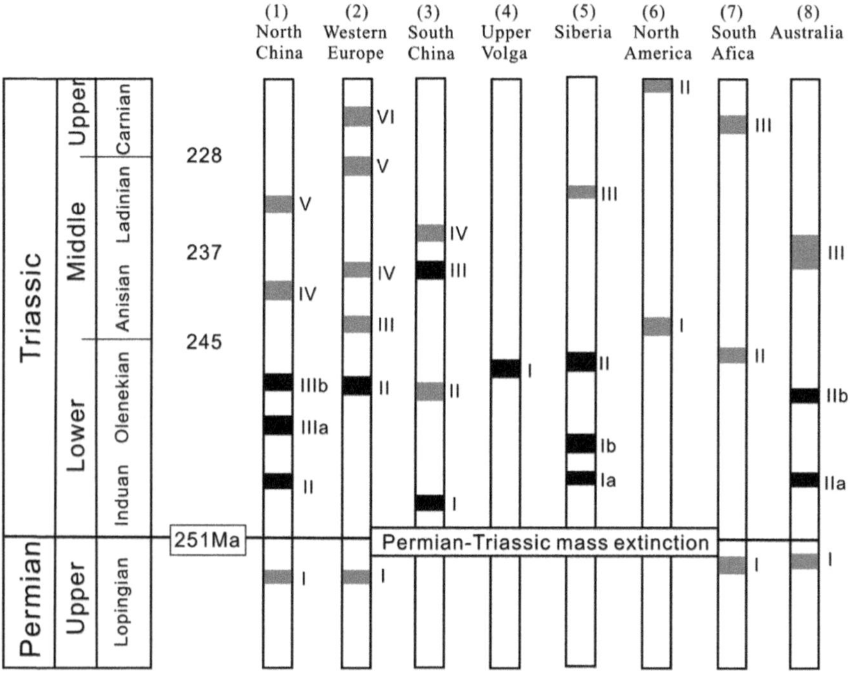

Fig. 7.4 Tentative simplified correlation table showing the distribution of the end-Permian and Triassicfloras in different regions of the world (slightly modified according to Grauvogel-Stamm and Ash 2005)

Triassic *Tongchuanophyllum* stage with two sub-zones (the *Isoetes* sub-zone and the *Scytophyllum* sub-zone), which is characterized by the broad-leaved pteridosperm *Tongchuanophyllum* and the sphenopsids *Equisetites* and *Neocalamites* [Fig. 7.4 (1).IV–V]. This two stage's recovery is considered as one model of Triassic land-flora from the end-Permian life crisis (Grauvogel-Stamm and Ash 2005).

7.4.2 *Triassic Land-Plant Recovery in Europe*

Kerp (2000) recently pointed out that the plant megafossil record from the Permian through the lower Middle Triassic of Europe is poorly known, and seems to preclude study of the Triassic land-plant recovery in this area. Similarly Meyen (1987) had recognized that a detailed reconstruction of the Permian–Triassic transition was impossible. However, a new analysis by Grauvogel-Stamm and Ash (2005), has given the Triassic land-plant recovery also went through two stages: a long survival interval in the Early Triassic, similarly dominated by the lycopsid *Pleuromeia* being the oldest Triassic land flora represented in Europe [Fig. 7.4, (2).II]; and the early Middle Triassic floras correspond to the recovery interval itself, with the early Anisian flora of northeastern France [Fig. 7.4, (2).III] (Grauvogel-Stamm 1978) representing the first phase and the Upper Anisian flora of Italy [Fig. 7.4, (2).IV] (Broglio et al. 2002) corresponding to the second phase.

In many other areas, the Uppermost Permian and Lower Triassic are represented by marine facies. However, in some areas continental facies with fossil plants are intercalated with marine deposits making possible to determine which of the recovery stages the plant record belongs to. In most areas discussed below (Siberia, North America, South Africa and Australia, respectively Fig. 7.4, (5), (6), (7) and (8), the recovery seems to have been completed during the Middle Triassic as well as in Europe and North China, South China.

7.4.3 *Triassic Land-Plant Recovery in South China*

1. **The plant assemblages of the Lower and Middle Triassic in South China**

In South China, Early and Middle Triassic strata are mostly marine facies. Fossil plants appeared only in Early Triassic Kayitou Formation of western Guizhou and eastern Yunnan, and Lingwen Formation of Qiongxian, Hainan Province (Fig. 5.2); the Middle Triassic Badong and Huangmaqing Formations, the middle part of the Yangtze River (Fig. 5.3), respectively Induan, Olenekian, Anisian and Ladinian. They consist of four assemblages, with the lowermost being Induan and the uppermost being Ladinian [Fig. 7.4, (3).I, II, III, IV].

In studying the recovery of the land plant in South China, the previous workers discussed a few. The main reason is the absence of a whole Early Triassic plant assemblage. With the research of fossil plants from western Guizhou and eastern Yunnan in the recent some years, the early Early Triassic plant assemblage is established in the thesis. Consequently, it is possible to discuss the land plant recovery in this area.

Black = Lycopsid-dominated floras; grey = more or less diversified floras. Please note that the columns and the plant-bearing beds are not at scale. For the source data, see the references in the relevant sections.

(1) North China: I-Late Permian (Sunjiakou Formation): *Ullmannia* zone; II-III: Induan-Olenekian: *Pleuromeia*-dominated floras (II-Induan: Liujiakou Fm. = *P. jiaochengensis* subzone; III - Olenekian: Heshanggou Fm.; III a = *P. epicharis* subzone; III b: *P. sternbergii* subzone); IV - Anisian (Ermaying Fm.): *Tongchuanophyllum* zone including the *Isoetes* subzone at the base and the *Scytophyllum* subzone above; V - Ladinian (Tongchuan Fm.) with *Danaeopsis*. **(2) Western Europe:** I-Late Permian (Lopingian): Zechstein flora, Val Gardena flora, N. Italy = conifer-dominated floras; II-Olenekian (Middle Buntsandstein), Germany = *Pleuromeia*-dominated flora; III-Early Anisian ('Grès à *Voltzia*' Fm., Upper Buntsandstein), NE France: conifer-dominated flora; IV - Upper Anisian (Dont Fm.), N Italy: conifer/cycadophyte/fern-dominated flora; V-Late Ladinian (Lower Keuper), Germany = sphenopsid-dominated flora (Wengen Fm.), N. Italy = *Voltzia*-dominated flora; VI - Middle Carnian (Schilfsandstein), Germany = sphenopsid-dominated flora. **(3) South China:** I- Induan (Kayitou Fm.) *Annalepis*- dominated flora. II-Olenekian (Lingwen Fm.) conifer-dominated flora; III Anisian (Lower Badong Fm.): *Pleuromeia-Annalepis*-dominated flora; IV-Ladinian (Upper Badong Fm.): *Annalepis*-dominated flora. **(4) Upper Volga (Russia):** I - Olenekian (Rybinsk Fm.): *Pleuromeia*-dominated flora. **(5) Siberia (Russia):** I a, I b - Induan: *Pleuromeia* and *Tomiostrobus* (sub-genus of *Annalepis*) dominated-flora; II-Upper Olenekian: *Pleuromeia*-dominated flora, Taymyr, Verkoyanye; III-Ladinian flora. **(6) North America:** I-Early Anisian (Moenkopi Fm.): flora with *Donwelliacaulis*; II-Late Carnian flora (Chinle Fm. with the Shinarump Member at the base and the overlying Petrified Forest Member, which is Late Carnian - Early Norian). **(7) South Africa (Gondwana):** I-Late Permian: *Glossopteris*-dominated flora; II-Late Olenekian (Burgersdorp Fm.): diversified flora; III-Early-Middle Carnian (Molteno Fm.): rich and diversified flora. **(8) Australia (Gondwana):** I-Late Permian: *Glossopteris*-dominated flora; II a, II b - Induan + Olenekian: lycopsid-dominated floras; III -Middle Triassic, (Basin Creek Fm. in the Nymboida Coal Measures, Clarence-Morton basin).

The Induan assemblage is dominated by the lycopsids *Annalepis* [Fig. 7.4, (3).I], but it also includes some elements of Peltaspermales and a few relicts of Late Permian Cathaysian flora. The feature of the Olenekian *Neuropteridium-Voltzia* assemblage (Zhou and Li 1979) [Fig. 7.4, (3).II] is that the conifers highlighting *Albertia* and *Voltzia*is dominating, the horsetails and *Neuropteridium* are quite also numerous, while Lycopsids are very rare, whose diversifications can be compared with those of the Buntsandstein flora of Europe.The Anisian assemblage [Fig. 7.4, (3).III] (Meng et al. 2000) is dominated by the lycopsids *Pleuromeia* (many in life position) and *Annalepis*, but it also includes most of the taxa described in the 'Grès à *Voltzia*' flora of northeastern France and the Ermaying Formation of North China, which are both early Middle Triassic. The Ladinian assemblage (Meng et al. 2000) [Fig.7.4, (3).IV] that includes mainly *Annalepis*, *Equisetites*, *Neocalamites*, *Scytophyllum*, *Voltzia*, *Taeniopteris* is said to resemble the upper part of the Ermaying Formation of North China.

2. **Taphocoenoses modes of the Lower and Middle Triassic in South China**

In the case of the Lower and Middle Triassic in South China, a total number of about 12 plant-bearing biostromes (Zhou and Li 1979; Meng et al. 2000; the author et al.) are the basis for the analysis resulting in enumeration of four types of plant-taphocoenoses based on plant preservation.

Autochthonous burial: Whole plants vertically buried in life position, or occurrences in one biostrome of a suite of isolated organs that have naturally detached, permitting the whole plant to be reconstructed.**Allochthonously burial over short distances (similar to parautochthonous):** Isolated organs, such as a leaf or a frond, that are completely buried after being shed and having undergone some transport within the habitat of growth.**Allochthonous burial:** partiallypreserved isolated organs that have been transported from the growth habitat to the place of deposition.**Fragmentation burial:** A variety of plant fragments or debris that have undergone transported, that can be barely identified at the genus level.

(1) **The *Annalepis -Peltaspermum* and the relicts of gigantopterids assemblage (Induan)**

The characterstics of this assemblage will be described in the following Chap. 9, genus *Annalepis* is the dominant element. This assemblage also is called the opportunistic pioneer lycopsid- dominated flora. All the taphocoenoses are restricted to grey-greenish or grey-yellowish muddy siltstone or mudstone representing deposits of a lagoon (Zhao Quanming, 2003, PhD thesis). This assemblage is dominated by *Annalepis*-taphocoenosis in allochthonously burial over a long distance: the *Annalepis* having sporadic sporophylls, megaspores in situ, microspores and suspicious strobile and root (Fig. 7.5). The part relicts in the top Xuanwei Formation (e.g. Bed 69 of the Chahe section) are in allochthonous burial. The pinnules relatively are complete, the second rachis of some fossil plants were preserved. Other plants, which are preserved as very small, or reworked? fragments, might be transported for a very long distance, belonging to the fragmentation burial.

(2) **The *Neuropteridium-Voltzia*assemblage (Olenekian)**

According to the data of Zhou and Li (1979), and Meng et al. (1992), a total of 38 species belonging to 28 genera have been described in the following Chap. 9, fragments of conifer *Voltzia* and *Albertia* are the dominant elements. The others are broken fragments. It shows that the Lingwen flora is an allochthonous burial.

(3) **The *Pleuromeia marginulata-Annalepiszeilleri* assemblage (Anisian)**

In South China, the Middle Triassic Badong Formation, consisting of a set of red clastic deposits intercalated with carbonate rock in the middle part, is widespread in

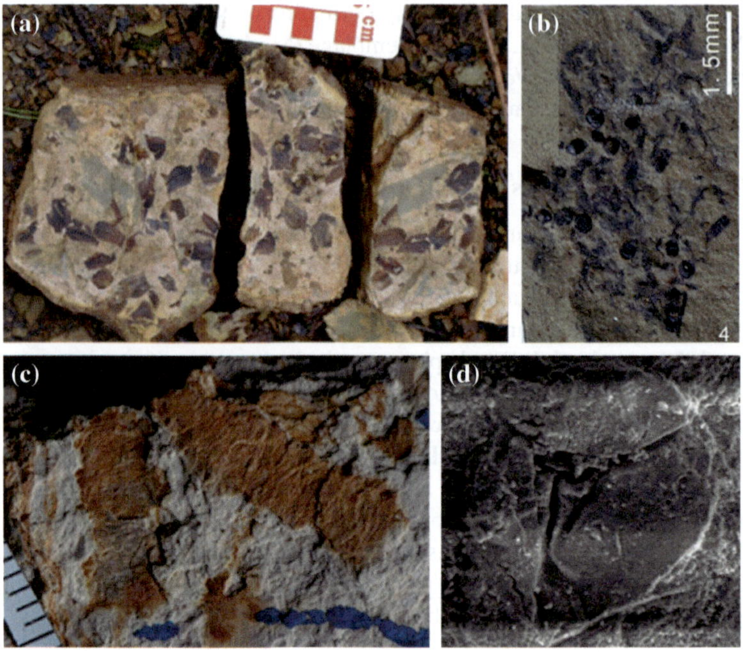

Fig. 7.5 Showing the *Annalepis*-taphocoenosis in allochthonously burial over a long distance. **a** sporadic sporophylls **b** possibly strobile **c** possibly roots **d** megaspore in situ

the Yangtze Gorge area, with a total thickness of about 1000 m. The formation may be evidently subdivided into 5 lithologic members in its named locality. The *Pleuromeia-Annalepis* flora was mainly found in the gray-green argillaceous siltstones or the siltstone lens in Members 2 and 5, but a few of specimens were collected from the sandy mudstone in Members 1 and 4.

The assemblage occurred in Members 1–2 of the Badong Formationis rich in composition, containing about 31 species belonging to 18 genera. The main elements are *Pleuromeia marginulata, P. sanxiaensis, P. hunanensis, Annalepis zeilleri, A. brevicystis* etc. Of them, the *Pleuromeia* is generally well preserved, with many complete plants, with some stems almost vertically preserved in sediments or some appendixes showing a growing state almost perpendicular to the bedding plane. The phenomenon mentioned above indicates that it may be autochthonous taphonocoenose. The *Annalepis* has both sporadic sporophylls and rather perfect cones. Presumably, it has ever been transported for a short distance, but basically is sub-autochthonous taphonocoenose. All other plants, which are preserved as fragments, might be transported for a long distance. So the assemblage represented a mixed plant taphonocoenose.

(4) **The *Annalepis latiloba-Scytophyllum* assemblage (Ladinian)**

The assemblage is represented by the plants from the Members 4–5 of the Bedong Formation.

The main elements of the assemblage include *Annaleplslatiloba*, *Equisetites arenceus*, *Neocalamites* sp., *Scytophyllum* sp., *Sphenozamites* sp., *Sinoctenis pulcella*, *Voltzia* sp., *Taeniopteris* sp., etc. The present assemblage shows rather simple genera and species and the presence without *Pleuromeia*, whereas the evident increase of *Scytophyllum* in amount and the occurrence having *Sphenozamites* and *Sinoctenis* which are commonly found from Late Triassic in South China.

Comparison with the *Pleuromeia marginulata-Annalepiszeilleri* assemblage,the *Annalepis*-taphocoenosis also is allochthonous burial over short distances. The *Scytophyllum*- taphocoenosis may be allochthonous burial.

3. **The process of vegetational recovery of the Lower and Middle Triassic in South China**

In South China, the Late Permian vegetation was composed mainly of Equisetales, Filicopsida, Pteridosperms and Gigantopteridales. The subsequent terminal Permian mass extinction destroyed almost all the Permian vegetation here, including most of the characteristic Paleozoic species. Only a few relicts at the generic level have survived the severe event. The present paleobotanical analysis suggests that in South China the subsequent Early and Middle Triassic vegetational recovery had two main stages.

(1) **Expansion of the vegetation stage in Early Triassic**

The synthetically data at the species level from megaplants in South China (including Fujian, Jiangxi, Guangdong, Jiangsu, Hunan, Sichuang, western Guizhou and eastern Yunnan) shows that the total species number passed from 203 known species (data from Li 1995) in the lower Upper Permian (Lungtan Formation and the lower part of the Xuanwei Formation) to 131 known species in the late Upper Permian (Dalong Formation and the upper part of the Xuanwei Formation) (data from Li 1995 and this thesis), 19known species in the early Lower Triassic (Kayitou F. = Induan) and 38taxa in the late Lower Triassic (Lingwen F. = Olenekian) This degressive characteristic is similar to that of North China (Fig. 7.6).

A. At the beginning of the Triassic, the landscape is assumed to have been rather barren. Shortly after, a few small monospecific herb-like *Annalepis* association in nearshore region of western Guizhou and eastern Yunnan, and *Peltaspermum* association intermittently occurred.

The *Annalepis*, including 4 known species, is a monogenetic association, and it likely grew in marginal area of nearshore. There was more or less influence by the break. The sporophylls of this genus are lanceolate, with an acute apex, truncate base and smooth lower surface. The sporangia are slender stick-shaped, attached closely to the central part of the ventral surface of the sporophyll, with a warped anterior apex of the sporophyll formed in water possibly because of its weight. All the features show that *Annalepis* possess a drift ecological character in water

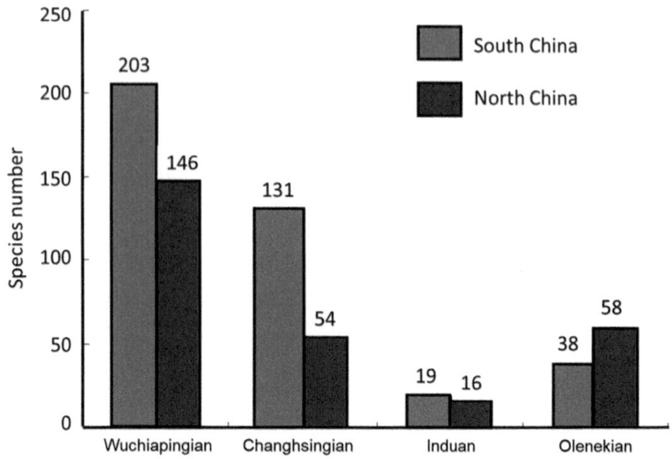

Fig. 7.6 Correlation of paleofloral species level between South and North China during Late Permian and Early Triassic

(Meng et al. 2000). In addition, the macrospores of *Annalepis* are relatively hard, with a smooth surface and larger specific gravity than other plants. These features are advantageous to the precipitation of the macrospores, and development of new plants in coastal environment (Meng et al. 2000).

The reduced and shortened pinna and sterile organs (?) of *Peltaspermum*, broken fragments of the Gigantopterids may be considered to represent another ecotype in the stage. They belong to fragmentation burial association. This taphocoenoses in the Kayitou Formation occur in 2.5m thickness, monocyclic sandy mudstone bed. No gymnosperm remains appeared.

The red-beds began to appear in the upper Kayitou and Dongchuan Formations (Olenekian stage), no fossil occurred.

B. In the second half of the early Triassic, the largest portions of muddy red-beds of the Lingwen Formation in Qionghai, Hainan province, and also Dongchuan Formation in western Guizhou and eastern Yunnan, indicates that dry climate had begun. The fossil plants only occur in gray, fine-grained sandstone overlying the red-rocks in Lingwen Formation. In the Qionghai taphoflora, fragments of conifer are numerous, but some articulates have been detected, but their body is small and the internodes shorten (Zhou and Li 1979). It appears that the plants were not preserved in situ. During this period, Gymnosperm (e.g. *Albertia* and *Voltzia*) and pteridosperm (e.g. *Neuropteridium*) began to be flourishing. These indicate that the mesophytic plants on upland and on shore could further develop.

The *Annalepis*-dominated flora of western Guizhou and eastern Yunnan is similar to the second zone (Liujiagou Formation) of North China [see Fig. 7.4, (1).II], which is also lycopsid *Pleuromeia*-dominated. In Gondwana, the herbaceous lycopsids (*Cylostrobus*, *Skilliostrobus*), which had been only a small part of the Permian land flora, expanded suddenly with the advent of the Triassic and became dominant during the Early Triassic, but then declined and virtually disappeared by the end of the Early Triassic. The Voltzialean conifer *Voltziopsis* followed the general trend of the herbaceous lycopsids. It became dominant in the Early Triassic land flora of Gondwana, after which it declined and became again common in the Late Triassic. The Early Triassic paleoflora in South China is correlated to that of the Gondwana. Therefore it can be similarly considered to represent the first stage of the Triassic land-plant recovery in South China.

(2) **A changeover: the occurrence of riparian vegetation in shore-lagoon**

The *Pleuromeia-Annalepis*flora consists of two assemblages, with the lowermost being Anisian and the overlying being Ladinian (Meng et al. 2000) [Fig. 7.4, (3).III and (3).IV]. The Anisian assemblage is dominated by the lycopsids *Pleuromeia* (many in life position) and *Annalepis*, but it also includes most of the taxa described in the 'Grès à *Voltzia*' flora of northeastern France and the Ermaying Formation of North China, which are both early Middle Triassic (Grauvogel-Stamm 1978) and (Wang 1996).

Based upon Meng's research (2000), all the *Pleuromeia* and *Annalepis* are distributed in shore-lagoon environment in the Yangtze valley. The Pleuromeia, representing a small type of monogenetic association, is composed of 2 species of dwarf bushes. As to the ecology of *Pleuromeia*, one indicated that it was initially considered to be a kind of xerophyte in desert (Magdefrau 1931, 1956) or halophile (Clarke and Hannon 1971). The other thought that it might grow in a relatively flat shoreline area (Kon'on 1973), frequently influenced by waves and tides. The *Annalepis*, representing also a monogenetic association, was a small type of aquatic herb (Meng et al. 2000). Similarly, it likely grew in the marginal area of nearshore with *Pleuromeia*'s shrub.

The fact that *Pleuromeia* was still present in the Anisian in South China, whereas in North China and Europe it disappeared or is very rare, may be explained by differences in palaeogeography and climate. According to recent palaeogeographic reconstructions (Ziegler et al. 2003), South China was at the equator in the Early Triassic, whereas North China and Europe were close to the northern tropic at that time, latitude that is characterized by arid climates (Fig. 7.7). South China reached the tropics only in the Middle Triassic which would explain why *Pleuromeia* proliferated only at that period (Fig. 7.8).

The Ladinian assemblage that includes mainly *Annalepis*, *Equisetites*, *Neocalamites*, *Scytophyllum*, *Voltzia*, *Taeniopteris* is said to resemble the upper part of the Ermaying Formation of North China. However, it also resembles the Ladinian floras of Siberia and Germany.

The *Pleuromeia-Annalepis* flora of South China closely resembles the coeval flora of North China and Europe. So this flora represents the second stage of the Triassic land-plant recovery in South China. Finally, in latest Middle Triassic, the process of vegetational recovery in South China had been completed and fully vegetated landscapes had again appeared there. It is the best evidence that Late Triassic coal-bearing strata appeared again in the whole South China.

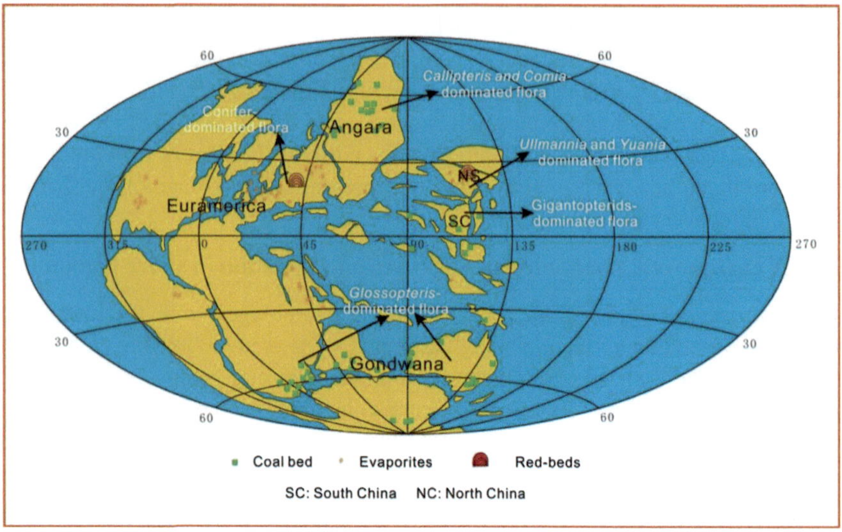

Fig. 7.7 Simply the distribution of the floras in different regions and paleogeographic reconstruction during Late Permian Changhsingian

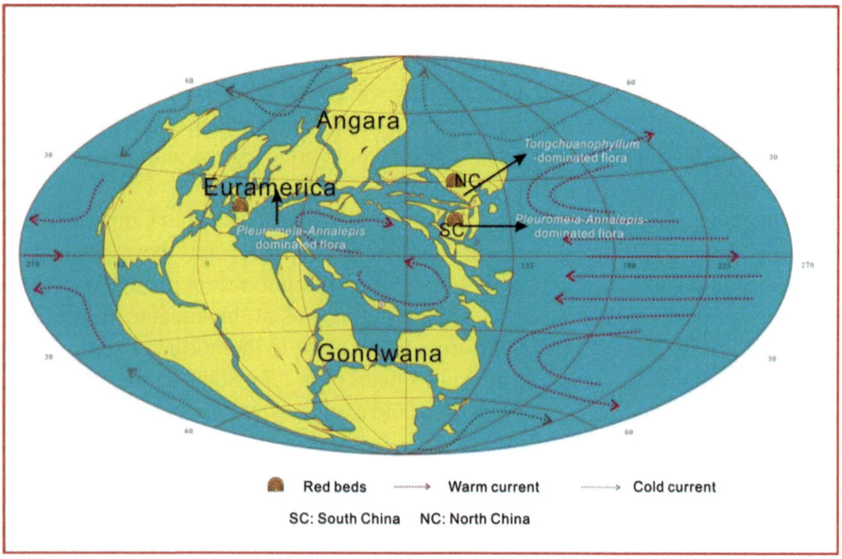

Fig 7.8 Simply the distribution of the floras in different regions and paleogeographic reconstruction during Middle Triassic Anisian

The above analysis clarifies that Permian–Triassic transitional Xuanwei flora in western Guizhou and eastern Yunnan is consistent in its transformation with quality of the global flora. Part of the Permian prosperous elements survived the mass extinction as the residual elements and continued to Early Triassic. While the new type of lycopsids appeared in a sudden around the P–T boundary at the beginning of Triassic and soon became predominant. Its recovery depends on the new types in the survivalinterval rather than the Paleozoic surviving flora, such as *Gigantopteris*. Specifically, *Annalepis* and *Peltaspermum* are mainly responsible for the recovery. This character is close to the recovery mechanism of foraminifera of the same date in South China (Tong 2004).

References

Balme BE (1970) Palynology of Permian-Tiassic strata in the Salt Range and Surghar, West Pakistan. In: Kummel B, Teichert C (eds) Stratigraphic boundary problems: Permian and Triassic of West Pakistan. University of Kansas, Special Publication 4, pp 306–453

Broglio LC, Fugagnioli A, Van Konijnenburg-Van Cittert J, Kustatscher E, Posenato R, Wachtler (2002) Anisian macroflora from the northern Dolomites (Monte Prà della Vacca/Kühwiesenkopf, Braies): a first report. Riv Ital Paleontogr Stratigr 108:1-390

Chen ZQ, Kaiho K, George AD (2005) Survival strategies of brachiopod faunas from the end-Permian mass extinction. Palaeogeography, Palaeoclimatology, Palaeoecology 224:232–269

Clarke LD, Hannon NJ (1971) The mangrove swamp and salt marsh communities of the Sydney district IV: the significance of species interaction. J Ecol 535–553

Dobruskina IA (1987) Phytogeography of Eurasia during the early Triassic. Palaeogeography, Palaeoclimatology, Palaeoecology 58:75–86

Erwin DH (1996) Understanding biotic recoveries: Extinction, survival, and preservation during the end-Permian mass extinction. In: Jablonski D, Erwin DH, Lipps JH (eds) Evolutionary Paleobiology. University of Chicago Press, Chicago, pp 398–418

Erwin DH (2001) Lessons from the past: biotic recoveries from mass extinctions. Proc Natl Acad Sci USA 98(10):5399–5403

Eshet Y, Rampino MR, Visscher H (1995) Fungal event and palynological record of ecological crisis and recovery across the Permian-Triassic boundary. Geology 23:967–970

Fang ZJ (2004) The Permian-Triassic boundary crisis: patterns of extinction, collapse of various ecosystems, and their causes. In: Rong JY, Fang ZJ (eds) Mass extinction and recovery, evidences from the Palaeozoic and Triassic of South China. University of Science and Technology of China Press, Hefei, pp 785–928 (in Chinese with English summary)

Fielding CR, Frank TD, Mcloughlin S, Vajda V, Mays C, Tevyaw AP, Winguth A, Winguth C, Nicoll RS, Bocking M, Crowley JL (2019) Age and pattern of the southern high-latitude continental End-Permian extinction constrained by multiproxy analysis. Nat Commun 10:385

Grauvogel-Stamm L (1978) La flore du Grès à Voltzia (Buntsandstein supérieur) des Vosges du Nord (France). Morphologie, anatomie, interprétations phylogénique et paléogéographique. Sci Géol Bull 50:1–225

Grauvogel-Stamm L, Ash SR (2005) Recovery of the Triassic land flora from the End-Permian life crisis. CR Palevol 4(6–7):593–608

He XL, Liang DS, Sheng SZ (1996) The Permian Floras from Jiangxi, China. China University of Minings Press, Xuzhou, pp 1–201

Kauffman EG, Erwin DH (1994) Biotic recoveries from mass extinction: initial meetings. Episodes 17:68–73

Kerp H (2000) The modernization of landscapes during the Late Paleozoic–Early Mesozoic. In: Gastaldo RA, DiMichele WA (eds) Phanerozoic terrestrial ecosystems. Paleontol Soc Pap 6, pp 9–113
Knoll AH (1984) Patterns of extinction in the fossil record of vascular plants. In: Nitecki MH (ed) Extinctions. University of Chicago Press, Chicago, pp 21–68
Kon'no E (1973) New species of *Pleuromeia* and *Neocalamites* from the Upper Scythian bed in the Kitakami Massif, Japan. Sci Rep Tohoku Univ, 2nd Ser (Geol) 43: 97–115
Li XX (ed) (1995) Fossil floras in China through the geological ages (English edition). Guangdong Science and Technology Press, Guangzhou, pp 1–695
Looy CV, Brugman WA, Dilcher DL, Visscher H (1999) The delayed resurgence of equatorial forests after the Permian-Triassic ecologic crisis. Proc Natl Acad Sci 96(24):13857–13862
Looy CV, Twitchett RJ, Dilcher DL, Van Konijnenburg-Van Cittert JH, Visscher H (2001) Life in the end-Permian dead zone. Proc Natl Acad Sci 98(14):7879-7883
Mägdefrau K (1931) Zur morphologie und phylogenetischen Bedeutung der fossilen Pflanzengattung *Pleuromeia*
Mägdefrau K (1956) Paläobiologie der Pflanzen. G. Fischer VEB, Jena, 444p
Mcloughlin S, Lindstrom S, Drinnan AN (1997) Gondawana floristic and sedimentological trends during the Permian-Triassic transition: new evidence from the Amery Group, northern Prince Charles Mountains East Antarctica. Antarct Sci 9(3):281–298
McManus HA, Taylor EL, Taylor TN, Collinson JW (2002) A petrified Glossopteris flora from Collinson Ridge, central transnatarctic Mountains: late Permian or Early Triassic? Rev Palaeobot Palynol 120:233–246
Meng FS, Zhang Z., Niu ZJ, Chen DY (2000) Primitive Lycopsid flora in the Yangtze Valley of China and systematics and evolution of Isoetales. Hunan Science and Technology Press, Hunan 1–87
Meyen SV (1987) Fundamentals of palaeobotany. Chapman and Hall, London
Meyen SF (1973) The Permian-Triassic boundary and its relation to the Paleophyte-Mesophyte floral boundary. In: Logan A, Hills LV (eds) The Permian-Triassic systems and their mutual boundary. Memoir of Canadian Society of Petroleum Geologists 2, pp 662–667
Mogutcheva NK (1996) Evolutionary stages of Triassic flora evolution in Siberia (Angarida). The Palaeobotanists 45:329–333
Ouyang S (1982) Upper Permian and Lower Triassic palynomophs from eastern Yunnan. Can J Earth Sci 19:68–80
Ouyang S, Li ZP (1980) The micro-floral in the Kayitou Bed of Fuyuan, Yunnan and its stratigraphic and paleobotanic significance. In: Nanjing Institute of Geology and Palaeontology, Academica Sinica (ed) Late Permian coal-bearing strata and palaeontological fauna in western Guizhou and eastern Yunnan. Science Press, Beijing, pp 123–194 (in Chinese)
Ouyang S (1986) Palynology of Upper Permian and Lower Triassic strata of Fuyuan district, Eastern Yunnan. Science Press, Beijing, 122 p (in Chinese with English Summary)
Pal PK, Ghosh AK (1997) Megafloral zonation of Permian-Triassic sequence in the Kamthi Formation, Talcher Coalfield, Orissa. The Palaeobotanists 46:81–87
Pant DD, Pant R (1987) Some Glossopteris leaves from Indian Triassic beds. Palaeontographica B205:165–178
Peng Y, Shi GR (2009) Life crises on land across the Permian-Triassic boundary in South China. Global Planet Change 65:155–165
Peng YQ, Yu JX, Gao YQ, Yang FQ (2006) Palynological assemblages of non-marine rocks at the Permian-Triassic boundary, western Guizhou and eastern Yunnan, South China. J Asian Earth Sci 28:291–305
Ricklefs RE, Buffetaut E, Hallam A, Hsu KH, Jablonski D, Kauffman EG, Legendre S, Martin P, McLaren DJ, Myers N, Traverse A (1990) Biotic systems and diversity – Report of working Group 4, Interlaken Workshop for Past Global changes. Palaeogeogr Palaeoclimatol Palaeoecol 82(1990):159–168

Rong JY, Fang ZJ (2004) Comparative analysis of the three major Palaeozoic mass extinctions and their subsequent recoveries in South China. In: Rong JY, Fang ZJ (eds) Mass extinction and recovery, evidences from the Palaeozoic and Triassic of South China. University of Science and Technology of China Press, Hefei, pp 931–1018 (in Chinese with English summary)

Shah SC (1987) Permian-Triassic boundary in the Peninsula. Palaeobotanist 36:58

Sheng JZ, Chen CZ, Wang YG, Rui L, Liao ZT, Bando Y, Nakamura K (1984) Permian-Triassic boundary in middle and eastern Tethys. Summary of Science Department of Hokkaido University 21(1):133–181

Tong JN (2004) Evolution of Foraminiferid Groups through the Palaeozoic-Mesozoic transition in South China. In: Rong JY, Fang ZJ (eds) Mass extinction and recovery, evidences from the Palaeozoic and Triassic of South China. University of Science and Technology of China Press, Hefei, pp 701–718 (in Chinese with English summary)

Traverse A (1988) Plant evolution dances to a different beat: plant and animal evolutionary mechanisms compared. Hist Biol 1:277–302

Twitchett RJ, Looy CV, Morante R, Visscher H, Wignall PB (2001) Rapid and synchronous collapse of marine and terrestrial ecosystems during the end-Permian biotic crisis. Geology 29:351–354

Utting J, Piasecki A (1995) Palynology of the Permian of northern continents: a review. In: Scholle PA, Peryt TM, Ulmer-Scholl DS (eds) The Permian of northern Pangea, 1: paleogeography, paleoclimate, stratigraphy. Springer-Verlag, Heidelberg, pp 236–262

Vajda V, McLoughlin S, Mays C, Frank TD, Fielding CR, Tevywa A, Lehsten V, Bocking M, Nicoll RS (2020) End-Permian (252 Mya) deforestation, wildfires and flooding - An ancient biotic crisis with lessons for the present. Earth Planet Sci Lett 529:115875

Visscher H, Brinkhuis H, Dilcher DL, Elsik WC, Eshet Y, Looy CV, Rampino MR, Traverse A (1996) The terminal paleozoic fungal event: evidence of terrestrial ecosystem destabilization and collapse. Proc Natl Acad Sci USA 93(5):2155–2158

Wang ZQ (1989) Permian gigantic palaeobotanical events in North China. Acta Palaeont Sinica 28(3):314–337 (in Chinese with English abstract)

Wang ZQ (1993) Evolutionary ecosystem of Permian–Triassic redbeds in North China: a historical record of global desertification. In: Lucas SG, Morales M (eds) The Nonmarine Triassic, New Mexico Ms Nat Hist Sci Bull 3:471–476

Wang ZQ (1996) Recovery of vegetation from the terminal Permian mass extinction in North China. Rev Palaeobot Palynol 91:121–142

Wang SY, Yin HF (2001) Study on terrestrial Permian-Triassic boundary in eastern Yunnan and western Guizhou. China University of Geosciences Press, Wuhan, 88p (in Chinese with English abstract)

Xie SC, Pancost R, Yin H, Wang H, Evershed R (2005) Two episodes of microbial change coupled with Permo/Triassic faunal mass extinction. Nature 434:494–497

Yao ZQ, Xu JT, Zhen ZG, Mo ZG (1980) Late Permian biostratigraphy and the Permian-Triassic boundary in Western Guizhou and Eastern Yunnan. In: Nanjing Institute of Geology and Palaeontology (ed) Stratigraphy and palaeontology of Late Permain coal-bearing formations in Western Guizhou and Eastern Yunnan. Science Press, Beijing, pp 1–69 (In Chinese)

Yu JX, Peng YQ, Zhang SX, Yang FQ, Zhao QM, Huang QS (2007) Terrestrial events across the Permian- Triassic boundary along the Yunnan-Guizhou border, SW China. Global Planet Change 55:193–208

Yu JX, Broutin J, Chen ZQ, Shi X, Li H, Chu DL, Huang QS (2015) Vegetation changeover across the Permian-Triassic boundary in Southwest China extinction, survival, recovery and palaeoclimate: a critical review. Earth-Sci Rev 149:203–224

Zhao XG, Mo ZG, Zhang SZ, Yao ZQ (1980) Late Permian fossil floras of Western Guizhou and Eastern Yunnan. In: Nanjing Institute of Geology and Palaeontology, Academica Sinica (ed) Late Permian coal-bearing strata and palaeontological fauna in Western Guizhou and Eastern Yunnan. Science Press, Beijing, pp 70–122 (in Chinese)

Zhou ZY, Li BX (1979) A preliminary study of the Early Triassic plants from the Qionghai District Hainan Island. Acta Palaeotol Sin 18(5):444–466 (In Chinese with English abstract)
Ziegler AM, Eshel G, Rees PM, Rothfus TA, Rowley D, Sunderlin D (2003) Tracing the tropics across land and sea: Permian to present. Lethaia 36:227–254

Chapter 8
Plant and Environment Co-evolution in Permian-Triassic Transition

Jianxin Yu, Xiao Shi, Zhen Xu, Hui Li, and Zongsheng Lu

The South and North China blocks were both located near the palaeo-equator in the Northern Hemisphere during the early-to-middle Permian. After the Late Guadalupian, the North China block drifted successively further northward to fnally collide with the Subangara terrane during the Lopingian (Late Permian)–Triassic transition (Fig. 8.1a). The terrestrial Lopingian–Early Triassic strata outcropping in China are much more extensive in the northern area, while only a few outcrop profiles on the Kangdian massif margins in the southern area exist (Fig. 8.1b). Permian-Triassic palaeobotanical research has, therefore, focused on both regions.The studies of Lopingian foras in China are since a long time closely related to the coal explorations leading to the characterization of an original "Cathaysian palaeoflora" with endemic plants including the iconic "gigantopterids". The first occurrence of Cathaysian floral elements was found in North and Northwest China characterized by the "*Neuropteris ovate-Cathaysiodendron nanpiaoense*" assemblage from the middle–upper part of the Taiyuan Formation. This endemic Cathaysian flora might have an Eastern origin from a "Procathaysian"flora, which perhaps existed as early as in Mississippian (Early Carboniferous) in China (Li 1995). From the middle of the Cisuralian,

J. Yu (✉) · Z. Xu · Z. Lu
School of Earth Sciences, China University of Geosciences, Wuhan, P.R. China
e-mail: yujianxin@cug.edu.cn

Z. Lu
e-mail: zshlu@cug.edu.cn

X. Shi
College of Earth Sciences, Jilin University, Changchun, Jilin, P.R. China
e-mail: xiaoshi@jlu.edu.cn

H. Li
Jiangxi Key Laboratory for Mass Spectrometry and Instrumentation, East China University of Technology, Nanchang, P.R. China
e-mail: huili_201860188@ecut.edu.cn

J. Yu et al. (eds.), *Plants and Palynomorphs around the Permian-Triassic Boundary of South China*, New Records of the Great Dying in South China,
https://doi.org/10.1007/978-981-19-1492-8_8

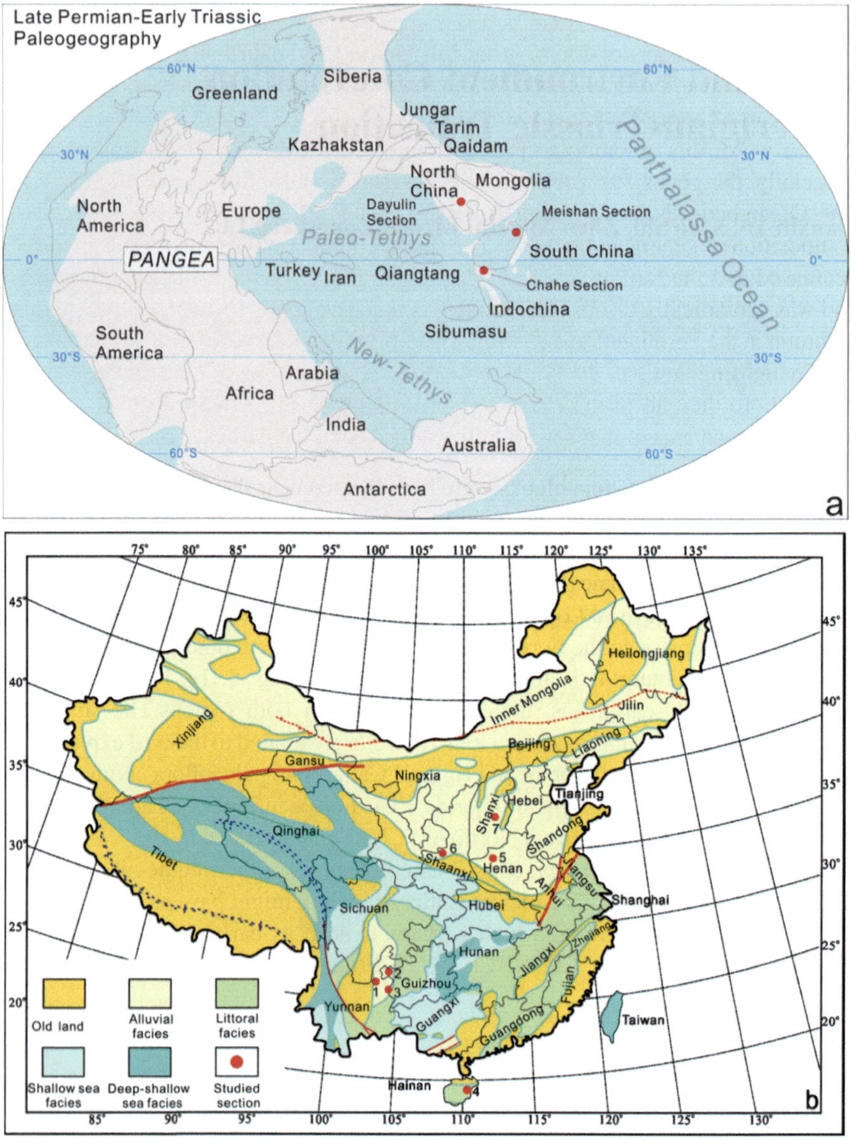

Fig. 8.1 **a** Palaeogeographic reconstruction of the Pangea "A" for Late Permian–Early Triassic showing the locations of South China and North China (modifed after Muttoni et al. 2009). **b** Palaeogeographic reconstruction of South China and North China during Late Permian Changhsingian and locations of the main sections cited in this paper (modifed after Wang 1985). (1) Lubei Section, Huize County, Yunnan Province; (2) Chahe Section, Weining County, Guizhou Province; (3) Chinahe Section, Xuanwei City, Yunnan Province; (4) Lingwen Section, Qionghai City, Hainan Province; (5) Dayulin Section, Yiyang City, Henan Province; (6) Zishiya Section, Linyou County, Shaanxi Province; (7) Guozigou Section, Pingyao County, Shanxi Province

this northern Cathaysian flora spread southwards as attested by its discovery in the Yangtze Platform of the South China Block, occurring then progressively in almost all of the Permian terranes of China. Some Cathaysian floral elements even migrated to the Arabian plate, in the North Gondwanan domain, from the Guadalupian (Middle Permian) to the Lopingian. The Guadalupian to Lopingian palaeofloras on the Arabian Plate were composed of a mixture of Gondwanan, Cathaysian and Euramerican elements (Berthelin et al. 2003).

8.1 Palaeoclimate Changes Over the P-Tr Transition in South China

8.1.1 The Late Permian Flora in South China: A "Stable" Homogeneous Cathaysian Assemblage

During the entire Permian time, the South China Block remained located within the equatorial sub-equatorial belt near the palaeoequator (Shen et al. 2013). This was under a stable, warm and humid ever-wet climate, favouring the persistence of large wetland areas, owing to the proximity to oceanic moisture, at a geological time scale. These features means that this area allows us to observe evolutionary modifications of terrestrial plants lineages not mainly driven by climatic changes.

As stated by Greb et al. (2006): *"the fossil record of wetland documents unique and long persistent floras and faunas with wetland habitats spawning or at least preserving novel evolutionary characteristics and, at other times, acting as a refugium"*. South China acted precisely in both functions: refugia for some Carboniferous-type peat-forming plant assemblages (including lycopsids, cordaites and tree ferns), and centre for emergence of new peat-forming swamp species (e.g., gigantopterids) and evolutionaryappearances of new "advanced models" of conifer vegetative leafy shoots.

Is the Lopingian of South China one of the "cradles" from where "modern" Mesozoic forms of conifers originated and spread out through the Pangea during the Mesozoic and Tertiary, diversifying into various lineages? This could explain recently discovered Lopingian fossils that include conifers with decussate leaves previously believed to appear first in the Late Triassic. This research is currently in progress.

8.1.2 Transition Between the Lopingian and Earliest Triassic Floras

8.1.2.1 Lopingian Changhsingian "*Gigantopteris*" Flora

Thefossil plants collected inthe Changhsingian deposits ofthe South China are abundant and diverse (Peng et al. 2009; Yu et al. 2015) comprising 105 species from 39 genera. This Changhsingian floral assemblageis dominated by species of *Gigantopteris* and *Gigantonoclea*, both genera containing 10 species and that having long been considered, together with *Lobatannularia,* as endemic to the Cathaysian floras (Table 8.1). Mostcommonspecies include *Lepidodendron oculus-felis* (Fig. 8.2a), *Lobatannulariamultifolia* (Fig. 8.2c), *Gigantonoclea lagrelii* (Fig. 8.2e), *Gigantopteris dictyophylloides* (Fig. 8.2f), and *Rhipidopsis lobulata* (Fig. 8.2g), and *Stigmariaficoides.* Other elements such as *Annularia shirakii* (Fig. 8.2b), *Lobatannularia cathaysiana, Paracalamites stenocostatus, Fascipteris hallei* (Fig. 8.2d), *Pecopteris marginata, Compsopteris contracta, Rhipidopsis panii* and *Lepidodendron acutangulum* are also commonly present (Li 1995). Due to taphonomic parameters fossil remains of conifers, surprisingly absent in the terrestrial deposits of the Lopingian Xuanwei Fm., often co-occur with marine fossils in the marine coeval deposits of South China. However, despite the large number of collected samples, the generic diversity of fossil conifers is noticeably low. Only four species in three genera were discovered, namely: *Pseudoullmannia frumentarioides* He et al. (1996), *Szecladia multinervia* Yao et al. (2000), and *Anshuncladus* sp. 1and sp. 2 ("tentative description" in Li et al. 2019). To the Permian–Triassic Boundary, the Cathaysian floras was in decline, except that some relics persisted into the earliest Triassic in South China.

8.1.2.2 Earliest Triassic *Lepacyclotes* (=*Annalepis*)-*Peltaspermum*floral Assemblage with Rare Relicts of *Gigantopteris* Flora After the Mass Extinction

The taxa of the "post crisis" earliest Triassic *Lepacyclotes* (=*Annalepis*)-*Peltaspermum* floral assemblage in South China are distributed in the lower Kayitou Fm. and include 18 species from 14 genera (Yu et al. 2015). Of these, 6 species from 4 genera were discoveredinterrestrial facies sections, while 16 species from 12 genera were found interrestrial-marine transitional facies sections. Dispersed and in situ megaspores of *Lepacyclotes* (=*Annalepis*) were also extracted from the same strata. This *Lepacyclotes* (=*Annalepis*)-*Peltaspermum* assemblage yieldied-Lycopsida: *Lepacyclotes* (=*Annalepis*) *zeilleri* (Fig. 8.3i, o) *Lepacyclotes* (=*Annalepis*) *latiloba* (Fig. 8.3j, m, n) *Lepacyclotes* (=*Annalepis*)*angusta* (Fig. 8.3k, l) *Lepacyclotes* (=*Annalepis*) *furongqiaoensis* and *Lepacyclotes* (=*Annalepis*) *brevicystis*–Sphenopsida: *Paracalamites stenocostatus*-Pteridophyta: *Pecopteris* sp. and ?*Sphenopteris* sp., –Pteridosperms: *Peltaspermum* cf. *martinsii,P. lobulatum, Peltaspermum* spp.

Table 8.1 Late Permian and Early Triassic fossil plants in South China and North China

Area	South China		North China	
Stage Names	Wuchiapingian	Changhsingian	Wuchiapingian	Changhsingian
Lycopsida				
Lepidodendron oculus-felis Abbado	X	X	X	
Cathaysiodendron acutangulum Halle	X	X	X	
Sphenopsida				
Sphenophyllum sino-coreanum Yabe	X	X	X	X
Annularia pingloensis Sze	X	X	X	
Annularia shirakii Kawasaki	X	X	X	
Lobatannularia multifolia Kon' no et Asama	X	X	X	
Lobatannularia spatulata (Kodaira) Kawasaki	X	X	X	
Schizoneura manchurensis Kon' no	X	X		
Noeggerathiopsida				
Tingia carbonica (Schenk) Halle	X		X	
Tingia trilobata Stockmans et Mathieu	X		X	
Plagiozamites linearis Yabe et Oishi	X			
Plagiozamites oblongifolius Halle	X	X		
Yuania chinensis Zhu et Du	X		X	
Yuania gigantea Zhu et Du	X		X	
Yuania magnifolia Wang et Wang				X
Filices				
Pecopterids				
Pecopteris marginata Gu et Zhi	X		X	
Pecopteris nervosa Halle	X		X	
Pecopteris norinii Halle	X	X	X	

(continued)

Table 8.1 (continued)

Area	South China		North China	
Stage Names	Wuchiapingian	Changhsingian	Wuchiapingian	Changhsingian
Pecopteris orientalis Schenk	X	X	X	X
Fascipteris densata Gu et Zhi	X	X	X	
Fascipteris hallei Kawasaki	X	X	X	
Fascipteris sinensis Gu et Zhi	X	X	X	
Fascipteris stena Gu et Zhi	X	X	X	
Szea (*Cladophlebis*) *sinensis* Yao et Taylor	X	X		
Pteridospermopsida				
Alethopterids				
Protoblechnum contractum Gu et Zhi	X	X	X	
Callipterids				
Callipteris lobulata Wang et Wang			X	X
Callipteris papillosa Wang et Wang				X
Gigantopterids				
Gigantonoclea acuminatiloba Shimakura	X			
Gigantonoclea guizhouensis Gu et Zhi	X	X		
Gigantopteris dictyophylloides Gu et Zhi	X	X	X	
Gigantopteris nicotianaefolia Schenk ex Potonié	X	X	X	
Taeniopterids				
Taeniopteris densissima Gu et Zhi	X	X	X	
Taeniopteris multinervis Weiss	X	X	X	
Taeniopteris tingii Halle	X		X	X
Cycadopsida				

(continued)

Table 8.1 (continued)

Area	South China		North China	
Stage Names	Wuchiapingian	Changhsingian	Wuchiapingian	Changhsingian
Pterophyllum eratum Gu et Zhi	X		X	
Ginkgopsida				
Rhipidopsis panii Chow	X	X	X	
Coniferae				
Walchia bipinnata Gu et Zhi	X		X	X
Ullmannia bronnii Göppert	X	X	X	X
Ullmannia frumantaria (Schlotheim) Göppert	X		X	X
Szecladia multinervia Yao, Rothway et Liu		X	X	X
Pseudoullmannia frumentarioides He, Liang et Sheng	X			
Pseudoullmannia bronnioides He, Liang et Sheng	X			
Pseudovoltzia liebeana (Geinitz) Florin				X
Quadrocladus solmsii (Gothan et Nagalhard) Schweitzer				X
Quadrocladus heterodermus Wang et Wang				X
Plantae incertae sedis		X		
Otofolium polymorphum Gu et Zhi	X	X		
Otofolium ovatum Gu et Zhi	X		X	
Psygmophyllum multipartitum Halle	X		X	

–Ginkgopsida: ?*Sphenobaiera* sp., – gigantopterids:*Gigantonoclea* sp. and *Gigantopteris* sp. Thisearliest Triassic floral assemblage is dominated overall by both *Lepacyclotes* (=*Annalepis*) and *Peltaspermum*, which occur in the lower part of the Kayitou Fm. in South China. Rare, small fragments of *Gigantopteris* and *Gigantoclea* were found in 2–3 horizons in the upper part of theterrestrial-marine transitional facies of the Kayitou Fm. This may indicate that some gigantopterids remains were

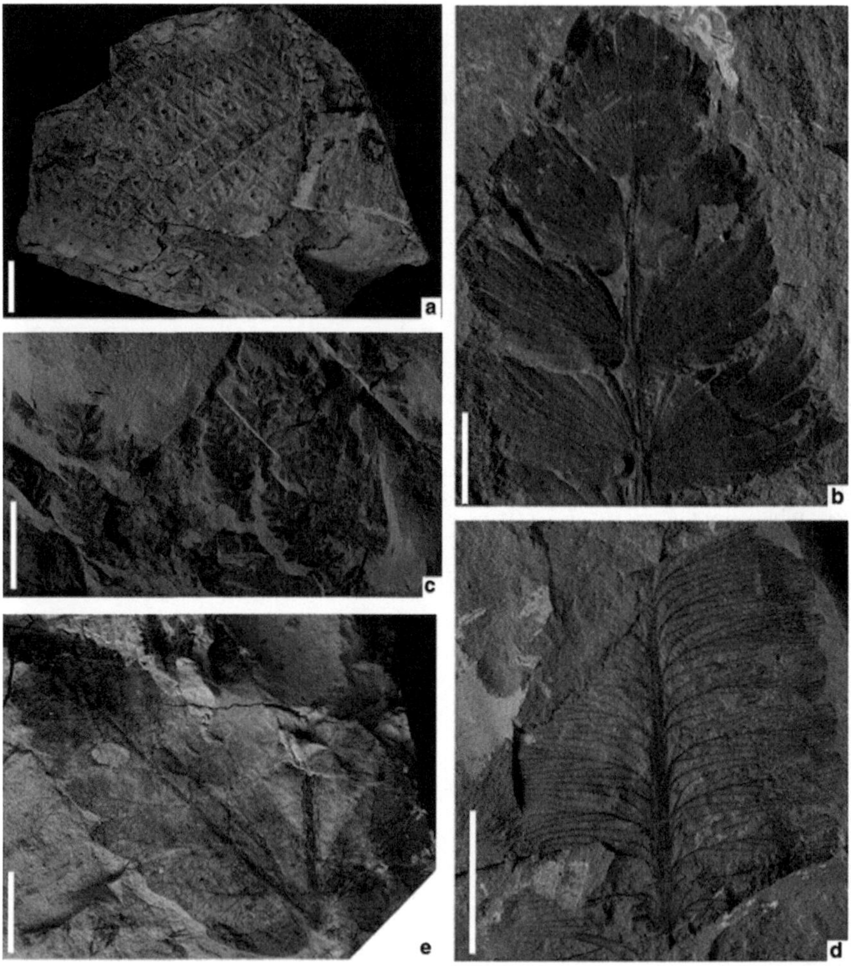

Fig. 8.2 The representative taxonomic groups of the Cathaysian flora of Late Permian Changhsingian upper Xuanwei Formation in South China. **a** *Lepidodendron oculus-felis (Abb) Zeiller* – Sample Number: YXC– (-9)-2; Locality: Bed 10, Chinahe Section, Xuanwei City, Yunnan Province; **b** Lobatannularia multifolia Kon'no et Asama Sample Number: YXC– (-1)-8; Locality: Bed 2, Chinahe Section, Xuanwei City, Yunnan Province; **c** *Annularia shirakii Kaw* Sample Number: CNH–0–18; Locality: Bed 2, Chinahe Section, Xuanwei City, Yunnan Province; **d** *Fascipteris hallei* Kaw Sample Number: CNH–0–9; Locality: Bed 2, Chinahe Section, Xuanwei City, Yunnan Province; **e** *Gigantonoclea lagrelii* (Halle) Koidz Sample Number: YXC– (-4)-1; Locality: Bed 5, Chinahe Section, Xuanwei City, Yunnan Province; **f***Gigantopteris dictyophylloides Gu et Zhi* Sample Number: YXC– (-24)-2; Locality: Bed 25, Chinahe Section, Xuanwei City, Yunnan Province; **g** *Rhipidopsis lobulatumMo* Sample Number: YXC– (-1)-13; Locality: Bed 2, Chinahe Section, Xuanwei City, Yunnan Province. All scale bars = 10 mm

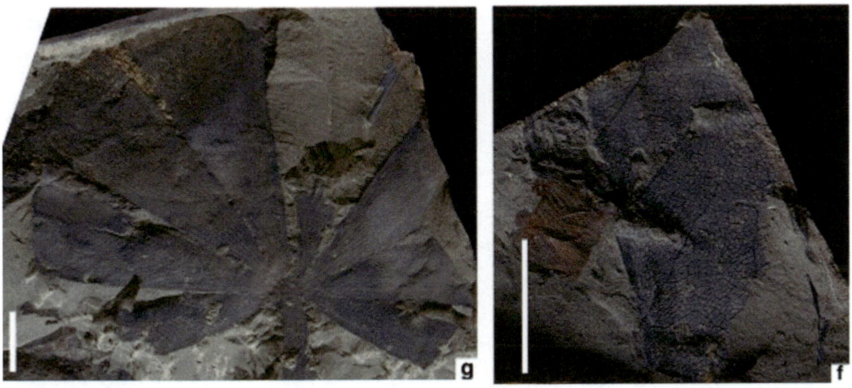

Fig. 8.2 (continued)

Fig. 8.3 The representative taxonomic groups of Early Triassic Kayitou and Feixianguan Formations in South China.**a** *Neocalamites* sp. Sample Number: YHL–5–41; Locality: Bed 5, Lubei Section, Huize County, YunnanProvince; **b** *Peltaspermum* sp. 1Sample Number: YHL–7–35; Locality: Bed 7, Lubei Section, Huize County, YunnanProvince; **c** *Equisetites acanthodon* Sze Sample Number: YHL–7–32; Locality: Bed 7, Lubei Section, Huize County, YunnanProvince; **d, e, g** -*Voltzia* spp. Sample Number: YHL–7–29, YHL–7–56 and YHL–7–28; Locality: Bed 7, Lubei Section, Huize County, Yunnan Province; **f** *Voltzia heterophylla* BrongniartSample Number: YHL–7–2; Locality: Bed 7, Lubei Section, Huize County, Yunnan Province; **h** Dwarf shoot of *Voltzia* sp.Sample Number: YHL–7–38; Locality: Bed 7, Lubei Section, Huize County, Yunnan Province; **i, o***Lepacyclotes*(=*Annalepis) zeilleriFliche*Sample Number: YXC– (-25)-16, YXC– (-25)-5A; Locality: Bed 26, Chinahe Section, Xuanwei City, Yunnan Province; **j, m, n** *Lepacyclotes*(=*Annalepis) latiloba Meng*Sample Number: YXC– (-25)-1, YXC– (-25)-23, YXC– (-25)-2A; Locality: Bed 26, Chinahe Section, Xuanwei City, Yunnan Province; **k, l** *Lepacyclotes*(=*Annalepis) angusta Meng* Sample Numbers: YXC– (-25)-14, YXC– (-25)-20; Locality: Bed 26, Chinahe Section, Xuanwei City, Yunnan Province. Figs. **a, c–g** scale bar = 10 mm; Figs **b, h–o**: scale bar = 5 mm

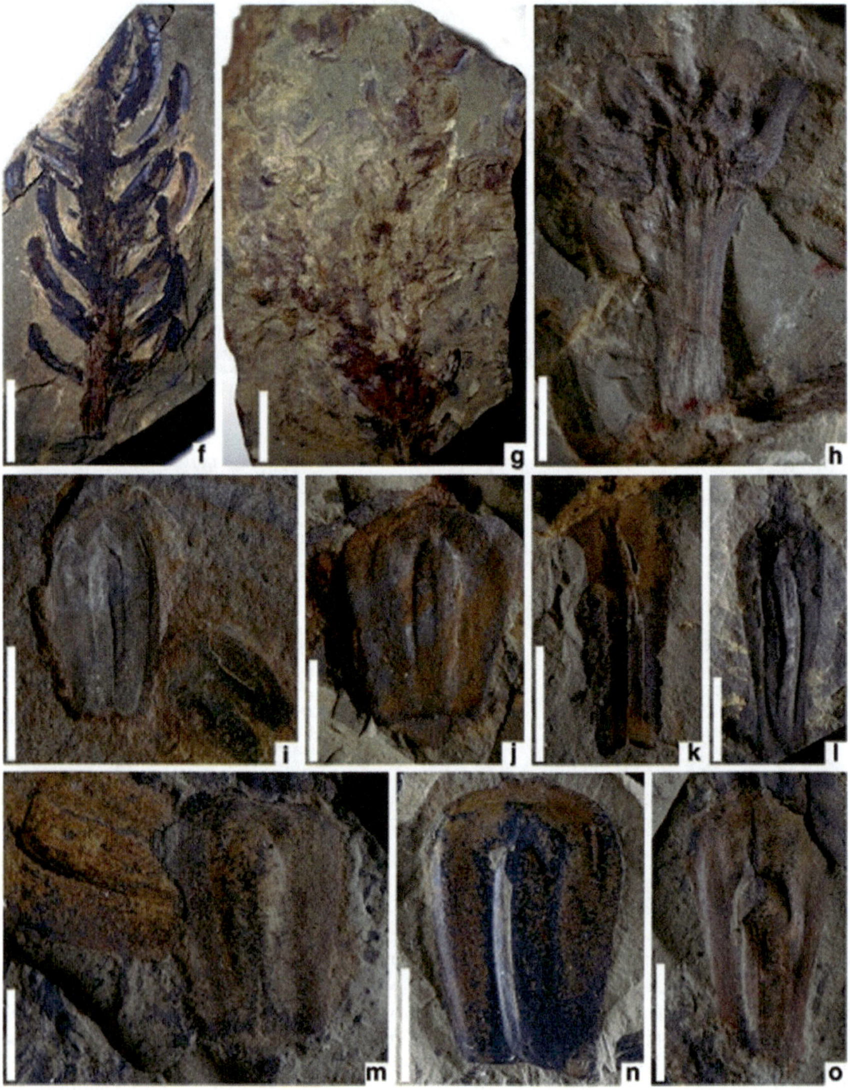

Fig. 8.3 (continued)

"paleophytic" survivors in Triassic depositsafter the PTB mass extinction. Moreover, numerous fragments of vegetative pinnae and female reproductive organs, related to Peltaspermales, were collected in the same 2–3 Early Triassiclayers. It is also true that Pteridophytes were very prosperous in the Carboniferous and Permian in South China (Li 1995). The occurrence of vegetative and reproductive organs of *Peltaspermum* in the Kayitou Fm. shows that this genus far from only survive the PTB mass extinction played also a crucial role in reconstructing evolutionary new lineages during the Paleozoic–Mesozoic transition. In South and North China, the *Peltaspermum* assemblage was also reported until themiddle Triassic and disappeared before the late Middle Triassic (Wang and Wang 1990). *Lepacyclotes* (=*Annalepis*), as a pioneering lycopsid genus initiating the recovery of land plants (Yu et al. 2010), was found in association with a diverse marine fauna from the basal Kayitou Formation in South China.

8.1.3 Late Early Triassic **Neuropteridium-Voltzia** *Floral Assemblage*

This assemblage is represented by the plants from the Lingwen Fm. of Hainan Provinceand the upper part of the Feixianguan Formation of Yunnan Province. This "post-recovery" paleofloraisplentiful, containing about 38 species belonging to 28 genera (Zhou and Li 1979; Meng 1992; Feng et al. 2018), namely: *Pleuromeia* sp., *Equisetites qionghaiensis*, *E. acanthodon.* (Fig. 8.3b) *Neocalamites* sp. (Fig. 8.3a) *Phyllotheca bella*, *P. marginans*, *Schizoneura* sp., cf. *Asterotheca szeiana*, *Todites shensiensis*, *Neuropteridium marginatum*, *Peltaspermum* sp. (Fig. 8.3c) *Vittaephyllum* sp., cf. *Ctenozamites cycadea*, *Leuthardtia ovalis*, *Qionghaia carnosa*, *Sphenobaiera* sp., *Yuccites ensiformis*, *Albertia latifolia*, *Albertia elliptica*, *Albertia* cf. *speciosa*, *Voltzia heterophylla* (Fig. 8.3f) *Voltzia. weismannii*, *Voltzia.* cf. *koenenüi*, *Voltzia* sp. (Fig. 8.3d, e, g) cf. *Glyptolepis longibracteata*, *Pagiophyllum* sp., *Brachyphyllum* sp., *Masculostrobus*? *prolatus*, *Taeniopteris mironervis*, *Taeniopteris costiformis* and *Taeniopteris hainanensis.* The assemblage is dominated by conifers highlighting representatives of *Albertia* and *Voltzia* (Fig. 8.3d, e, f, g) whose diversifications can be compared with those that are observedinthe Middle "Buntsandstein" flora of Europe. Moreover, the horsetails and *Neuropteridium* are also quite numerous, while lycopsids are very rare. Generally, this palaeoflora of the Heshanggou Fm. of North China can be analysed as a "MiddleBuntsandstein-like" European flora. Therefore, the age of the assemblage is likely late Early Triassic. In the late Early Triassic, the largest portions of muddy red beds inthe Lingwen Fm. (Qionghai region, Hainan province) and also in theupper Feixianguan Formation (Western Guizhou and eastern Yunnan) indicates the installation of a drier climate. Fossil plant remains are only foundin gray, fine-grained sandstone overlying the red rocks of the Lingwen Formation and the upper Feixianguan Formation. During this period, conifers (e.g. *Albertia* and *Voltzia*) and pteridosperms (e.g. *Peltaspermum*) began to flourish. Thisindicates that mesophytic new plants both in upland and nearshore environments could further develop.

8.2 Palaeoclimate Changes Over the P-Tr Transition in North China

8.2.1 *The Lopingian Flora in North China: A Subangaran-Euramerican Flora with Cathaysian Relicts*

Although the first Cathaysian plants appeared during the basalmost Permian and flourished during the Cisuralian-Guadalupian to become a highly diverse "Cathaysian paleoflora" in the North China block, from the late Guadalupian to the Lopingian, most of these Cathaysian elements gradually disappeared in the North China block and were progressively invaded by Subangaran and Euramerican elements. This dramatic process is due to the progressive aridification resulting from the northward drifting of this block, leading finally to a suture of the block with the "Subangaran realm".

The decline and final disappearance of the Cathaysian flora in North China are due chiefly to the fact that the Sino–Korean–Tarim (North China) Block (Fig. 8.1a) drifted from its Cisuralian to Guadalupian sub-equatorial location to the north. This northward drifting gradually made this block closer to the Subangara plate, allowing a progressive climate change from the Guadalupian to Lopingian: the hot and dry Euramerican-type climate migrated from West to East until it encompassed the Subangara and North China Block (Li 1995). As the North China blocks moved to the north with its Cathaysian vegetation cover, Angaran and Cathaysian floral elements started to mix mainly during the late Guadalupian in eastern Xinjiang and the Qilian Mountains of Gansu Province. Finally, in the Lopingian, the Subangaran flora increasingly prevailed in the north North China Block and Northwest China intermingling with some surviving Cathaysian plant relics. In the middle of the North China block, many representative members of the Cathaysian flora were replaced by Zechstein migrating conifers from Western Europe in a Lopingian "Shihchienfeng flora" (Sunjiagou Fm.- Fig. 8.4), in which only a small number of Cathaysian elements of relatively drought-enduring habit, e.g., *Yuania* and *Taeniopteris taiyuanensis* survived (Shi 2016).

8.2.2 *Lopingian* **Ullmannia** *Aff.* **Bronnii-Yuania Magnifolia** *Plant Assemblage with Rare Gigantopteris Flora Relicts*

The Lopingian fossil plants in North China remained little known due tothe dominance of unusual and almost barren "red beds" (Fig. 8.4). Fortunately, we recently discovered some new rich fossiliferous localities named *"Dayulin Section in Henan Province, Zishiya Section in Shaanxi Province and so on"* in the uppermost grey intercalation of the Sunjiagou Fm. (Sects. 5–7 in Figs. 8.1b and 8.4). Those new

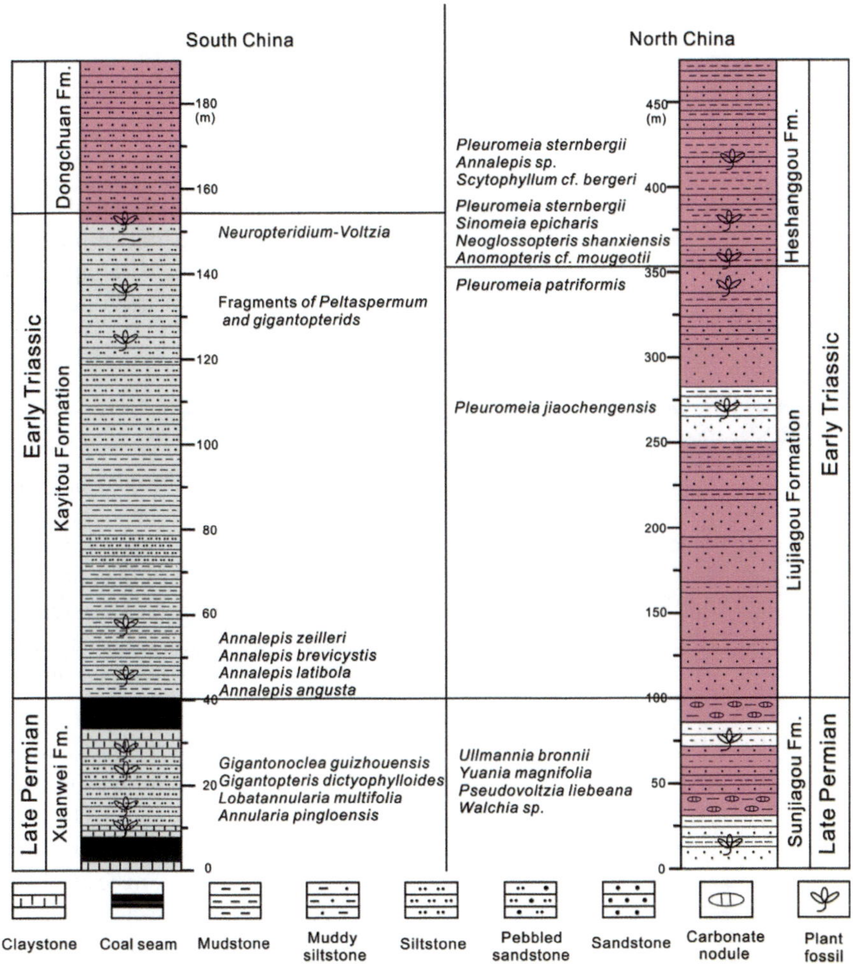

Fig. 8.4 Correlating palaeofloral keyelements and sedimentary characteristics between South China and North China during Permian–Triassic transition (in pink: purple-reddish muddy siltstones and thin sandstones deposits)

discovered fossil plants can be integrated in the *"Ullmannia* aff. *bronnii* - *Yuania magnifolia"* assemblage, defined by Wang Ziqiang and others (Wang 1983, 1985; Wang and Wang 1986; Pott et al. 2010).

This newly full assemblage mainly include *Yuania magnifolia*, *Yuania* spp., *Sphenopteris* sp., *Pecopteris* cf. *arcuata*, *Pecopteris* spp., Sphenopsids (Fig. 8.5a) *Callipteris lobulota*, *Germaropteris martinsii*, *Tatarina* cf. *sinuosa*, *Ptetophyllum* cf. *striatum* (Fig. 8.5c), *Sphenobaiera micronervis*, *Walchia* ? sp., *Pseudovoltzia liebeana* (Fig. 8.5 d, e, f) *Pseudovoltzia* spp. (Fig. 8.5g, h) *Ullmannia bronnii*, *Ullmannia frumentaria*, *Ullmannia* sp. (Fig. 8.5i) *Szecladia multinervia* (Fig. 8.5k,

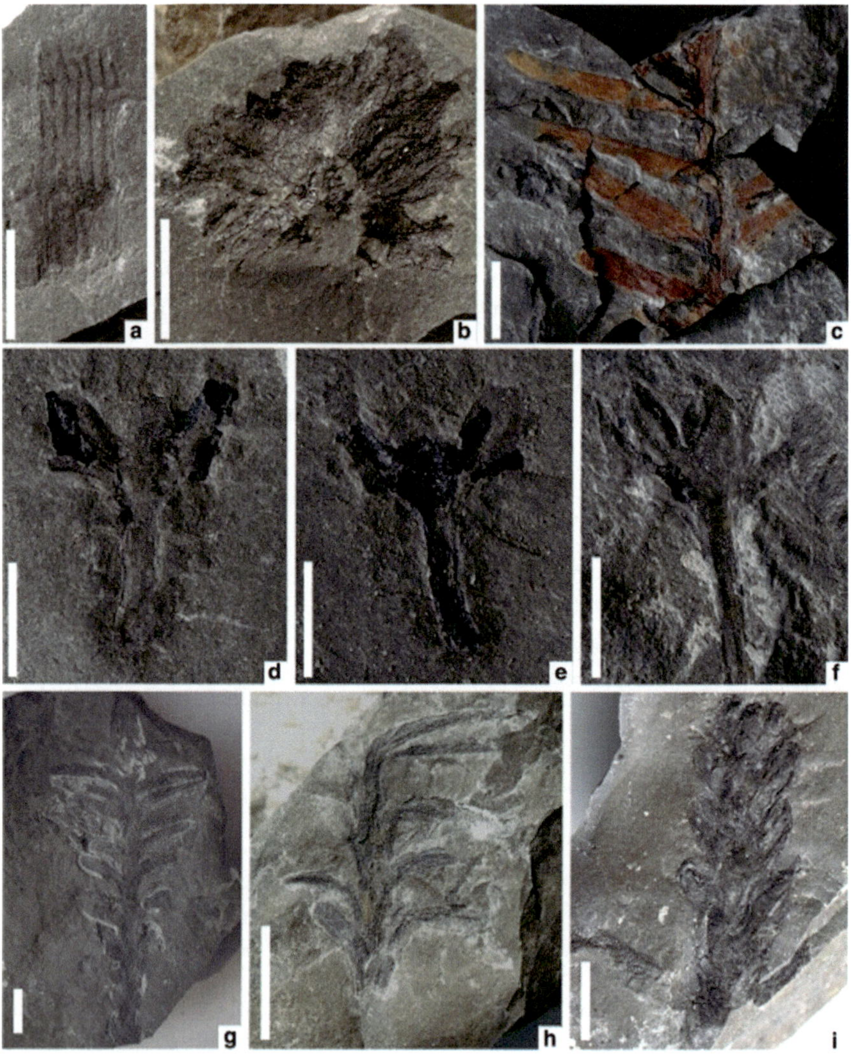

Fig. 8.5 The representative taxonomic groups of the Late Permian Changhsingian Sunjiagou Formation in North China. **a** Sphenopsid stem,Sample Number: ZSY-1; Locality: the Sunjiagou Formation, Zishiya Section, Linyou County, Shaanxi Province; **b** *Peltaspermum* sp. 2,Sample Number: ZSY-2; Locality: the Sunjiagou Formation, Zishiya Section, Linyou County, Shaanxi Province; **c***Pterophyllum* cf. *striatum*, Sample Number: DYL–1–7; Locality: Bed 1, Dayulin Section, Yiyang County, Henan Province; **d**, **e**, **f**Dwarf shoots of *Pseudovoltzia liebeana* (Geinitz) Florin Sample Number: DYL–7–7, DYL–7–8 and DYL-7–9; Locality: Bed 7, Dayulin Section, Yiyang County, Henan Province; **g**, **h** Vegetative shoots of *Pseudovoltzia* spp.,Sample Number: ZSY-3 and ZSY-4; Locality: the Sunjiagou Formation, Zishiya Section, Linyou County, Shaanxi Province; **I** *Ullmannia* sp.,Sample Number: ZSY-5; Locality: the Sunjiagou Formation, Zishiya Section, Linyou County, Shaanxi Province; **j**, **m***Feysia* ? spp., Sample Number: ZSY-6 and ZSY-9; Locality: the Sunjiagou Formation, Zishiya Section, Linyou County, Shaanxi Province; **k**, **l** *Szecladia multinervia* Yao, Rothwell et Liu Sample Number: ZSY-7 and ZSY-8; Locality: the Sunjiagou Formation, Zishiya Section, Linyou County, Shaanxi Province. Figs. a, c, g, h, k, l, m: scale bar = 10 mm; Figs b, d-f, i, j: scale bar = 5 mm;

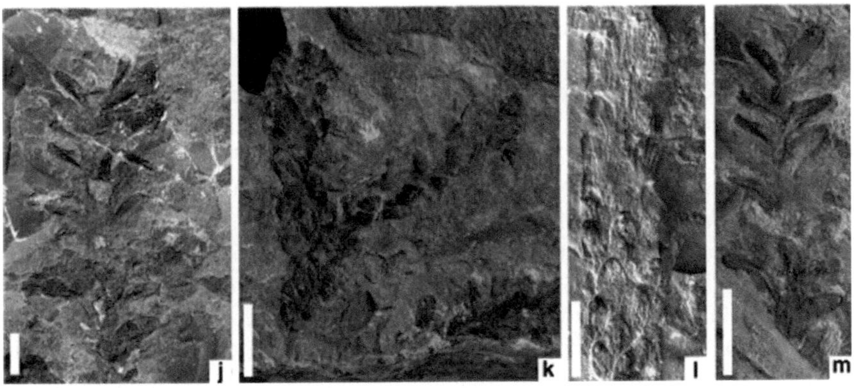

Fig. 8.5 (continued)

l)?*Ortiseia leonardii*, ?*Quadrocladus* spp., ?*Feysia* spp. (Fig. 8.5j, m) *Esterella* sp., *Lesleya anastomosis*, *Scytopyllum sunjiagouense*, *Peltaspermum dafungshanense*, *Peltaspermum* sp. 2 (Fig. 8.5b) *Taeniopteris liulinensis*, *T. Taiyuanensis.*

With only a few Cathaysian relict elements (most of the gigantopterids, *Lobatannularia, Fascipteris*, Lepidodendrales) all disappeared and with incoming Euramerican and Subangaran elements, this assemblage is globally confined to the Changhsingian (Late Lopingian) stage except a few species that appeared as early as the Carboniferous (*Sphenopteris* spp.) and rare examples where others extended into the Mesozoic. Due to environmental constraints, this palaeoflora became dominated by "xerophytic" plants and devoid of "peat-formingplants", conversely to what is observed in coeval South China deposits.

It is important to note that the generic and specific composition of the Lopingian palaeoflora in North China includes almost all the key "xerophytic" forms known in the latest Permian "Zechstein flora" of Western Europe, even if they did not become numerically dominant. Therefore, the Changhsingian Sunjiagou palaeoflora, totally devoid of peatforming plants due to its forming under too dry conditions, is obviously distinct from the coeval Changhsingian "Gigantopteris" flora of the South China Xuanwei Formation, still represented by peat-forming wetland vegetation.

Nevertheless, the occurrence of the conifer *Ullmannia bronnii* in both South China (Western Guizhou and Eastern Yunnan) and North China domains indicates that some floristic exchanges remained possible even during the Late Permian for more "xrophytic" elements.

8.2.3 Early Triassic Pleuromeia Jiaochengensis *Assemblage*

The *Pleuromeia jiaochengensis* assemblage is based on the fossil plants in the Liujiagou Fm. (Wang 1989) (Fig. 8.4). The 200 to 480 m thick Liujiagou Fm. is composed of a set of greyish purple and purple feldspathic sandstones intercalated with sandy shale and argillite, occasionally with lenses of greyish green feldspathic sandstones. Most of its fossil plants content are autochthonous.16 species from 12 genera have been identified (Wang and Wang 1982; 1989), including *Pleuromeia altinis*, *P. jiaochengensis*, *P. pateriformis*, *Macrostachya gracilis*, *Phyllotheca yusheensis*, *Scolopendrites* (*Crematopteris*) *brevipinnata*, *C. Scolopendrites* (*Crematopteris*) *circinalis*, *Peltaspermum lobulatum*, *P.calycinum*, *Voltzia* sp., *Willsiostrobus* sp.. This assemblage is dominated by the Lycopsids (characterized by *Pleuromeia*) in association with horsetails (highlighting *Macrostachya gracilis* and *Phyllotheca yusheensis*) that are also important elements of the Lower-Middle Buntsandstein paleoflora of Western Europe. Floristically, the assemblage differs strongly from the underlying *Ullmannia bronnii-Yuania magnifolia* assemblage of the Lopingian Sunjiagou Formation (Wang and Wang 1986), while it is closer to the overlying *Pleuromeia sternbergii* assemblage of the Heshanggou Formation (Fig. 8.4). Therefore, this palaeofloral assemblage is likely to be of Early Triassic age, perhaps deposited during the middle to late Induan Stage.

8.3 Analysis of Paleoclimate Difference Between South and North China Blocks During the Permian–Triassic Transition

8.3.1 Different Paleogeographic Locations

Above mentioned, the updated palaeogeographic reconstructions (Ziegler et al. 2003; Yin et al. 2013) show that the South China block was located nearby the equatorial zone during the Permian-Triassic transition, whereas the North China block and Europe were close to the northern tropic latitude characterized by arid climates (Fig. 8.1). Therein, the South China block had a warm and humid climate condition favorable to the development of coal-forming swamps (Golonka et al. 1994; Yin et al. 1999; Fluteau et al. 2001). Nevertheless, the periods of significant coal accumulations, linked with terrestrial environment, varied in different areas depending on various episodes of marine transgression-regression throughout the Permian times. In Zhejiang, Fujian and northeastern Guangdong Provinces, eastern part of South China, coal measures deposited only during the Capitanian (late Middle Permian) to Wuchiapingian (early Late Permian) (Fig. 8.6), whereas the Changhsingian deposits were fully marine facies due to the late Wuchiapingian transgression prevailling in those areas (Chen et al. 1998). In Hunan, Jiangxi, northern Guangdong, and eastern Guangxi Provinces, central part of South China, coal measures were deposited only during the Wuchiapingian when terrestrial-marine transitional deposits overlaid the marine Middle Permian rocks (Fig. 8.6). In contrast, the coverage of the Changhsingian coal measures shrank to the southern part of the Upper Yangtze Platform (Li and Yao 1980; Yu et al. 2007) due to the succeeding transgression that generated

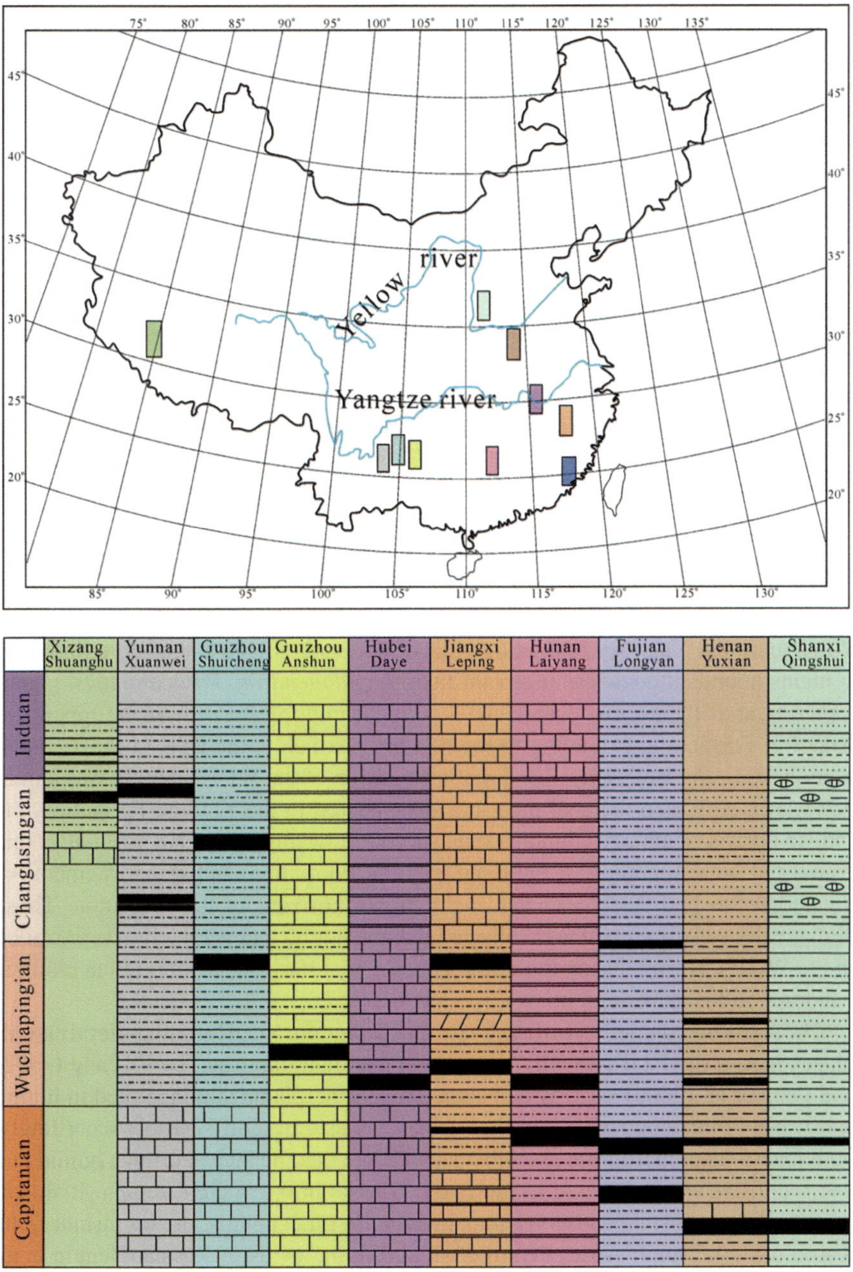

Fig. 8.6 Showing temporal appearance of the Late Permian coal measures in South China (after Li and Yao 1980)

massive carbonate platforms and reef settings cross the entire South China block in that time (Chen et al. 1998). In contrast, the coal measures were continuously deposited through the early Late Permian until the latest Permian in WGEY (i.e., the Tucheng section), in which the uppermost Xuanwei Formation coal seams represent the youngest coal deposits of the Permianin South China. These coal measures were createdby accumulation of the tropical *Gigantopteris* flora in swampy areas. A general picture of distributions of coal measures shows that the Permian coal-bearing strata tended to migrate from east to west throughout the Permian cross the entire South China due to the westward migration of coal forming-swamp habitats (Fig. 8.6).

Conversely in North China, far from the seaduring the Late Permian, the coal-forming environments decreased strongly due to a severe warming giving rise to a hot and arid climatic condition. At the same time Euramerican plants, especially meso-xerophytic woody scrubs, dominated by *Ullmannia*-like conifers, migrated to this region.

8.3.2 The Paleoflora Reflecting Paleoclimate

The uppermost Xuanwei Formation palaeofloral assemblage of latest Permian age contains abundant coal-forming plant taxa: *Lepidodendron*, *Paracalamites*, gigantopterids, and other forms. Coal seams occurred in multiple horizons in the uppermost Xuanwei Formation, extending to the horizon just below the PTB in the Tucheng section.Thus, the latest Permian climatic condition of the WGEY region was humid and warm. Although the coal measures disappeared in the Kayitou Formation, the Induan floral assemblage contains some surviving swampy plant *Paracalamites* that co-existed with *Annalepis* at the basal Kayitou Formation, and some fragments of gigantopterids and Peltaspermales at the higher horizons of the formation. Those climatically sensitive plant taxa reflected that the humid and warm climate prevailed in the WGEY region during the Early Triassic, just like the latest Permian climatic condition in the same region.

The plant-based inference of the P-Tr climates is reinforced by palynofloral assemblages across the PTB in WGEY. Palynological data published by Ouyang (1982) and Peng et al. (2006) showed that some Palaeozoic plants that inhabited in humid and warm climate persisted into the earliest Triassic time in WGEY. Accordingly, both macro- and microfloral assemblages indicate that the latest Permian humid and warm climate were inherited to the Early Triassic in the WGEY region. It should be noted that climatic conditions in many Late Permian continents, for instance, the Karoo basin of South Africa (Ward et al. 2000), the Urals of Russia (Benton et al. 2004), the Junggar basin of Xinjiang, northwest China (Yang et al. 2010) underwent a dramatic change from the latest Permian warm, humid climate to the Early Triassic dry, hot weather condition. In contrast, the warm, humid climatic condition prevailed over the P-Tr transition in WGEY, which may have provided suitable climatic condition for some Permian plant relicts to survive the PTB biocrisis and the rise of some Mesozoic pioneers in Early Triassic in WGEY.

8.3.3 Conclusions

In South China, during the Changhsingian, a "*Gigantopteris*" flora representing a tropical rainforest climate flourished. This rainforest floral association disappeared rapidly due to the impact of the Permian–Triassic worldwide mass extinction events.

During the Earliest Triassic recovery, pioneering transitional *Lepacyclotes* (= *Annalepis*)-*Peltaspermum* floral assemblage appeared including rare surviving relicts of *Gigantopteris* flora, coeval to the progressive disappearance of the tropical rainforest climate. And after a long time with almost barren red-bed deposits under a probable dry climate, in the Late-Early Triassic, a more innovative and stable *Neuropteridium-Voltzia* floral assemblage emerged only in few scattered areas that may indicate the setup of an extremely heterogeneous terrestrial environment.

Conversely, in North China, the Changhsingian *Ullmannia* aff. *Bronnii-Yuaniamagnifolia* floral assemblage may correspond to the arid climate episode. But following the Permian–Triassic mass extinction processes, a sedimentation of red-bed deposits testifying the establishment of a dryer climate in North China that persisted much longer than in South China. During the Early Triassic, new few fossil plants (e.g. *Pleuromeia, Macrostachya, Phyllotheca* etc.) assigned to a "*Pleuromeia* floral assemblage" may indicate some increase in humidity. Overall, the vegetation cover suffered severe impacts during the Permian–Triassic mass extinction both in North and South China blocks. But in North China due to extensive plate tectonic movement to a higher latitude palaeogeographic position, the post-crisis vegetal covers became drastically different to the South China one due to the progressive incoming of Subangaran plants. It seems well to show that the impacts of the mass extinction event on floras in the tropical South China are less strong than in North China. As a matter of fact, the post-crisis recovery of the vegetal cover in the tropical South China block was faster than in the more or less "dry-temperate" North China block as witnessed by the much more diverse and disparate earliest Triassic plant fossil assemblage there.

Finally, at the geological timescale, the combination of speciation processes and cross migrations with the adjacent territories led the Late Permian and Triassic floristic assemblages to diverge more and more between the Northern and Southern "Cathaysian" blocks.

These divergent trajectories were driven by an up to 35° counter-clockwise rotation of the Pangea (between 260 and 230 Ma.) around a Euler pole located in Northernmost South America (Upper Permian to Middle Triassic) that triggered a deep reorganization of the whole Tethyan realm (Riel et al. 2018). In this tectonic context, the South China blocks remained in a sub-equatorial location, while the North China block drifted up the 30°North.These different palaeogeographic histories led to the paradox that most of the Cathaysian plant taxa started to disappear as early as in the Middle Permian on the North China block, "cradle" on which the first Cathaysian plants appeared around the Carboniferous-Permian transition, migrating afterwards to the South China. The last "refugium" for these plants until the latest Permian was then the South China block.

8.4 Plant and Environment Co-evolution

Plants have a bidirectional relationship with the environment. Plant evolution and distributions are strongly influenced by prevailing environmental conditions and plants alter the environments in which they grow through photosynthesis and consequences of their growth on water and biogeochemical cycles as well as altering physical environments (Vajda et al. 2020). To evaluate the palaeoenvironmental influences of the floral changes evaluated here, a timescale for environmental events has been compiled through the End Permian to Middle Triassic, concentrating on the EPPC, the Early Induan survival stage, the late Induan species richness recovery, and the Olenekian abundance recovery. This is divided into two parts: the Changhsingian to Early Grisbachian phase set against the timescale of the marine Meishan section from which precise zircon and marine animal ages have been determined, and the Induan to Anisian interval with lower stratigraphic resolution due to lack of precisely dated events in terrestrial facies (Fig. 8.7).

8.4.1 The End Permian Plant Crisis and Marine Ecosystem Collapse

The EPPC is composed of the plant species decline happened before the marine extinction event and the following abundance drop before the PTB (Fig. 8.7). It is set against the long-term backdrop of global climate drying and habitat loss for spore producing plants that negatively affected plant species richness and abundance following the formation of Pangaea from the Middle Permian (Kidder and Worsley (2003); Roscher et al. 2011; Benton et al. 2014; Blomenkemper et al. 2018). As outlined in Fig. 19, mechanisms of plant-environment interactions are intricately linked from which many concurrent changes to plant species richness and abundance as well as environmental factors combined to make the latest Permian a time of significant upheaval. Lower plant abundances would have reduced CO_2 uptake via photosynthesis leading to increased atmospheric CO_2 concentrations and global warming. Worsening of drought conditions appears to be evidenced by gradually increasing wildfire frequency evidenced by charcoal concentrations (Shen et al. 2011; Yan et al. 2019; Chu et al. 2020). Contemporaneous volcanic eruptions and the degassing of CO_2, CH_4 and other greenhouse gasses contributed to climbing

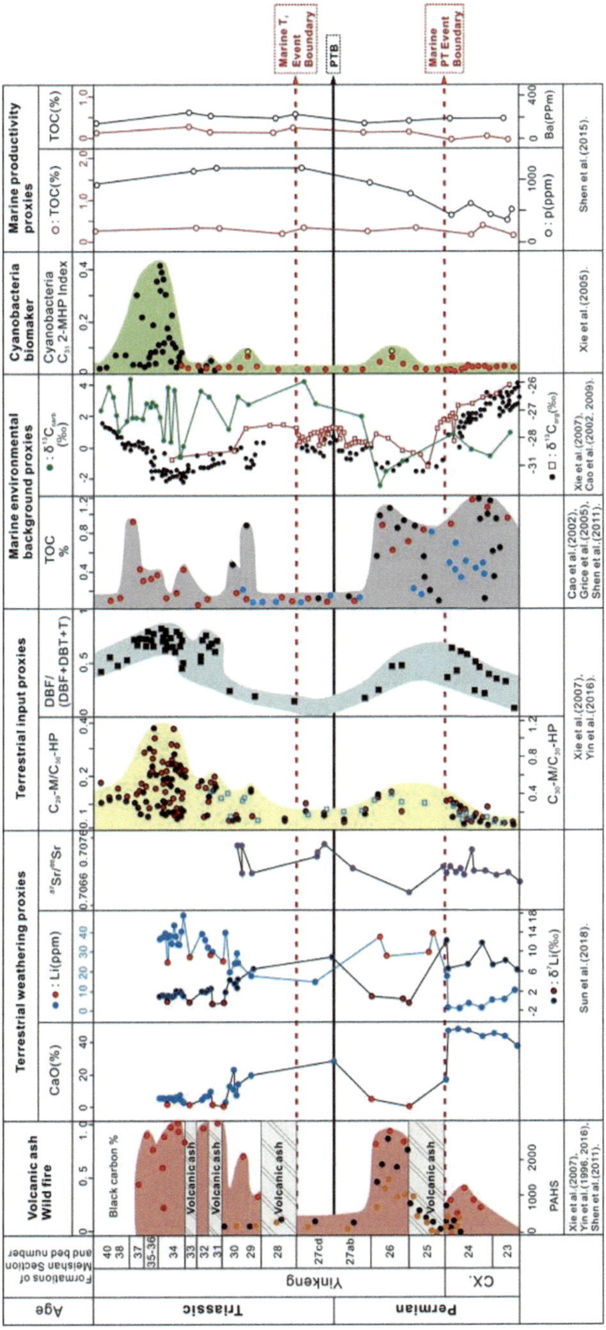

Fig. 8.7 High-resolution comparison between terrestrial events and marine feedback from the End-Permian GSSP Meishan Section Bed 23 to Early Triassic Bed 40. C. X. = Changhsingian; modified from Yin et al. (2016)

global temperature (Sun et al. 2012; Retallack 2013; Benton et al. 2014; Burgess et al. 2017). In addition to greenhouse gasses, volcanic activity also produced toxic gasses that led to acid rain and depletion of the ozone layer resulting in increased UVB radiation (Benca et al. 2018; Black et al. 2018). Acid rain reduced plant growth and along with increased UVB radiation had negative effects on plant reproduction through increased malformations and sterility (Visscher et al. 2004; Benca et al. 2018). Plant rooting system are significant in soil fixation so the loss of terrestrial ecosystems in the EPPC could have resulted in anomalous soil erosion and siltation with large amounts of terrestrial debris flowing into water system (Shen et al. 2019; Vajda et al. 2020). Abnormal terrestrial inputs into water systems could result in eutrophication from elevated nutrient inputs and lead to blooms of plankton such as *cyanobacteria* (Algeo et al. 2011; Shen et al. 2015; Kaiho et al. 2016). Furthermore, oxidation and decomposition of terrestrial debris comsued the oxygen in the water system and may have triggered anoxia in nearshore settings, then the euxinia (Algeo et al. 2011; Kaiho et al. 2016). However, in South West China during the EPPC, dysoxia rather than anoxia occurred in shallow marine facies and a "death by siltation" mechanism for shallow marine ecosystems following the loss of terrestrial vegetation is not supported as carbonates rather than surges in clastic sediments were deposited (Wignall et al. 2020). While these may have been apparent in other areas globally, in South West China clastic sediment were likely trapped in alluvial settings by base level rise and marine transgression (Wignall et al. 2020). Moreover, rising temperatures in reducing gas solubility in water would have worsened freshwater and oceanic anoxia and help accelerate the ecosystem collapse in these settings. In this model, feedback circles exist in which reduction of plant abundance is key to deteriorating environmental conditions, and deteriorating environmental condition are key to terrestrial ecosystem loss.

8.4.2 *The Early Induan Survival and Marine Ecosystem Delayed Recovery*

Terrestrial input proxies suggest that soil erosion increased during the earliest Induan (Grisbachian) and was likely related to the delayed plant abundance recovery (Fig. 8.8). The instability of plant fixation was coincident with enhancement of chemical weathering, which appears to be related to the anomalous terrestrial debris entering water systems and the increase of marine productivity coming from the explosive propagation of plankton (Algeo et al. 2010, 2011; Shen et al. 2015; Xu et al. 2017). The flora at this time was dominated by the herbaceous lycopods *Tomiostrobus* and *Pleuromeia* which havesmall and shallow rooting systems (Retallack et al. 1975; Yu et al. 2010) that appear to have been insufficient to effectively bind soils; these plants were restricted to near shore settings (Yu et al. 2010) and together with their low abundance infers that terrestrial plant communities at this time were unable to slow down soil erosion (Algeo et al. 2011; Boyce et al. 2016). Furthermore, carbon

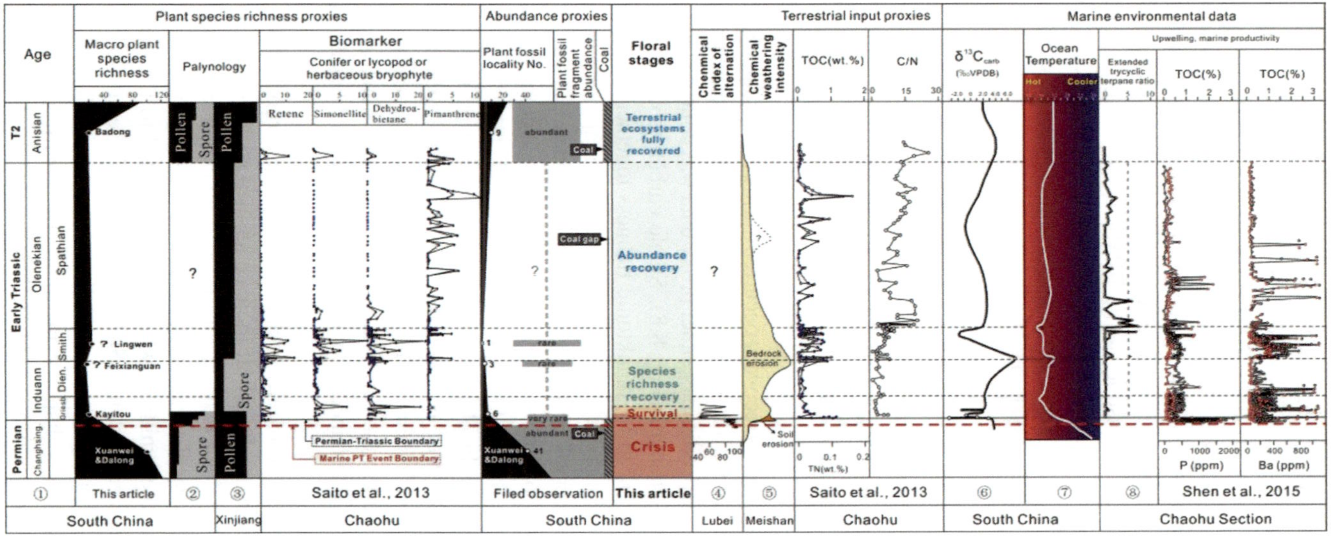

Fig. 8.8 Comparison between floral change pattern proxies from macro plant fossils, coal, palynology, biomarker and environment event such as terrestrial input, marine feedback, marine temperature through the End-Permian Changhsingian to Middle Triassic Anisian. (1) Stratal data from Burgess et al. (2014); (2) Palynology data from Changhsingian to Induan of South China from Zhang et al. (2004), Yu et al. (2008), Ou yang et al. (2007) and Middle Triassic Anisian of South Chin from Qu et al. (1990); (3) Palynology data from Changhsingian to Anisian from Qu et al. (1986); (4) Chemical Index of Alternation (CIA) from Xu et al. (2017); (5) Chemical weathering rate from Algeo et al. (2011); (6) Carbonate Carbon Isotope from Burgess et al. (2014); (7) Oceanic temperature from Sun et al. (2012); (8) Extended trycyclic terpane ratio data from Saito et al. (2013)

fixation efficiency was limited based on limited species richness and abundance of plant populations, with this unable to contribute to environmental improvement (Boyce et al. 2016). The plant crisis continued into the earliest Induan (Grisbachian) as a vicious cycle of feedback mechanisms (Fig. 8.8) attested by the reappearance of cyanobacteria booms (Xie et al. 2005, 2007) as well as steadily rising temperature (Sun et al. 2012). All the available evidence indicates that the prosperity of the opportunist lycopods together with the Cathaysian floral refugial elements failed to define either species richness or abundance recovery, or a successful attempt for ecological recovery. The second episode of the EPME occurred shortly after the Permian Triassic Boundary and was restricted to marine ecosystems (Marine T_1 Event Boundary in Fig. 8.7), from which protracted instability on land continuously delayed the recovery of marine ecosystems.

8.4.3 Distinct Episodes of Plant Species Richness and Abundance Recovery and Their Impacts in Ecosystem Stabilization

From our data plant species richness recovery occurred during the Induan while abundance recovery possibly occurred from the Olenekian. To identify the environmental impact of the plant two recovery episodes, other proxy data was collected to indicate the stages of environmental changes (Fig. 8.8). Stable carbon isotopes of marine carbonate rocks can be influenced by planktonic richness together with terrestrial plant inputs (Sun et al. 2012; Shen et al. 2015). There were two major marine primary productivity enhancement interval, respectively the Griesbachian and the Smithian substages. While the corresponding carbonate stable carbon isotope values only had one positive peak in the Induan-Olenekian boundary indicated the prosperous of terrestrial producer-most likely plantsrather than the marine primary producer plankton (Sun et al. 2012; Saito et al. 2014; Burgess et al. 2014). Ocean temperatures had almost synchronous changes with carbonate rock stable carbon isotope ratios, decreasing slightly during the early stage of plant species richness recovery but rebounding during plant abundance recovery, and turned to stable subsequently (Sun et al. 2012).

Proxies of marine primary productivity such as total organic carbon (TOC), phosphorus (P), and biogenic barium (Ba_{bio}) all indicate an increase of marine primary productivity after the EPPC which may have resulted from enhancement of terrestrial chemical as well as physical weathering. Marine primary productivity came to be relatively stable after the Smithian which may be related to the reestablishment of stability in terrestrial ecosystems (Shen et al. 2015). The extended trycyclic terpane (ETR) ratio is also a proxy that could indicate ocean upwelling and marine productivity if the value is greater than 4 (Saito et al. 2013), with increased marine upwelling due to trade wind intensification indirectly indicating the decrease of atmospheric pCO_2 (Saito et al. 2013). The ETR data from the Chaohu section near the

GSSP Meishan Section reveals a strong positive peak near the Smithian-Spathian (S–S) boundary which indicates that the ocean and climate states are switching from elevated atmospheric pCO_2 level in the end-Permian to a decreased level during S–S. Furthermore, data from Retallack (2011) and Galfettial (2007) show the drawdown of atmosphere CO_2 is related to enhanced organic carbon burial, with values in the Chaohu section indicating that ecosystem productivity is much higher than the crisis and is possibly a sign of ecosystem recovery after the S–S (Beerling and Woodward 2001).

The evidence amassed above indicates that separate plant species richness and abundance recovery episodes had different effects on environmental stability and thus had different patterns recorded in sedimentary strata. It is the plant abundance recovery that was efficient in helping the reestablishment of ecosystems together with increased environmental stability, while the initial species richness recovery had much less significant effects on terrestrial environments.

References

Algeo TJ, Twichett RJ (2010) Anomalous early Triassic sediment fluxes due to elevated weathering rates. J Earth Sci 38(11):1023–1026

Algeo TJ, Chen ZQ, Fraiser ML, Twitchett RJ (2011) Terrestrial–marine teleconnections in the collapse and rebuilding of early Triassic marine ecosystems. Palaeogeogr Palaeoclimatol Palaeoecol 308:1–11

Beerling D, Woodward FI (2001) Vegetation and the terrestrial carbon cycle: modelling the first 400 million years. Cambridge University Press, pp 1–405

Benca JP, Duijnstee I, Looy CV (2018) UVB–induced forest sterility: implications of Ozone shield failure in Earth's largest extinction. Sci Adv 4(2):e1700618

Benton MJ, Newell AJ (2014) Impacts of global warming on Permian-Triassic terrestrial ecosystems. Gond Res 25(4):1308–1337

Benton MJ, Tverdokhlebov VP, Surkov MV (2004) Ecosystem remodelling among vertebrates at the Permian-Triassic boundary in Russia. Nature 432(7013):97–100

Berthelin M, Broutin J, Kerp H, Crasquin-Soleau S, Platel JP, Roger J (2003) The Oman Gharif mixed paleoflora: a useful tool for testing Permian Pangea reconstructions. Palaeogeogr Palaeoclimatol Palaeoecol 196(1):85–98

Black BA, Neely RR, Lamarque J, Elkins-Tanton LT, Kiehl JT, Shields CA, Mills MJ, Bardeen C (2018) Systemic swings in End-Permian climate from Siberian traps carbon and sulfur outgassing. Nat Geosci 11(12):949–954

Blomenkemper P, Kerp H, Hamad AA, Dimichele WA, Bomfleur B (2018) A hidden cradle of plant evolution in Permian tropical lowlands. Science 362:1414–1416

Boyce CK, Dimichele WA (2016) Arborescent lycopsid productivity and lifespan: constraining the possibilities. Rev Palaeobot Palynol 227:97–110

Burgess SD, Bowring S, Shen SZ (2014) High-precision timeline for earth's most severe extinction. Proc Natl Acad Sci USA 111(9):3316–3321

Burgess SD, Muirhead JD, Bowring SA (2017) Initial pulse of Siberian traps sills as the trigger of the End-Permian mass extinction. Nature Commun 8:164

Chen ZQ, Jin YG, Shi GR (1998) Permian transgression-regression sequences and sea-level changes of South China. Proc R Soc Victoria 110:345–368

Chu DL, Grasby SE, Song HJ, Corso JD, Wang Y, Mather TA, Wu Y, Song HY, Shu WC, Tong JN, Wignall PB (2020) Ecological disturbance in tropical peatlands prior to marine Permian-Triassic mass extinction. Geology 48:288–292
Feng Z, Wei HB, Guo Y, Bomfleur B (2018) A conifer-dominated Early Triassic flora from Southwest China. Sci Bull 63:1462–1463
Fluteau F, Besse J, Broutin J, Ramstein G (2001) The Late Permian climate: what can be inferred from climate modelling concerning Pangea scenarios and Hercynian range altitude? Palaeogeogr Palaeoclimatol Palaeoecol 167:39–71
Golonka J, Ross JI, Scotese CR (1994) Phanerozoic paleogeographic and paleoclimatic modeling maps. In: Embry AF et al (eds) Pangea: global environments and resources. Canadian Society of Petroleum Geologists Memoir 17:1–47
Greb SF, Dimichele WA, Gastaldo RA (2006) Evolution and importance of wetlands in earth history. In: Greb SF, DiMichele WA (eds) Wetlands through time. Geological Society of America (special Paper) 399, pp 1–40
He XL, Liang DS, Sheng SZ (1996) The permian floras from Jiangxi, China. China University of Minings Press, Xuzhou, pp 1–201
Kaiho K, Saito R, Ito K, Miyaji T, Chen ZQ (2016) Effects of soil erosion and anoxic-euxinic ocean in the Permian-Triassic marine crisis. Heliyon 2(8):137
Kidder DL, Worsley TR (2003) Causes and consequences of extreme Permo-Triassic warming to globally equable climate and relation to the Permo-Triassic extinction and recovery. Palaeogeogr. Palaeoclimatol. Palaeoecol 203:207–237
Li XX (ed) (1995) Fossil floras in China through the geological ages (English edition). Guangdong Science and Technology Press, Guangzhou, pp 1–695
Li XX, Yao ZQ (1980) Permian coal-bearing formations in South China. J Stratigr 4:241–255 (In Chinese)
Li H, Yu JX, McElwain JC, Yiotis C, Chen ZQ (2019) Reconstruction of atmospheric CO_2 concentration during the late Changhsingian based on fossil conifers from the Dalong Formation in South China. Palaeogeogr Palaeoclimatol Palaeoecol 519:37–48
Meng FS (1992) Fossil plants. In: Yichang Institute of Geology and Mineral Resources et al (eds) Geology of hainan island, 1. Stratigraphy and Palaeology. Geological Publishing House, Beijing, pp 166–170, 175–183 (in Chinese)
Muttoni G, Mattei M, Balini M, Zanchi A, Gaetani M, Berra F (2009) The drift history of Iran from the Ordovician to the Triassic. Geol Soc London (special Publications) 312(1):7–29
Ouyang S (1982) Upper Permian and Lower Triassic palynomophs from Eastern Yunnan. Can J Earth Sci 19:68–80
Ouyang S, Zhu HC (2007) Query the assumption of "End-Permian Fungal Spike Event", with special reference to the permo-triassic transitional palynofloras. Acta Palaeontologica Sinica 46(4):394–410
Peng Y, Shi GR (2009) Life crises on land across the Permian-Triassic boundary in South China. Global Planet Change 65:155–165
Peng YQ, Yu JX, Gao YQ, Yang FQ (2006) Palynological assemblages of non-marine rocks at the Permian-Triassic boundary, western Guizhou and eastern Yunnan, South China. J Asian Earth Sci 28:291–305
Pott C, McLoughlin S, Lindström A (2010) Late Palaeozoic foliage from China displays affinities to Cycadales rather than to Bennettitales necessitating a re−evaluation of the Palaeozoic Pterophyllum species. Acta Palaeontol Pol 55(1):157–168
Qu LF (1990) Palynological assemblages of Middle and Late Triassic in Sangzhi, Hunan, and their stratigraphical significance. J Stratigr Paleotol 23:81–95 (In Chinese with English abstract)
Qu LF, Wang Z (1986) Triassic spores and pollen. In: Permian and Triassic strata and fossil assemblages in the Dalongkou area of Jimsar, Xinjiang. Geol. Mem., Series 2(3). Geological Publishing House, Beijing, pp 113–173 (in Chinese with English summary)
Retallack GJ (1975) The life and time of a Triasic lycopod. Alcheringa 1:3–29

Retallack GJ (2011) Exceptional fossil preservation during CO_2 greenhouse crises? Palaeogeogr Palaeoclimatol Palaeoecol 307:59–74

Retallack GJ (2013) Permian and Triassic greenhouse crises. Gondwana Res 24:90–103

Riel N, Jaillard E, Martelat J-E, Guillot S, Braun J (2018) Permian-Triassic Tethyan realm reorganization: Implications for the outward Pangea margin. J S Am Earth Sci 81:78–86

Roscher M, Stordal F, Svensen H (2011) The effect of global warming and global cooling on the distribution of the latest Permian climate zones. Palaeogeogr Palaeoclimatol Palaeoecol 309:186–200

Saito R, Kaiho K, Oba M, Takahashi S, Chen ZQ, Tong J (2013) A terrestrial vegetation turnover in the middle of the Early Triassic. Global Planet Change 105:152–159

Saito R, Oba M, Kaiho K, Schaeffer P, Adam P, Takahashi S, Nara FE, Chen ZQ, Tong JN, Tsuchiya N (2014) Extreme euxinia just prior to the Middle Triassic biotic recovery from the latest Permian mass extinction. Org Geochem 73:113–122

Shen WJ, Sun YG, Lin YT, Liu DH, Chai PX (2011) Evidence for wildfire in the Meishan Section and implications for Permian-Triassic events. Geochim Cosmochim Acta 27:1992–2006

Shen J, Algeo TJ, Hu Q, Xu GZ, Zhou L, Feng QL (2013) Volcanism in South China during the Late Permian and its relationship to marine ecosystem and environmental changes. Global Planet Change 105:121–134

Shen J, Schoepfer SD, Feng Q, Song HY (2015) Marine productivity changes during the End-Permian crisis and Early Triassic recovery. Earth-Sci Rev 149:136–162

Shen J, Chen J, Algeo TJ, Yuan SL, Feng Q, Yu J, Zhou L, O'Connell B, Planavsky NJ (2019) Evidence for a prolonged Permian-Triassic Extinction interval from global marine mercury records. Nat Commun 10:1563

Shi X (2016) Fossil woods and environmental changes during the Permian–Triassic transition in Northwest China. PhD thesis China University of Geosciences–Wuhan/Université Pierre et Marie Curie-Paris 6, France, 182 p

Sun YD, Joachimski MM, Wignall PB, Yan CB, Chen YL, Jiang HS, Wang LN, Lai XL (2012) Lethally hot temperatures during the Early Triassic greenhouse. Science 338(6105):366–370

Vajda V, McLoughlin S, Mays C, Frank TD, Fielding CR, Tevywa A, Lehsten V, Bocking M, Nicoll RS (2020) End-Permian (252 Mya) deforestation, wildfires and flooding - An ancient biotic crisis with lessons for the present. Earth Planet Sci Lett 529:115875

Visscher H, Looy CV, Collinson ME, Brinkhuis H, van Konijnenburg van Cittert JH, Kürschner WM, Sephton MA (2004) Environmental mutagenesis during the End-Permian ecological crisis. Proc Natl Acad Sci USA 101(35):12952–12956

Wang ZQ (1983) New materials of fossil plants from Shiqianfeng Group in N China. Bull Geol Soc Tianjing 1(2):27–80 (in Chinese)

Wang ZQ, Wang LX (1982) A new species of the lycopsid Pleuromeia from the Early Triassic of Shanxi China and its ecology. Palaeontology 25(1):215–225

Wang ZQ (1985) Palaeovegetation and plate tectonics: palaeophyto geography of North China during Permian and Triassic times. Palaeogeography, Palaeoclimate, Palaeoecology, 49(1):25–45

Wang ZQ, Wang LX (1986) Late Permian fossil plants from the lower part of the Shiqianfeng (Shihchienfeng) Group in North China. Bull Tianjin Instit Geol Mineral Resour 15:1–80 (in Chinese with English abstract)

Wang ZQ (1989) Permian gigantic palaeobotanical events in North China. Acta Palaeont Sinica, 28(3):314–337

Wang ZQ, Wang LX (1990) Late Early Triassic fossil plants from upper part of the Shiqianfeng Group in North China. Shanxi Geol 14(1):23–40 (in Chinese with English abstract)

Ward PD, Montgomery DR, Smith RMH (2000) Altered river morphology in South Africa related to the Permian-Triassic extinction. Science 289:1740–1743

Wignall PB, Chu D, Hilton JM, Dal Corso J, Wu Y, Wang Y, Atkinson J, Tong J (2020) Death in the shallows: The record of Permo-Triassic mass extinction in paralic settings, southwest China. Global Planet Change 189(2020):103176

Xie SC, Pancost R, Yin H, Wang H, Evershed R (2005) Two episodes of microbial change coupled with Permo/Triassic faunal mass extinction. Nature 434:494–497

Xie SC, Pancost RD, Huang JH, Wignall PB, Yu JX, Tang X, Chen L, Huang XY, Lai XL (2007) Changes in the global carbon cycle occurred as two episodes during the Permian-Triassic crisis. Geology 35:1083–1086

Xu GZ, Feng QL, Deconinck JF, Shen J, Zhao TY, Young AL (2017) High-resolution clay mineral and major elemental characterization of a Permian-Triassic terrestrial succession in southwestern China: diagenetic and paleoclimatic/paleoenvironmental significance. Palaeogeogr Palaeoclimatol Palaeoecol 481:77–93

Yan Z, Shao L, Glasspool LJ, Wang J, Wang X, Wang H (2019) Frequent and intense fires in the final coals of the Paleozoic indicate elevated atmospheric oxygen levels at the onset of the End-Permian Mass Extinction Event. Int J Coal Geol 207:75–83

Yang W, Feng Q, Liu YQ, Tabor N, Miggins D, Crowley J, Lin JY, Thomas S (2010) Depositional environments and cyclo- and chronostratigraphy of Uppermost carboniferous-Lower triassic fluvial-lacustrine deposits, Southern bogda mountains, NW China–a terrestrial paleoclimatic record of Mid-Latitude NE Pangea. Global Planet Change 73:15–113

Yao ZQ, Liu LJ, Rothwell WG et al (2000) Szecladia new genus, a Late Permian conifer with multiveined leaves from South China. J Paleontol 74:524–531

Yin HF, Wu SB, Du YS, Peng YQ (1999) South China defined as part of Tethyan archipelagic ocean system. Earth Sci J China Univ Geosci 24:1–12 (in Chinese with English abstract)

Yin HF, Song HJ (2013) Mass extinction and Pangea integration during the Paleozoic-Mesozoic transition. Sci China: Earth Sci 56:1791–1803 (In Chinese with English abstract)

Yu JX, Peng YQ, Zhang SX, Yang FQ, Zhao QM, Huang QS (2007) Terrestrial events across the Permian- Triassic boundary along the Yunnan-Guizhou border, SW China. Global Planet Change 55:193–208

Yu JX, Broutin J, Huang QS, Grauvogel-Stamm L (2010) *Annalepis*, a pioneering lycopsid genus in the recovery of the Triassic land flora in South China. C R Palevol 9:479–486

Yu JX, Broutin J, Chen ZQ, Shi X, Li H, Chu DL, Huang QS (2015) Vegetation changeover across the Permian-Triassic boundary in Southwest China extinction, survival, recovery and palaeoclimate: a critical review. Earth-Sci Rev 149:203–224

Yu JX (2008) Floras (macro- and microfloras) and evolutionary dynamics across the Permian–Triassic boundary along Guizhou and Yunnan border, South China. PhD Thesis, University Geosciences China Wuhan–University Pierre et Marie Curie, Paris 6, 220 p

Zhang KX, Yu JX, Lin QX, Jing YL, Chen B (2004) Palynological assemblage in section d of Meishan, Changxing, Zhejiang and its significance of global correlation. Earth Sci J China Univ Geosci 29(3):253–262

Zhou ZY, Li BX (1979) A preliminary study of the Early Triassic plants from the Qionghai District Hainan Island. Acta Palaeotol Sin 18(5):444–466 (In Chinese with English abstract)

Ziegler AM, Eshel G, Rees PM, Rothfus TA, Rowley D, Sunderlin D (2003) Tracing the tropics across land and sea: Permian to present. Lethaia 36:227–254

Chapter 9
Systematic Palaeontology: Part 1 Plant Fossils

Jianxin Yu, Zhen Xu, Hui Li, and Zongsheng Lu

The classification of plant fossils adopted herein is generally consistent with that of extant plants (Table 9.1), according to the degree of plant differentiation, anatomical structure, nutritional mode, reproductive and life history types. However, the classification of plant fossils has its particularity. Due to the large body of most plants and the scattered preservation of various organs in the stratum, it brings some difficulties to the natural classification of some plant fossils, so it is often supplemented by artificial morphological classification.

The naming of plant fossils also follows Linnaeus's double name method and is named in Latin. In order to facilitate the understanding of the classification system and name meaning of plant fossils, the following table shows the classification system and corresponding word endings adopted by the simplified version of plant fossils, as well as the standard format when writing the names of genera and species of plant fossils (Table 9.2), in which the names of genera and species are named by the discoverer without word endings.

All the specimens described in this paper are preserved in the Laboratory of Geobiology, Faculty of Earth Sciences, China University of Geosciences, Wuhan, People's Republic of China, with the prefix of GWC, GWZ, YXC, YXM, GPT, CLC, GAXM,

J. Yu (✉) · Z. Xu · Z. Lu
School of Earth Sciences, China University of Geosciences, Wuhan, P.R. China
e-mail: yujianxin@cug.edu.cn

Z. Xu
e-mail: janexu_cug@qq.com

Z. Lu
e-mail: zshlu@cug.edu.cn

H. Li
Jiangxi Key Laboratory for Mass Spectrometry and Instrumentation, East China University of Technology, Nanchang, P.R. China
e-mail: huili_201860188@ecut.edu.cn

J. Yu et al. (eds.), *Plants and Palynomorphs around the Permian-Triassic Boundary of South China*, New Records of the Great Dying in South China,
https://doi.org/10.1007/978-981-19-1492-8_9

Table 9.1 Classification system of plants

Bryophyta	Early Paleozoic — Rec.	
Protopteridophyta	S-D	Ferns
Lycophyta	D — Rec.	Ferns
Sphenophyta(Arthrophyta)	D — Rec.	Ferns
Pteridophyta	D — Rec.	Ferns
Progymnospermophyta	$D_{2\text{-}3}$ — P	Gymnosperms
Pteridospermophyta	D_3 — K_1	Gymnosperms
Cycadophyta	D_3 — Rec.	Gymnosperms
Ginkgophyta	P — Rec.	Gymnosperms
Coniferophyta: Cordaitopsida	D_3 — T_1	Gymnosperms
Coniferophyta: Coniferopsida	C_2 — Rec.	Gymnosperms
Gnetophyta	T — Rec.	Gymnosperms
Anthophyta(Angiospermae): Dicotyledones, Monocotyledones	(J_3?) – Rec.	Angiosperms

Table 9.2 Standard format to the names of genera and species of plant fossils

Classification System of Plants	Respective Suffix	Example
Kingdom		Plantae
Phylum	-phyta	Rhyniophyta
Subphylum	-phytina	
Class	-opsida	Rhyniopsida
Subclass	-idae	
Order	-ales	Rhyniales
Suborder	-ineae	
Family	-aceae	Rhyniaceae
Subfamily	-oideae	
Genus		*Rhynia*
Subgenus		
Species		*Aglaophyton (Rhynia) major*
Subspecies		

GHDSA and ZK. Abbreviations of localities (sections): GWC = Chahe section, Weining County, Guizhou Province; GWZ = Zhejue section, Weining County, Guizhou Province; YXC = Chinahe section, Xuanwei City, Yunnan Province; YXM = Mide section, Xuanwei City, Yunnan Province; GPT = Tucheng section, Panxian County, Guizhou Province; CLC = Chunlecun section, Hezhang County, Guizhou Province; GAXM = Xinmin section, Anshun City, Guizhou Province; GHDSA = Duanshan A section, Huishui County, Guizhou Province; ZK = Zuankong meaning drilling well), Fuyuan County, Yunnan Province.

Division Lycophyta

Class Lycopsida

Order Lepidocarpales

Family Lepidocarpaceae

Genus *Lepidodendron* Sternberg, 1820

Lepidodendron oculus-felis (Abbado) Zeiller, 1900 (Fig. 9.1-6)

1900 Lepidodendron oculus-felis (Abbado, 1900) Zeiller, in Ann. Mines Paris, 19:8, Pl.7

1963 Lepidodendron oculus-felis (Abbado, 1900) Zeiller, Paleontologica Sinica, Pl.8, Figs. 1–3; Pl.9, Figs. 1–2

1974 Lepidodendron oculus-felis, Paleozoic plants of China, p. 28, Pl.12, Fig. 5

Holotype:

Diagnosis (emended): Leaf cushion longitudinal rhombus, rhombus, rhomboid or lenticulated; length–width ratio is 2 to 1/2; crowded in range. Big leaf scar; the width slightly wider than the length; lenticulated to rhomboid; large vertex and base angle, sharp lateral angle; usually with lateral extensive line; situated in the middle or upper part of leaf cushion; sightly cat-eye-shaped; bundle scar wide "V" shape; lateral scar small, round, situated on the extension line of two neighbor lateral angle. Ligular pit situated on the top of leaf scar, occasionally concave. Leaf cushion usually flat; sometimes with faint ridge line and cross striation; occasionally small triagular pit on the top.

Material examined: GWC-21-15, GWC-21-17.

Occurrence: Upper Carboniferous Taiyuan Formation and Upper Permian Upper Shihezi Formation in North China, Gansu, Ningxia, Shaanxi, Liaoning; Early to Late Permian Qixia, Maokou and Longtan formaations in Fujian, Guangxi, Guangdong, Jiangsu and Xuanwei Formation in Western Guizhou and Eastern Yunnan.

Description (after Zeiller 1901). The cushions are crowed, rhomboidal in shape, about 1 cm broad by approximately 2 cm long. The leaf-scars are cats-eye-shaped, situated at a little above the middle of the cushions; upper and basal angles obtusely rounded; lateral angles acutely rounded. The broadly V-shaped vascular and two lateral cicatrices with round in shape, lie nearly one line. The ligule-scars are clear, round in shape, lies almost directly above the leaf-scars.

Discussion and Comparison. This species has a widespread geographical distribution in the Cathaysia-land and has often been described and illustrated. The impressions of the present specimens show the characteristic cats-eye-shaped leaf scars more or less distinctly.

Lepidodendron acutangulum (Halle) (Stockmans and Mathieu 1957) (Figs. 9.1-3, 9.1-4 and 9.1-11)

1900 Sigiillaria acutangula Abbado, in Palaeont. Sinica, Ser. A, Vol. I, 2 p. 5, Pl. II, Figs. 6–7

Fig. 9.1 1, 2 *Lepidodendron lepidophloides* Yao Locality: 1 Chahe section, Weining County, Guizhou Province Sample Number: GWC-17-1, 2 Chinahe section, Xuanwei City, Yunnan Province Sample Number: YXC-(-9)-3, 3, 4, 11 *Lepidodendron acutangulum* (Halle) Stockm. et Math. Locality: 3, 4 Mide section, Xuanwei City, Yunnan Province Sample Number: YXM (A)-43-1 11 Xinmin section, Anshun City, Guizhou Province Sample Number: GAXM11019, 5, 7 *Lepidodendron* sp. Locality: 5 Chinahe section, Xuanwei City, Yunnan Province Sample Number: YXC-(-9)-2, 7 Chahe section, Weining County, Guizhou Province Sample Number: GWC-21, 6 *Lepidodendron oculus-felis* (Abbado) Zeiller Locality: Chahe section, Weining County, Guizhou Province Sample Number: GWC-21-15, 8 *Lepidostrobophyllum* sp. Locality: Xinmin section, Anshun City, Guizhou Province Sample Number: GAXM11030, 9, 10 *Lepidostrobophyllum hastatum* (Lesq.) Chal Locality: Mide section, Xuanwei City, Yunnan Province Sample Number: YXC-(-6)-1A, YXC-(-6)-2A

1901 Lepidodendron oculus-felis (Abbado, 1900) Zeiller, in Ann. Mines Paris, 19:8, Pl.7

1957 Lepidodendron acutangula (Halle) Stockm. et Math., in Publ. Ass. et. Paleont. 32:56

1963 Lepidodendron oculus-felis (Abbado, 1900) Zeiller, Li, Paleontologica Sinica, Pl.15, Figs. 1–4

1974 Lepidodendron acutangula (Halle) Stockm. et Math., Li et al., Paleozoic plants of China, p.30, Pl.12, Figs. 1–4

Holotype:

Diagnosis (emended): Relatively big leaf cushion; rhomboid or lenticulated; densely spirally arranged. Leaf scar shape equals to the cushion, a little bit smaller; account for 2/3 area of leaf cushion; usually situates in the middle of leaf cushion; upper and basal angles wide, lateral angle sharp; the shape of leaf cushion resembles cat-eye. Big bundle scar, wide "V" shape. Lateral scar small and round. Leaf, bundle and lateral scars situated on the extension line of two neighbor lateral angle.

Material examined: YXM (A)-43-1.

Occurrence: Upper Permian Longtan Formation, in South China; occasionally in Upper Carboniferous Taiyuan and Shanxi formations, Lower Permian lower Shihezi Formation in North China.

Description (after Stockman and Mathieu). Leaf cushion and leaf scar transversely rhombic, with the lateral angle acute and elongated; leaf-scar 7 × 15 mm, cushions 9 × 22 mm; leaf-scar normally placed in the middle of the cushion. The margins of the leaf scars flatly arched, with somewhat concave sides; the lateral angles acute. The broadly V-shaped vascular and two lateral cicatrices with round in shape, lie in one line. The ligule-scars are clear, with a longitudinal cranny in shape, most directly above the leaf-scars. The surfaces of leaf cushions are smooth.

Discussion and Comparison. The shape of leaf cushion and leaf scar are similar to those of *Lepidodendron lepidophloides* (Yao et al. 1980), but they differ in that the lower margins of the leaf-scar and leaf cushions of *L. lepidophloides* are superposed.

Lepidodendron sp. aff. L. lepidophloides Yao, 1980 (Figs. 9.1-1, 9.1-2, 9.1-5 and 9.1-7)

1980 *Lepidodendron lepidophloides* Yao, p. 74, Pl.III, Figs. 1, 2, 6

Holotype: *Lepidodendron lepidophloides* Yao (1980), pl. 3, Fig. 1, 1a, 2, 6, deposited in the Nanjing Institute of Geology and Palaeontology, Chinese Academy of Sciences, impression fossil, from Xuanwei Formation (Lopingian), Panxian, Guizhou Province, Fuyuan and Xuanwei, Yunnan Formation, China.

Diagnosis: Leaf cushion spirally arranged, crowd, almost rhomboid. Leaf scar situates in the lower part of leaf cushion; upper edge nearly parallel to that of leaf cushion; cross leaf cushion's lower margin near lateral angle, or extension line cross in lateral leaf angle; lower margin coincide with leaf cushion lower margin; account 1/3 area of leaf cushion. Bundle scar situates in the middle of leaf scar, wide "V" shape; lateral scar round; three scar range in a line. Ligular pit situate above leaf scar's top vertex angle; lengthwise crack like.

Material examined: GWZ-15-1, GWZ-42-1, GWC-21-17, GPT-17-1, GPT-18-1.

Occurrence: Zhejue and Chahe sections, Weining County, Guizhou; Late Permian Xuanwei Formation. Tucheng section, Panxian County, Guizhou; Early Triassic Kayitou Formation.

Description: Leaf cushion is relatively large, counter-rhombic, 8 mm high and 1.5 mm wide, densely and spirally arranged. Shape of leaf scar similar to leaf cushion but slightly smaller, with upper and lower sides slightly round and two lateral sharp corners, both its lower side and leaf cushion are the superposition. The vascular is located the center of the leaf scar, wide "V"-shaped. The lateral cicatrices are round. They lied in the line. The ligule-scar is unclear in shape.

Discussion and Comparison: Following Yao (1980), that the leaf cushions of this species being counter-rhombic and the lower sides of the leaf scar and cushion being the superposition are the character of the genus *Lepidophloios*. But it is different from genus *Lepidophloios* by the area's proportion of the leaf scar to the leaf cushion and the location of the ligule-scar. The big leaf scar and the leaf ligule-scar being just above the top angle are the character of genus *Lepidodendron* in Cathaysia flora. A similar single leaf-cushion has been found in Beds 17 and 18 in Tucheng section (Induan).

Genus ***Lepidostrobophyllum*** Hirmer, 1927

Lepidostrobophyllum hastatum (Lesq.) Chal, 1974 (Figs. 9.1-9, 9.1-10)

1974 *Lepidostrobophyllum hastatum* (Lesq.) Chal, Paleozoic plants of China, p. 36, Pl. 17, Figs. 7–9

Holotype: *Lepidostrobophyllum hastatum* (Lesq.) Chal (1974), p. 36, Pl. 17, Figs. 7–9, deposited in the Nanjing Institute of Geology and Palaeontology, Chinese Academy of Sciences, a pith cast, from Xuanwei Formation (Changhsingian), South China.

Diagnosis: Upper part of sporophyll is 15–17 mm in length; sharp apex; lower margin extends to halberd hamulus or ear-shape; 7 mm in width; distinct mid vein. Lower part of sporophyll is 8–10 mm in length; narrowing downwards; distinct mid ridge.

Material examined: YXC-(-6)-1A.

Occurrence: Upper Carboniferous Tangshan and Kaiping formations in Hebei, North China; Chinahe Section, Upper Permian Xuanwei Formation, Yunnan, South China.

Description: Upper part of Sporophyll is nearly 8 mm, with sharp apex, inferior margin significantly extends to hamulus of the halberd or ear shape, 5 mm wide, midrib is not obvious; Lower part of sporophll is nearly 3 mm, narrows downward to triangular shape, midrib is obvious. Oval sporangium is preserved in the adaxial surface of sporophyll.

Discussion and Comparison: Lepidostrobophyllm is the sporophyll of Lepidodendrales, usually the upper part of it is lanceolate or ensiform, margin entire, single vein; the lower part is usually triangular shape; the adaxial surface of sporophyll ocassionally contains sporangium. This species is distinguished from others by typical halberd shape inferior margin and narrowed to triangular shape sporophyll lower part.

Genus ***Stigmaria*** Brongniart, 1822

Stigmaria ficoides (Sternb.) Brongniart, 1822 (Figs. 9.2-1, 9.2-2, 9.2-5 and 9.2-8)

1822 *Stigmaria ficoides* Brongniart, pp. 228, 239, Pl.2, Fig. 7

1927 Stigmaria ficoides, Halle, p. 181, Pl.49, Figs. 5, 11–12

1974 Stigmaria ficoides, Paleozoic plants of China, p. 37, Pl.21, Fig. 1

Holotype:

Diagnosis: Root scar round, umbilicate, spirally arranged. Surface flat, occasionally wrinkled.

Material examined: GWC-21-3, GWC-21-13, GWC-21-14, GWC-69-36; GWZ-36-2, GWZ-46-3, GWZ-47-2, GWZ-47-4, etc.

Occurrence: Lower Devonian to lower Upper Permian, all around the world.

Description: These specimens are found as often as any other fossil, especially in the "underclay" or "coal seam floor" immediately under coal seams from the four sections. Some impressions with the roots are still in position (Pl.5, Fig. 1). The root scars are small, about 2–3 mm in diameter, rather distant from each other in proportion to their size, and regularly placed. The scars are in all respects typical, having a black central dot of carbonaceous material corresponding to the vascular strand.

Discussion and Comparison:

Stigmaria rugulosa (Goth 1974) (Figs. 9.2-3, 9.2-10)

1974 *Stigmaria rugulosa* Goth, Paleozoic plants of China, p. 37, Pl.20, Figs. 6–7

Holotype: *Stigmaria rugulosa* Goth, 1974, p. 37, Pl.20, Figs. 6–7, deposited in the Nanjing Institute of Geology and Palaeontology, Chinese Academy of Sciences, a pith cast, from Xuanwei Formation (Changhsingian), South China.

Diagnosis: Root scar round, umbilicate. Wrinkled surface.

Material examined: GWZ-36-1, GWZ-46-5.

Occurrence: Upper Devonian Wutong Formation in Nanjing, South China; Lower Permian Shanxi Formation in Taiyuan, North China; Upper Permian Longtan Formation in Guangdong, South China; Carboniferous Wanshoushan Group in Yunnan,

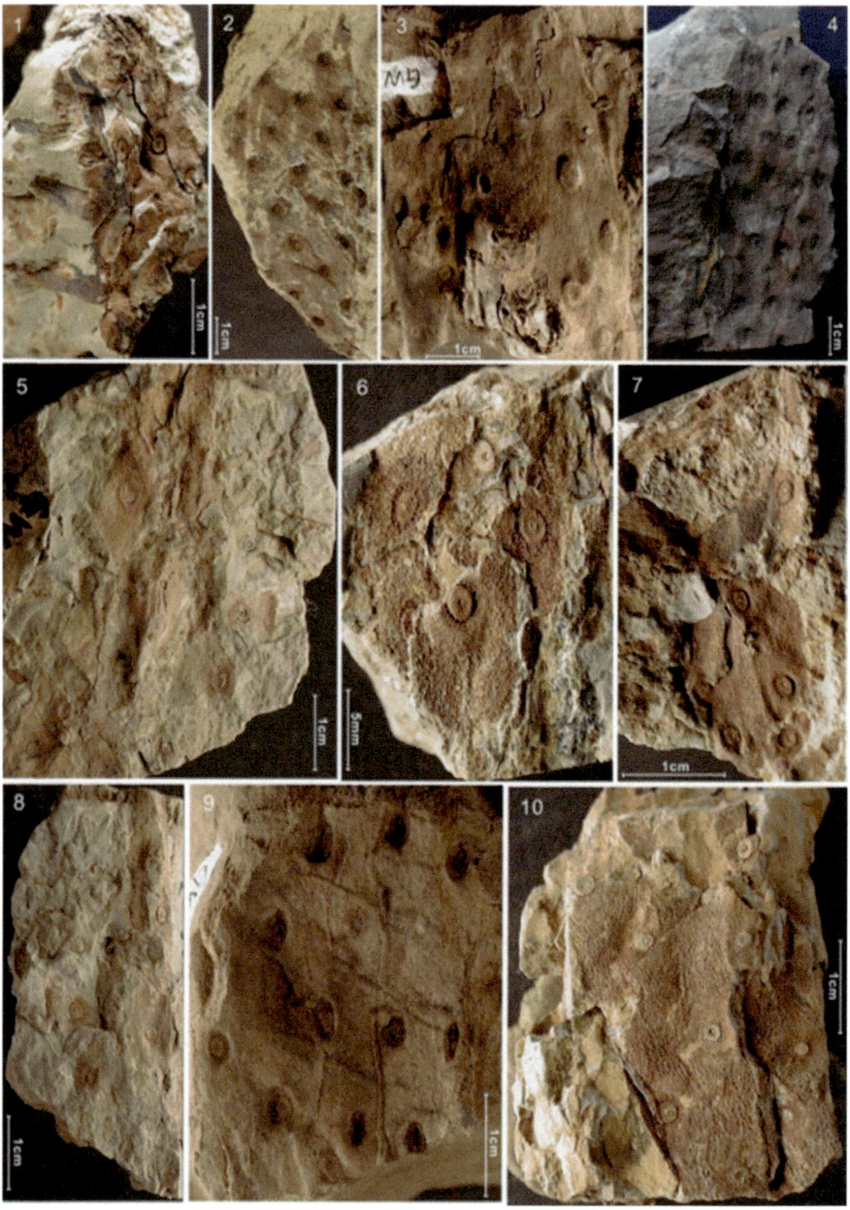

Fig. 9.2 1, 2, 5, 8 *Stigmaria ficoides* (Sternb.) Brongn. Locality: 1, 5, 8 Zhejue section, Weining County, Guizhou Province Sample Number: GWZ-47-2, GWZ-46-1, GWZ-46-3, 2 Chahe section, Weining County, Guizhou Province Sample Number: GWC-21-14, 3, 10 *Stigmaria rugulosa* Goth. Locality: Zhejue section, Weining County, Guizhou Province Sample Number: GWZ-36-1, GWZ-46-5, 4, 6, 7 *Stigmaria radiatopunctata* Goth. & Sze Locality: 4 Mide section, Xuanwei City, Yunnan Province Sample Number: CNH-8 + 0.5-1 6, 7 Zhejue section, Weining County, Guizhou Province Sample Number: GWZ-47-4, GWZ-46-7, 9 *Stigmaria* sp. Locality: Chahe section, Weining County, Guizhou Province

South China; Zhejue Section, Upper Permian Xuanwei Formation in Guizhou, South China.

Description: The specimen differs in having some longitudinally wave-shaped wrinkle on the surface of the compressions. Other character is the same as *Stigmaria ficoides*. Root scars round and umbilical.

Discussion and Comparison: The distinct feature of this species is that the surface of the compression has some longitudinally wave-shaped wrinkle, which is differ from the other species.

Stigmaria radiatopunctata (Goth and Sze 1974) (Figs. 9.2-4, 9.2-6 and 9.2-7)

1974 *Stigmaria radiatopunctata* Goth and Sze, Paleozoic plants of China, p. 37, Pl.20, Fig. 5

Holotype: *Stigmaria radiatopunctata* Goth and Sze, 1974, p. 37, Pl.20, Fig. 5, deposited in the Nanjing Institute of Geology and Palaeontology, Chinese Academy of Sciences, a pith cast, from Xuanwei Formation (Changhsingian), South China.

Diagnosis: Root scar oval, with margin. Radial convex rib on margin, like wheel. Longitudinal wrinkle on surface, countless dot between wrinkles.

Material examined: GWZ-46-7, GWZ-47-4.

Occurrence: Upper Devonian Wutong Formation in Jiangsu, South China; Zhejue Section Upper Permian Xuanwei Formation in Guizhou, South China.

Description: Root scar round or oblong, with costa. There are some radial and convex ribs on the costa, which look like a wheel. The surface between the root scars have longitudinally wave-shaped wrinkle.

Discussion and Comparison: Compared to *S. rugulosa*, the difference between the present specimens and *S. rugulosa* is that the latter have no costa in the root-scars and no radial ribs.

Order **Isoetales**

Family **Isoetaceae**

Genus *Lepacyclotes* (=*Annalepis*) Neuberg, 1936

Lepacyclotes (=*Annalepis*) *zeilleri* ((Fliche 1910) Yu emended, 2021) (Figs. 9.10-1 to 9.10-26)

2010 *Annalepis zeilleri*, Yu et al., Figs. 5, 1–12

2020 Tomiostrobus sinensis, Feng et al., Figs. 3–6

2021 Tomiostrobus (Annalepis) angusta, Yu et al.

Holotype: *Annalepis zeilleri*, Yu et al. 2010, Fig. 5, 1–12, deposited in China University of Geosciences (Wuhan), Main building, 5th floor, Basement NO. B104, a impression fossil, from Xuanwei Formation (Changhsingian), South China.

Diagnosis: (*Annalepis zeilleri*, emended after Fliche, Ye, Grauvogel-Stamm and Meng). Herbaceous; strobile close to spherical; diameter ranging from maximum 7 cm to minimum 3 cm; with spherical stem, 0.7–2.5 cm in diameter; without tropho-phyll. Sporophyll more or less cespitose; lanceolate, spatulate, oval, cuneate or listric; 1.6–3.2 cm in length; widest portion is in 1/4–1/3 from the top, 1.1–1.5 cm, sharp narrowing from here to the top and gently narrowing to the base; arrow apex, blunt stylus or blunt; base concave or truncature. Sporangium virgate; insert at the middle to lower and center of sporophyll adaxial surface; gradually extend from the base to the top; 0.7–2.6 cm in length, usually is the 2/3 of sporophyll, 0.3–0.7 cm in width; a longitudinal groove reaches the top from the base, with 5–6 trabecula on each side. A ligular scar and labellum situate slightly above the sporangium; ligular scar nearly round, 1 mm or more in diameter; labellum heart-round shape, 4 mm in length, 6 mm in width, with indistinct radial striae on the surface and mildly converge towards the top. Leaf apex grows out of the sporophyll phyllopodium, linear. Microspore equatorial outline is oval to wide-oval, 55–66.7 μm in diameter, 43.3–51.7 μm in height; monolote, reaching or close to equator; with narrow lip, 2 μm in width, constantly curve in the middle; extine is divided into inner and outer layers; the inner layer is slightly thicker than the outer layer; inner and outer layers are separated separation layer 8.3–16.7 μm in width; outer layer of extine approximately 1 μm in width, with coarse granular and spiny ornamentation on the surface; spine base 1–2 μm in width, 1–3.5 μm in length; spine apex sharp or blunt sharp. Megaspore unknown.

Feng et al. (2020) described that *Tomiostrobus sinensis* was heterosporous plant with helically arranged sporophylls, Smooth-margined sporophylls of various shapes, consisting of proximal fertile and distal sterile portions, proximal portion of the sporophyll bearing a single sporangium on the adaxial surface covered by a bell-shaped velum. A ligula is presented distal to the velum. The distal portion of the sporophyll gradually narrows from the proximal portion and has a long, tapering apex, which is commonly broken off leaving a frayed apex. Microsporophylls present on top of the protocorm, megasporophylls borne on the protocorm's mid- to lower parts. Megaspores of *Biharisporites* (Potonié 1956) Bharadwaj and Tiwari, 1970 type, trilete, with a pronounced alveolate wall.

Material examined: YXM (B)-19-1, YXM (B)-19-2, YXM (B)-21-27, GPT-17-4, GPT-17-3, GPT-18-2.

Occurrence: Early Triassic Kayitou Formation in western Guizhou and eastern Yunnan, South China.

Description: All the specimen are dispersed sporophyll; spatulate, cuneate or listric; 0.8–1.6 cm in length; widest portion is in 1/5–1/4 from the top, 0.5–0.8 cm, sharp narrowing from here to the top and gradually narrowing to the base; length–width ratio is around 2; arrow apex, usually broken and leave the base of leaf apex. Sporangium virgate;; stick close to the base and center of sporophyll adaxial surface; in most cases it graduallyy extend drom the base to the top, occasionally it narrows from the 2/3 of the top; 0.6–1.2 cm in length, 0.1–0.3 cm in width; a longitudinal ridge extend from the sporangium base to the top. Ligular scar tightly attached to the sporangium top, smaller than 0.5 mm in diameter; oval.

Discussion and Comparison: The samples of this species are most abundant in the base of the Kayitou Formation. Presently most specimens are similar to the holotype *A. zeilleri* named by Fliche in shape of sporophyll and sporangium. Whereas, it is much more smaller than the *Annalepis zeilleri*. It also does not have Middle Triassic *Annalepis* typical arrow like leaf apex, and the Early Triassic lycopods has thin and tapering apex which is usually broken. Feng edit the genus *Annalepis* to *Tomiostrobus* in 2020. He proposed all the dispersed sporophyll of various shape belong to same species *Tomiostrobus sinensis* because the different shape sporophyll were preserved together. Without detailed in situ spore and cuticle data, we still keep the morphotype species name *zeilleri* and accept the genus name *Tomiostrobus* for Early Triassic lycopods.

Tomiostrobus* (*Annalepis*) *brevicystis (Meng 1995) (Figs. 10.2-1 to 10.2-21)

1995 *Annalepis brevicystis*, Meng et al., p. 20, Pl.3, 8–13

1998 *Annalepis brevicystis*, Meng et al., p. 773, Pl.I, Figs. 1–11, 1d

1998 *Annalepis brevicystis*, Meng et al., p. 48, Pl.9, Figs. 8–21; Pl.13, Figs. 11–18

2021 Tomiostrobus (Annalepis) angusta, Yu et al.

Holotype: *Annalepis brevicystis*, Meng et al. 1995, p. 20, Pl.3, 8–13, deposited in the Wuhan Geological Survey Center (Yichang Base), China Geological Survey, Yichang, Hubei, impression fossil, from Badong Formation (Anisian), Xianfeng, Hubei Province and Fengjie, Sichuan Province, China.

Diagnosis (*Annalepis zeilleri* emended after Meng 1995). Herbaceous, unknown stem. Sporophyll lanceolate, oval-lanceolate, spatulate, triangle; 4.5–7.5 cm in length, widest part situates in the base, 1.8–3 cm in width, narrows from the base to the top; tapering apex, concave or wide arcuate base. Sporangium insert at base or lower part of sporophyll adaxial surface; long-oval or short virgate; length is 1/5 of the sporophyll or more, 1.4–2 cm in length, 0.5–0.8 in width; with a longitudinal groove in the middle. Above the sporangium is a leaf ligule scar and labellum; leaf ligular scar is suborbicular, 1–1.5 mm in diameter; labellum heart-round shape; vague. Leaf apex linear to ribbon-like; with a bundle scar in the middle, both sides are composed of lots of rectangle-like compartments; 1.5–1.7 cm in length, 0.3 cm in width. Leaf membrane on both side of sporophyll bulge to indistinct ridge; parallel to side edge. Megaspore equatorial outline oval, 450–512 μm in diameter; thick trilete, reaching equator; extine baculate ornamentation on the surface, ornamentation of proximal surface contact zone is thin and delicate, baculate ornamentation on distal surface and equatorial region is dense and thick; terminus of baculate ornamentation inflate. Microspore unknown.

Material examined: YXM (B)-21-7a, b; YXM (B)-21-9.

Occurrence: Early Triassic Kayitou Formation in western Guizhou and eastern Yunnan, South China.

Description: All specimen is dispersedly preserved. Sporophyll triangular or ovate; 1.4–2.1 cm in length, 1.2–1.5 cm in width; the widest part usually in the base, distal portion gradually narrowing since the base, base arcuate. Sporangium obovate or stick-shaped, attached closely to the lower or basal part of the adaxial surface of the sporophyll; about 1/4 of length of sporophyll length, 0.3–0.4 cm in length, 0.3–0.5 cm in width; with a longitudinal groove at the median. Ligular pit oblong or round. Leaf-tip linear, vascular bundle scar located in the middle of the leaf-tip.

Discussion and Comparison: The distinct feature of the present specimens is the short sporangium, no more than 1/3 length of the sporophyll, commonly located on the base of the sporophyll, which differs from other species. *T. brevicystis* is the secondary in abundance to genus *Tomiostrobus* (*Annalepis*) in the Early Triassic Kayitou Formation.

Tomiostrobus* (*Annalepis*) *angusta (Meng 1995) (Figs. 10.2-22, 10.2-23 and 10.2-24)

1995 *Annalepis angusta*, Meng et al., p. 19, Pl.2, 7–9.

2021 Tomiostrobus (Annalepis) angusta, Yu et al.

Holotype: *Annalepis angusta*, Meng et al. 1995, p. 20, Pl.3, 8–13, deposited in the Wuhan Geological Survey Center (Yichang Base), China Geological Survey, Yichang, Hubei, impression fossil, from Badong Formation (Anisian), Xianfeng, Hubei Province and Fengjie, Sichuan Province, China.

Diagnosis (*Annalepis angusta* emended after Meng, 1995): Herbaceous. Sporophyll narrow and small, densely spirally arranged; overlap each other; each sporophyll is composed of an inflated distal portion and a stipiform proximal portion which sharply at first then gradually narrows toward the base; 15–22 mm in length, proximal portion 6 mm in width, distal portion 1.5 mm in width; blunt top, truncature base. Sporangium thin virgate, tightly attached to the stipiform portion on the adaxial of sporophyll; 3/4 of the sporophyll in length, top portion 3 mm in width, base 1 mm in width, with a longitudinal groove in the middle. Leaf apex extent out of the sporophyll phyllopodium, linear, 5 mm in length, 0.5 mm in width; with indistinct bundle scar in the middle; compartment of neither side is not clear. Ambilateral leaf membrane of sporangium is thin, less than 0.5 mm in width. Mega- and Microspore are unknown.

Material examined: YXM (B)-19-23, YXM (B)-21-5, YXM (B)-21-1, etc.

Occurrence: Early Triassic Kayitou Formation in western Guizhou and eastern Yunnan, South China.

Description: Dispersed sporophyll, small and narrow, consisting of inflated and stalk-shaped parts, 1.3–1.7 cm long, the broadest up to 1.0 cm, the top obtuse, the base broken. Sporangium thin stick-shaped, attached closely to central part of the stalk-shaped part of the ventral surface of sporophyll, length about 3/4 of length of sporophyll, with a longitudinal groove at the median. Ligular pit is elliptic. Membrana leaf of both sides of sporangium is narrow, less than 1 mm in width.

Discussion and Comparison: This specimen is typical of stipiform sporophyll suddenly narrowing from 3/4 to 1/2 of the sporophyll, different from *T. zeilleri* and *T. brevicystis.*

***Tomiostrobus* (*Annalepis*) sp.** (Figs. 10.1-1 to 10.1-26; Figs. 10.2-25, 10.2-26)

Material examined: YXM(B)-21-37.

Occurrence: Early Triassic Kayitou Formation in western Guizhou and eastern Yunnan, South China.

Description: Sporophyll is rounded square to fan-shaped. The width is equal to or bigger than the length of the sporophyll. The width of it is almost the same in the whole fertile sporophyll. Sometimes the top part of the fertile portion is a little bit wider than the bottom. The sterile sporophyll-the leaf aepx is usally broken and only the base is preserved. The sporangium is in the longitudinal attached to the middle of the adaxial surface, rounded-rectangular. It has one vertical indistinct mid ridge, reaching from the bottom to the top of the sporangium, but not extending out of the sporangium.

Discussion and Comparison: The size and shape of those specimen is the transitional type of the *T. zeilleri* and *T. brevicystis,* which is rounder than the later two species. The leaf apex angle is also different from the *T. brevicystis*. Due to the lack of spore and cuticle data, those specimen is documented as on of the morphotype indeterminata species.

Division **Arthrophyta**

Class **Sphenopsida**

Order **Equisetales** Du Mortier, 1829

Family **Calamitaceae** Unger, 1840

Genus ***Paracalamites* Zalessky, 1927**

Type species: *Paracalamites striatus* Zalessky, 1927. Ufimian or Kazanian (Upper Permian) of the Daniko-Shor locality; Pechora Basin, The northeastern part of the European part of Russia.

Diagnosis (emended): a form genus, articulate stems and rhizomes, inner cast of the stems with opposite-in-node ribs. Most longitudinal rib and longitudinal groove pass straight through the node without alternation. A few ribs may occasionally alternate at the nodes.

Discussion and Comparison: *Paracalamites* is a form genus to which the isolated stem fragments from Lidgetton could be referred. However, there seems little to be gained by this, for there appears to be confusion as to the path of the ridges through the nodes in stems referred to it, for Rigby states: Zalessky erected this genus *Paracalamites* in 1932 to include Permian articulate stems and rhizomes having the ribs of the pith cast opposite at each node, as in *Asterocalamites*, not alternating. Later Rigby continues: "From *Paracalamites* sp. described below it is thought that ribs dichotomise and recombine with adjacent ribs at nodes rather than dichotomise and recombine with themselves as in *Asterocaiamites*. He then goes on to describe the specimen as follows: Ribbing is distinct in only a few places. It is opposite at nodes on both the inner and outer surfaces. There is some slight resemblance to Stellotheca robusta.

Because *Paracalamites striatus* Zalessky, 1927, is typified by a rather indistinctive pith cast, the name *Paracalamitina striata* Zalessky, 1934, a name typified by a specimen showing the characteristic outer surface with scars of leaves and branches is preferred. Genus *Paracalamitina* is emended to include stem remains with branch and/or leaf scars and oppositely positioned ribs. *Paracalamitina* differs from *Calamitina* Weiss, which has alternating ribs, and from *Paracalamites* Zalessky by the presence of shoot and/or leaf scars. *Paracalamites* Zalessky should be used as a formal genus for inner cast of the stems with opposite-in-node ribs.

Paracalamites stenocostatus (Gu and Zhi 1974) (Figs.)

1974 *Paracalamites stenocostatus* Gu and Zhi, pp. 52, pl.29, Figs. 1–4

Holotype: *Paracalamites stenocostatus* Gu and Zhi, 1974, pl. 4, Fig. 8, deposited in the Nanjing Institute of Geology and Palaeontology, Chinese Academy of Sciences, a pith cast, from Xuanwei Formation (Changhsingian), Panxian, Guizhou Province, China.

Diagnosis (emended): Consisting of node and internode, internodes usually longer than broad. Longitudinal ribs are straight and narrow, widths are usually less than 1 mm. blunt or acute end. seldomly visible small and long-ovate infranodal scar. Longitudinal grooves are shallow and slender. Most longitudinal rib and longitudinal groove pass straight through the node without alternation. A few ribs may occasionally alternate at the nodes.

Occurrence: Upper Permian and Lower Triassic, widely distributed in Changxing county of Zhejiang Province, Anxi county of Fujian Province, Mojiang county of Yunnan Province and Lianzhou county of Guangdong Province, western Guizhou Province (Gu and Zhi 1974; Zhao et al. 1980; Wan et al. 2011) in addition to Xuanwei county of Yunnan.

Description: This species has internodes that are usually longer than broad, up to 3.9 cm long and 2.4 cm wide, and the nodes are not contracted. The nodal lines are straight. The flattened ribs separated by shallow and thin groove are narrow, with blunt or acute end. A few ribs may occasionally alternate at the nodes, while others meet end to end. They seldom have visible ovate infranodal scar.

Discussion and Comparison: Compared with the *Paracalamites stenocostatus*, the most obvious feature of *Calamites cistii* (Brongniart) and *Mesocalamites cistiformis* (Stur) is the staggered arrangement of longitudinal ribs and longitudinal grooves at the node position. In general, the *Paracalamites* is characterised by the straight and narrow longitudinal rib, shallow and slender longitudinal groove, and most longitudinal rib and longitudinal groove pass straight through the node without alternation. A few ribs may occasionally alternate at the nodes.

Genus ***Annularia*** Sternberg, 1822

Type species: *Annularia spinulosa* Sternberg, 1822, from Sternberg's collection housed in the National Museum of Prague.

Diagnosis (emended): The lanceolate to obovate or spathulate shaped single-veined, dorsiventral, leaves of equal or unequal length, positioned in verticils radiating from the nodes with a narrow fused base forming a circular nodal disk. The articulated axes of all the three types of leafy shoots of the present collection show ridges and furrows. Robust leaf whorls borne on a slender stem, spreading out horizontally without cup like sheath. Probably have disk like leaf whorl; the leaves are united at the base up to more than 2/3 of their length. The leaves radiate from a shallow depression with a cuneate base, attachment directly at the nodes with little or no adherent leaf sheath; the leaf lamina is flat with an indistinct median vein.

Discussion: Comments on the morphology of *Annularia* and its taxonomic distinction *Annularia* von Sternberg, 1821 resembles *Asterophyllites* Brongniart, 1822. Both fossil-genera incorporate small Calamites-type axes bearing leaf-whorls (verticils) at each articulation (node), having single-veined leaves united at the base in each whorl. However, the leaves of *Annularia* are generally rigid and have different shapes (linear, lanceolate, oblanceolate and spathulate), with the broadest part in the middle of leaf. In contrast, leaves of *Asterophyllites* are considerably narrower with an acicular to filiform shape, and generally arranged in cup-shape whorls (e.g. Álvarez-Vázquez and Wagner 2016).

Annularia pingloensis (Gu and Zhi 1974) (Figs. 9.3-1 to 9.3-6)

1974 *Annularia pingloensis* (Sze); Gu and Zhi, p. 52, pl.29, Figs. 1–4

Holotype: *Annularia pingloensis* Gu and Zhi, 1974, pl. 29, Fig. 1–4, deposited in the Nanjing Institute of Geology and Palaeontology, Chinese Academy of Sciences, a impression fossil, from Heshan Formation (Lopingian), Pingle, Guangxi Province, China.

Diagnosis (emended). Leaves of a whorl usually less than 10 in number, linear and lanceolate in shape, with an acute apex, unequal in size, the undermost pair being the shortest, the other ones length equal, separately from each other. The midrib is distinct and runs into the apex. The impeller sometimes has upper blade defect.

Occurrence: Upper Permian and Lower Triassic. Xuanwei City, Yunnan Province, Weining County, Guizhou Province, Guangxi, China.

Description: The tri-branches are faint, and irregularly striated, 8 mm in width. The bi-branchs are slender and distichous, 2 mm in width. The length of the internodes is almost equal. Leaves of a whorl are usually less than 10 in number, linear and

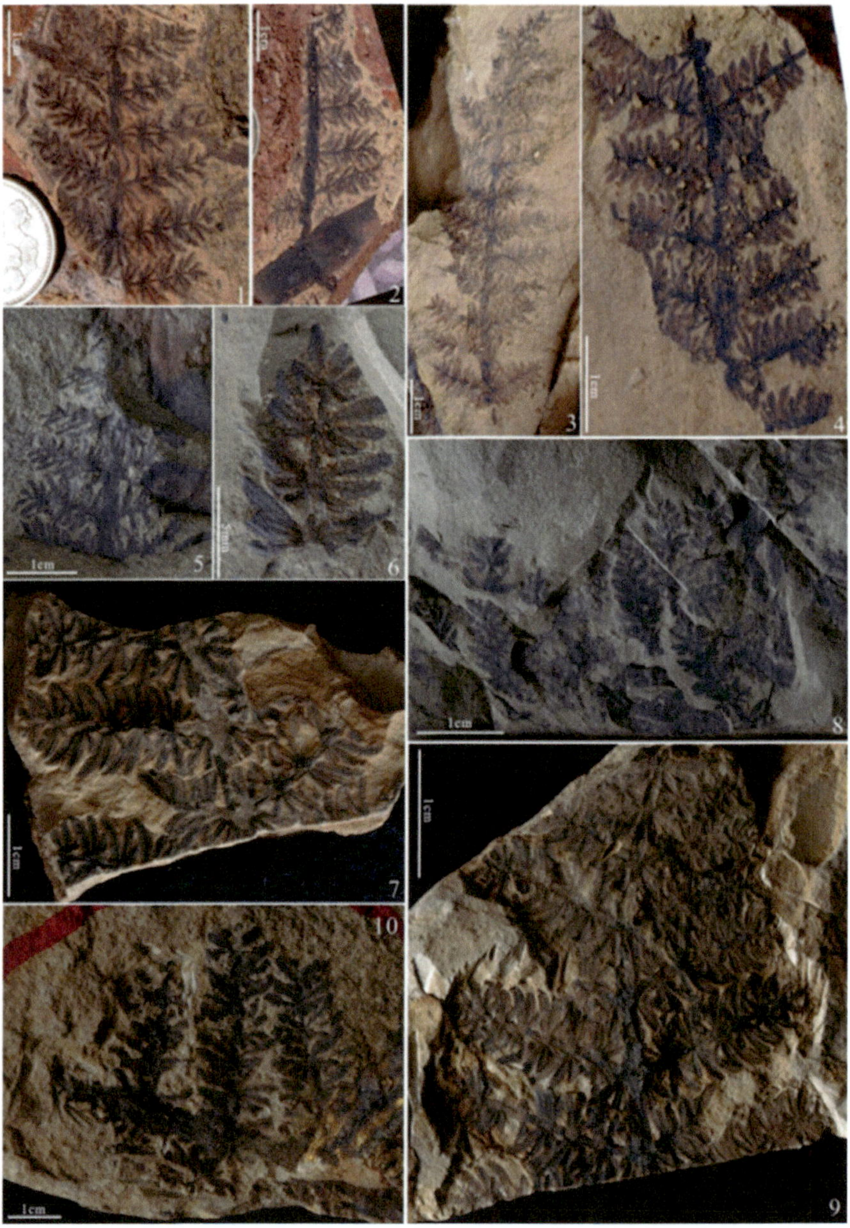

Fig. 9.3 1-6 *Annularia pingloensis* (Sze) Gu et Zhi Locality:1, 2 Mide section B, Xuanwei City, Yunnan Province Sample Number: YXM (B)-12-16, YXM (B)-12-17, 3, 4 Chahe section, Weining County, Guizhou Province Sample Number: GWC-18-97, GWC-18-98, 5, 6 Chinahe section, Xuanwei City, Yunnan Province Sample Number: YXC-(-1)-145A, YXC-(-1)-11, 7, 8, 9 *Annularia shirakii* Kawasaki Locality: 7, 9 Kele section, Hezhang County, Guizhou Province Sample Number: CLC-16-2, CLC-16-1, 8 Chinahe section, Xuanwei City, Yunnan Province Sample Number: CNH-0-18, 10 *Annularia hunanensis* Gu et Zhi Locality: Kele section, Hezhang County, Guizhou Province Sample Number: CLC-3-5

lanceolate in shape, with an acute apex, unequal in size, the uppermost pair being the longest and the other ones becoming smaller downward, separately from each other. The midrib is distinct and runs into the apex.

Discussion and Comparison: The difference between *Annularia shirakii* and *Annularia hunanensis* is that the internode shortening is not obvious and has midrib, and only the lowest pair is shorter.

Annularia shirakii (Figs. 9.3-7, 9.3-8, 9.3-9)

1927 *Annularia shirakii* Kawasaki, pp. 9, pl.XIV, Fig. 76

1927 *Annularia crassiuscula* Halle, pp. 34–36, pl.6, Figs. 10–11; pl.7, Figs. 10–13

1974 *Annularia shirakii* Kawasaki, Li and others, pp. 55, pl.31, Figs. 8–11

Holotype: *Annularia shirakii Kawasaki*, 1927, pl. XIV, Figs. 76, from Gaofangshan series (Lopingian), Panxian, Guizhou Province, North Korea.

Diagnosis (emended). All levels branche internodes length gradually shorten upward. Branch impeller at all levels sometimes with a lack of upper leaves, arranged in butterfly shape. Leaves shorter, on terminal branches up to 12 mm, usually shorter, lanceolate, slightly curved forward, widest in middle, apex rather pointed. The midrib is not obvious.

Occurrence: Upper Permian, in Guizhou, Yunnan, China.

Description: The bi-branch are slender, 1 mm in width. The lengths of the internodes gradually shorten upward. Leaves of a whorl are usually less than 10 in number, lanceolate in shape, with an acute apex, unequal in size, often slightly curved, widest in the middle of the leaf. The midrib is indistinct.

Discussion and Comparison: The main difference between this species and *Annularia papilioformis* Kaw is that the former has smaller leaves, closely arranged impellers, and appears in the middle and Late Permian (Paleozoic plants in China, 1974); the latter has not only larger leaves and loosely arranged impellers, but also appears in the early age, from the Late Carboniferous to the middle Permian (Gu and Zhi 1974). The leaves of *Annularia hunanensis* are smaller and the tip is round, the length difference between the upper and lower leaves is not obvious, and the leaves are not curved, while the leaves of *Annular shirakii* Kaw are larger and the tip is gradually pointed, and the upper leaves are obviously curved upward.

Genus ***Lobatannularia*** Kawasaki, 1927

Type species: *Lobatannularia inequifolia* Kawasaki, 1927 on the genus *Lobatannularia* Kawasaki 1927 from Permian beds in South Manchuria and Shanxi, China.

Diagnosis (Kon'no and Asama, 1950 emended): Shoot calamitina-type, slender, articulate; generally have unequal braches at node. Lateral leaf-whorl spreads out in one plane oblique or nearly parallel to the axis, being divided into two symmetrical

Fig. 9.4 1-6 *Lobatannularia multifolia* Kon'no et Asama Locality: 1, 4, 6 Chahe section, Weining County, Guizhou Province Sample Number: GWC-3-16, GWC-53-2, GWC-2-15, 2, 3 Chinahe section, Xuanwei City, Yunnan Province Sample Number: YXC-P_3x-9, CNH-3-0.5-1, 5 Tucheng section, Panxian County, Guizhou Province Sample Number: GPT-10-2, 7, 8, 9 *Lobatannularia cathaysiana* Yao Locality: 7 Chinahe section, Xuanwei City, Yunnan Province Sample Number: YXC-(-1)-8, 8, 9 Chahe section, Weining County, Guizhou Province Sample Number: GWC-18-29, GWC-21-4, 10-12 *Lobatannularia heianensis* (Kod.) Kawasaki Locality: 10 Chinahe section, Xuanwei City, Yunnan Province Sample Number: YXC-(-1)-40, 11 Chahe section, Weining County, Guizhou Province Sample Number: GWC-21-1, 12 Kele section, Hezhang County, Guizhou Province Sample Number: CLC-21-9-1, 13, 14 *Lobatannularia ensifolia* (Halle) Halle Locality: Chahe section, Weining County, Guizhou Sample Number: GWC-58-10, GWC-66-11, 15-17 *Schizoneura manchuriensis* Kon'no Locality: 15, 16 Chahe section, Weining County, Guizhou Sample Number: GWC-5-4, GWC-18-19, 17 Chinahe section, Xuanwei City, Yunnan Province Sample Number: YXC-(-1)-108 ▶

lobes; terminal one not divided into lobe, forming a single fan-shaped. Leaf-segments of each lobe roughly coherent along sutural lines, showing considerable difference in length, namely lowest one shortest and curved upward: leaf slender and uninerved, having both fan-shaped terminal whorl and special habit of branching to be called "Pseudo-dichotomous polypodial".

Discussion: *Annularia* has linear, oblanceolate or spatulate leaves and no lower leaf defect. Genus name used in Paleozoic *Lobatannularia* has two lobed impeller with more or less curved leaves. The lower leaves are obviously absent. The genus name is often used in Paleozoic and occasionally in Middle Triassic.

Lobatannularia cathaysiana (Yao, 1980) (Figs. 9.4-7, 9.4-8 and 9.4-9)

1927 *Annularia maxima* Halle, p. 13, pl.1, Figs. 8, 9

1927 *Annularia heianensis* (Kodaira), p. 27, pl.5, Figs. 13, 14

1930 *Annularia lingulatus*?, Chang, Geol. Kwangtung & Kwangsi, p. 2, pl.1, Figs. 1–3

1980 *Lobatannularia cathaysiana* Yao, pp. 76–77, pl.3, Fig. 5; pl.5, Figs. 1–6

Holotype: *Lobatannularia cathaysiana* Yao, 1980, pl.3, Fig. 5; pl.5, Figs. 1–6, deposited in the Nanjing Institute of Geology and Palaeontology, Chinese Academy of Sciences, a impression fossil, from Xuanwei Formation (Lopingian), Panxian, Guizhou Province, China.

Diagnosis (emended): Leaf whorled, 24–28 leaves in each whorl, being divided into two lobes. Leaves linear to counter-lanceolate, the apex is blunt. Leaves near to the proximal side bend falcate. The lateral leaves are longer than the upper and lower ones, and gradually broad towards the apex. The leaves are united for about two thirds or three fourths of their length. The leaves are very unequal in size, the smallest leaves at the lower portion of the branch, about one fifth of the length of the longest. 1–2 leaves near to the upper portion of the branch are straight and separated. The single midvein is wide and distinct.

Occurrence: Upper Permian, Guizhou, China.

Description: Leaf whorled, 24–28 leaves in each whorl, being divided into two lobes. Leaves linear to counter-lanceolate, the apex is blunt. Leaves near to the proximal side bend falcate. The lateral leaves are longer than the upper and lower ones, and gradually broad towards the apex. The leaves are united for about two thirds or three fourths of their length. The leaves are very unequal in size, the smallest leaves at the lower portion of the branch, about one fifth of the length of the longest. 1–2 leaves near to the upper portion of the branch are straight and separated. The single midvein is wide and distinct.

Discussion and Comparison: The main character of the specimens is that 1–2 leaves near the upper portion of the lobe are relatively straight and departed, which length is not more than one-half of the longest one. This species is different from *L. heianensis* by: (1) *L. cathaysiana* has more leaves than *L. heianensis* in a whorl, the latter is 14–26 leaves; (2) linear or lanceolate, the leaf's apex is acute or blunt, but shows a projecting prickle (mucronate), the latter is spoon-shaped, blunt-tipped; (3) The angle between the two lobes in this species is bigger than the latter having 55–120°.

Lobatannularia multifolia (Kon'no and Asama, 1950) (Figs. 9.4-1 to 9.4-6)

1924 *Schizoneura heianensis* Kodarra, p. 163, pl.23, Fig. 2

1950 *Lobatannualria multifolia* Kon'no and Asama, p. 27, pl.4, Figs. 9–12

Holotype: *Lobatannualria multifolia* Kon'no and Asama, pl.4, Figs. 9–12. Deposited in the Nanjing Institute of Geology and Palaeontology, Chinese Academy of Sciences, a impression fossil, from Xuanwei Formation (Lopingian), Weining County, Guizhou Province, China.

Diagnosis (emended): Ultimate branch about 1–2 mm wide; internode 5–7 mm long, distinctly divided into two lobes, each lobe oblique obovate with 18–20 leaves; leaves completely connected, linear lanceolate, slightly curved upward; apex obtuse or subacute, widest near the apex. Leaves are very unequal in size, those at the lower portion of the valve being the smallest, are about one fourth of the longest leaf at the middle portion of the valve. The leaf has a single midvein.

Occurrence: Upper Permian, Anhui, Sichuan, Guizhou. In western Guizhou and eastern Yunnan, China.

Description: The impeller is obviously divided, with 20 leaves per valve, bending upward; the leaf length is 5–35 mm, the widest near the top; the leaf tip is blunt to subacute, almost all united, with single vein; the upper and lower leaves are wide.

Discussion and Comparison: *Lobatannualria nampoensis* (Kawasaki 1925) Kim and Kimura, 1988 is stem slender. Leaf-whorl borne at node, divided into two symmetrical lobes, obliquely attached to the stem, but the terminal whorl not divided into lobes, forming a single fan-shaped whorl. Leaf-whorl varying in size according to the position of the plant; towards the distal part it becoming shorter and smaller. *Lobatannualria multifolia* Kon'no and Asama impeller is obviously divided, the widest near the top; the leaf tip is blunt to subacute, almost all united, with single vein; the upper and lower leaves are wide.

Lobatannularia heianensis ((Kod.) Kaw, 1927) (Figs. 9.4-10, 9.4-11, 9.4-12)

1925 *Schizoneura heianensis* Kodaira, Kod., p.163, Pl.XXIII, Figs. 3–4

Holotype: *Lobatannularia heianensis* (Kod.), Kaw., 1927, pl.VI, Fig. 23. Form Kaisen Coal-field, a impression fossil, Pingan serise (Lopingian), North Korea.

Diagnosis (emended): The impeller obviously divided, with 12 (14)–14 (16) leaves per valve; the leaves linear lanceolate to lanceolate; except for 2–3 leaves near the lower lobe, little difference in leaf length, the tip more pointed, with short tip; most of the leaves are united, only the top is 3–5 mm apart; with single vein; obviously lower lobe deficiency.

Occurrence: Upper Permian, Guizhou and Yunnan, China.

Description: Ultimate branch about 1.5 mm wide, distinctly bivalvate, each valve oblique obovate with 8–10 leaves; most portion of the leaf is connected, only distal 2–4 mm near leaf apex is separated; leaves linear oblanceolate, about 1.5–2 mm wide, apex abruptly round and bearing a mucro. The leaves are very unequal in size; the smallest leaves being at the lower portion of the lobe, are about one fourth of the length of the longest, which curved upwards.

Discussion and Comparison: This species can be compared with *Lobatannslaria multifolis* Konno et Asama, but the latter one is not The tip of the leaf blade is almost completely connected without separation. This species is also slightly different from *L. cf. heianensis* (Kod.) Kaw (Paleozoic plants in China, 1974, pp. 57–58, pl.32, Figs. 11–12) which has been found in northern China. In the latter, the leaves are not spoon shaped, the tip is sharp, and the number of leaves per valve is more.

Lobatannularia ensifolia (Halle 1927, 1928) (Figs. 9.4-13, 9.4-14)

1927 *Annularites ensifolia* Halle, Halle, pp. 20–26, Pl.1, Figs. 1–5, Pl.2, Figs. 1–2, pp. 3, pl.4, Figs. 1–3

1974 *Lobatannularia ensifolia* (Halle) Halle, Paleozoic plants of China, pp. 57, pl.33, Figs. 1–2; pl.34, Fig. 1

Holotype: *Lobatannularia ensifolia* (Halle) Halle, 1974, pl.33, Figs. 1–2; pl.34, Fig. 1, deposited in the Nanjing Institute of Geology and Palaeontology, Chinese Academy of Sciences, a impression fossil, from Shihezi Formation (Lopingian), Taiyuan, Shanxi Province, China.

Diagnosis (emended): The last two branches divided into the last branches in pseudo dichotomous form (a pair of forked last branches are produced on each node, and each of them has an undeveloped branchlet on the outside). The impeller divided into two lobes, each of which has about 7–10 leaves. Leaves lanceolate, up to 8 cm long, the shortest about 1 cm, 5 mm wide, apex acuminate, base separated or slightly united, with single vein. The lower blade of impeller short, and the blade near the lower blade the shortest. The top impeller round to wide oval.

Occurrence: Upper Permian, Sichuan, Guizhou and Yunnan, China.

Description: The leaves number about 18 in a whorl (Pl.12, Fig. 1), more or less falcate, often confluent for a short distance, narrow lanceolate in shape, apex not preserved. The leaves are unequal in size, the smallest one at lower portion of the lobe, the largest being 3 or 4 times as long as the shortest. The midrib of the leaves is not very distinct.

Discussion and Comparison: *L. lingulata* (Halle 1927, 1928) impeller is obviously divided into 8–10 leaves per valve, the leaves are obviously upward curved and completely separated. The widest part is at the top, apically acutely needled; with single vein; the lower leaves are obviously absent. *Lobatannularia ensifolia* (Halle) Halle, 1974 each of which has about 7–10 leaves. Leaves is lanceolate and apex acuminate, which base separated or slightly united, with single vein. The lower blade of impeller is short, and the blade near the lower blade the shortest.

Genus ***Schizoneura*** Schimper et Mougeot, 1844

Type species: *Schizoneura paradoxa* Schimper & Mougeot, 1844, nom. cons. (Zijlstra et al. 2007).

Diagnosis (emended): Stems divided into long internodes and usually swollen nodes; nodes of main stems bearing whorls of leaves or branches; branch nodes bearing whorls of long, linear, single-veined leaves; immature whorls with leaves entirely fused along the lateral margins, forming a cylindrical sheath; mature whorls with leaves laterally fused into two or more strap-shaped, lanceolate or ovate leaf-sheath segments; in a few specimens, individual whorls with leaves free to their bases.

Discussion: *Schizoneura* was established by Schimper and Mougeot in 1844 based on specimens from the Early Triassic strata in France. The leaf stem of the leaf leaf is divided into segments and internodes, and the longitudinal rib and longitudinal groove are straight through the nodes. The leaves are slender and long than internode, each node is often 10–22, with single vein, and the edges are combined with two petals, and the stem is hugged on both sides of the node. The reproductive organs of the plant are spiked. The typical species of the three genera are not difficult to distinguish, but some transitional types of specimens can not be distinguished from each other. Generally, they are classified into the genus according to which typical species they are close to He et al. (1996).

Schizoneura manchuriensis (Kon' no, 1942) (Figs. 9.4-15, 9.4-16 and 9.4-17)

1960 *Schizoneura manchuriensis,* Kon' no, pp. 163–188, pl.16–20

1974 *Schizoneura manchuriensis,* Kon' no, pp.60, pl.34, Fig. 11; pl.35, Figs. 1–3

Holotype: *Schizoneura manchuriensis* Kon'no, 1942. pl.34, Fig. 11; pl.35, Figs. 1–3. from the *Gigantopteris nicotianaefolia*-bearing Formation in Penchihu Coal-field, Northeastern China.

Diagnosis (emended): The stem has nodes and internodes, and the internodes at the lower end of the branch may be the longest. The impeller consists of 8–16 leaves. The leaf sheath divided into two petals, opposite in shape, and inserted on the node in the form of half embracing stem. The leaf petals are oblong to lanceolate, 10–20 mm wide, 30–55 mm long, irregularly dehiscent or only apically dehiscent; the immature leaves at the top of the stem are connected with each other to form a roughly cylindrical closed sheath. The leaf blade is linear, about 60 mm long and only 20 mm wide. The tip is sharp and connected with each other, with obvious suture line and single vein, accounting for about 1/2 of the leaf width. The reproductive organs are terminal, spicate, and consist of a round of sterile bracts with closed sheaths and 6–7 rounds of cyst stalks. There are 20–23 cyst stalks in each round. There are six sporangia in each sporangium stalk.

Occurrence: Upper Permian, Guizhou and Yunnan, China.

Description: Leaves number 8–12 in each whorl, linear to lanceolate. They are symmetrically divided into two parts (or half whorls), which are oblong in shape, look like the opposite. Each part is 5–10 mm wide and 28–30 mm long, containing of 4–6 leaves, each other irregularly split or united.

Discussion and Comparison: This species is similar to some species of the genus *Lobatannularia*, especially to the specimen of *Lobatannularia multifolia*, but the latter has more leaves, different length and slightly curved forward.

Division **Pteridophyta**

Class **Marattiales et filicales incertae sedis**

Genus ***Sphenopteris*** (Brongn.) Sternberg 1825

Sphenopteris cf. ***tenuis*** (Schenk) Halle emended, 1927 (Fig. 9.6-1, 1a)

1927 *Sphenopteris tenuis*, Halle, p. 58, Pl.15, Figs. 15–17.

1974 *Sphenopteris tenuis*, Paleozoic plants of China, p. 76, Pl.46, Figs. 5–6; Pl.47, Figs.1–7.

2009 *Sphenopteris tenuis*, Seyfullah and Hilton, 2009, pp. 205, Figs. 23–26.

Lectotype: Specimen S138382 comprising fertile frond as illustrated by Halle 1929 pl. 1, Figs. 1–4.

Syntype: Specimen S138062 comprising vegetative foliage as illustrated by Halle (1927) pl. 15, Fig. 17.

Diagnosis (Seyfullah and Hilton, 2009): Frond tripinnate or quadripinnate. Penultimate pinnae subopposite or alternate, triangular to lanceolate, with smooth, winged rachis. Ultimate pinnae alternate or subopposite, with winged rachis, ovate to lanceolate; pinnate, pinnatifid or serrate. Well-developed pinnules, forming an angle of about 30 with the parent rachis, lanceolate to oblanceolate, acute, tapering towards a rather broad decurrent base; pinnules in distal part of the frond and pinnae entire, confluent towards the extremities; becoming lobes of the ultimate pinnae; pinnules in the proximal part dissected into lobes, becoming more prominent towards frond base and eventually forming free pinnules. Fine but distinct median vein; secondary veins few, of almost the same strength as the median vein, at an acute angle to the midvein, in the larger lobed pinnules commonly bifurcating. Ovules borne on unmodified pinnules; attached to the base of the lateral pinnules, or to pinna rachis. Ovules ovoid or narrowly ovoid, 4–5 mm long and 2–2.5 mm wide, rounded at the base, with a hollow, tubular apex.

Material examined: GPT-10-4.

Occurrence: Tucheng section, Panxian County, Guizhou; Late Permian Xuanwei Formation.

Description: The rachis of the penultimate pinnae is smooth and winged. The ultimate pinnae are lanceolate; pinnate, pinnatifid and the pinnules are lanceolate to oblangceolate, tapering downwards but attached by a broad decurrent base.

Discussion and Comparison: The specimen in Figs. 6.1, 6.1a, 6.2 shows the characteristic rachis and pinnules of *S. tenuis*. But the venations of the present specimen are unclear, it may be well belong to *S.* cf. *tenuis*.

Genus ***Pecopteris*** Brongniant, 1822

Type Species: *Pecopteris penniformis* Brongniart, 1822, Middle Pennsylvanian (Upper Carboniferous); Dudweiler, Germany; Muséum national d'histoire naturelle (France).

Diagnosis (emended): Multiple pinnate frond; rachises surface smooth or with fine longitudinal lines, or scales, tumors, thorns and other attachments, some even with abnormal pinnules. Pinnules on either side or adaxial side of the rachises, usually linguiform, elliptic, or rectangular; entirely attached in the ultimate rachises or slightly contracted at the base; alternate or subopposite, touching or separated each other. Pinnules sides more or less parallel, margin generally entire, occasionally undulate or lobed. Pinnate venation, the midrib distinct usually, and the lateral veins not difurcating or several times.

Discussion: Genus *Pecopteris* has a widespread geographical distribution in the Cathaysia-land and has often been described. *Pecopteris* has traditionally been used as the name of a form fossil-genus of fern and fern-like fossil pinnate frond with small, dentate pinnules, regardless of their reproductive organs (Brongniart 1822). With expanding knowledge of the fructifications attached to these fronds, paleobotanists are trying to improve the classification with details of the sporangia. Most of the species which sporangia found show characters that seemed to suggest they had marattialean affinities and were attributed to various fossil–genera based on detailed differences in their sporangial structures.

Pecopteris gracilenta (Gu and Zhi 1974) (Fig. 9.5-10)

1974 *Pecopteris gracilenta* Gu and Zhi, Paleozoic plants of China, p. 95–96, Pl.65, Figs. 1–4

Holotype: *Pecopteris gracilenta* Gu and Zhi, 1974, p. 95–96, Pl.65, 1–4, deposited in the Nanjing Institute of Geology and Palaeontology, Chinese Academy of Sciences, impression fossil, from Upper Shihezi Formation (Guadalupian), Taiyuan, Shanxi Province, China.

Diagnosis (emended): Frond fragment at least tripinnate. Ultimate rachises thin, Pinnules thin lamina, with apices obtuse. The midrib not distinct usually, as slender as the lateral veins, and decurrent at the base; the lateral veins sparse and slender, not difurcating or once.

◂**Fig. 9.5** 1 *Pecopteris (Asterotheca)* cf. *hemitelioides* Brongniort Locality: Chinahe section, Xuanwei City, Yunnan Province Sample Number: YXC-(-1)-44, 2-5 *Pecopteris (Asterotheca) guizhouensis* Zhang Locality: 2, 4 Chinahe section, Xuanwei City, Yunnan Province Sample Number: YXC-(-1)-98A, YXC-(-1)-175, 3 Zhejue section, Weining County, Guizhou Province Sample Number: GWZ-38-10, 5 Chahe section, Weining County, Guizhou Province Sample Number: GWC-58-3, 6-9 *Pecopteris (Asterotheca) orientalis* (Schenk) Potonie Locality: 6, 7, 9 Chahe section, Weining County, Guizhou Province Sample Number: GWC-18-59, GWC-58-31, GWC-18-61, 8 Chinahe section, Xuanwei City, Yunnan Province Sample Number: YXC-(-24)-1A, Fig. 10 *Pecopteris gracilenta* Gu et Zhi Locality: Chinahe section, Xuanwei City, Yunnan Province Sample Number: YXC-(-1)-126, 11, 12 *Pecopteris lativenosa* Halle Locality: 11 Chinahe section, Xuanwei City, Yunnan Province Sample Number: YXC-(-1)-15, 12 Chahe section, Weining County, Guizhou Province Sample Number: GWC-58–20, 13 *Pecopteris taiyuanensis* Halle Locality: Chinahe section, Xuanwei City, Yunnan Province Sample Number: YXC-(-6)-4, 14, 15 *Pecopteris sahnii* Hsu Locality: 14 Kele section, Hezhang County, Guizhou Province Sample Number: CLC-29–11-1, 15 Chahe section, Weining County, Guizhou Province Sample Number: GWC-58-28, 16, 17 *Pecopteris longifolioides* Zhang Locality: 16 Kele section, Hezhang County, Guizhou Province Sample Number: CLC-29-1-1, 17 Chahe section, Weining County, Guizhou Province Sample Number: GWC-18-2, 18 *Pecopteris marginata* Gu et Zhi Locality: Chahe section, Weining County, Guizhou Province Sample Number: GWC-58-14a, 19 *Pecopteris echinata* Gu et Zhi Locality: Chahe section, Weining County, Guizhou Province Sample Number: GWC-58-7b

Material examined: GWC-5-2, GWC-58-1.

Occurrence: Chahe section, Weining County, Guizhou; Late Permian Xuanwei Formation,

Description: The pinnules are well characterized by the very distinct and ascending venation. Ultimate rachis is very slender. The characteristic thin lamina is very distinct. The pinnules are ovate to oblong, forming an angle of about 45° with the pinna rachis. The apex of the pinnules is obtuse. The midrib is slender, slightly decurrent, attaining to the apex. The lateral veins are rare, arising at a narrow angle, bifurcating once or no; the first basal lateral veins at both distal and proximal sides of the midrib are very long. The first lateral veins at the proximal pinnules are given off from the decurrent part of the midrib.

Discussion and Comparison: The distinct feature of this species is that thin lamina and slender rachis and veins, which are differ from the other species of genus *Pecopteris*.

Pecopteris echinata (Gu and Zhi 1974) (Fig. 9.5-19)

1974 *Pecopteris echinata* Gu and Zhi, Paleozoic plants of China, p. 94; Pl.62, Figs. 12–13; Pl.63, Figs. 1–7

Holotype: *Pecopteris echinata* Gu and Zhi, p. 94; Pl.62, Figs. 12–13; Pl.63, Figs. 1–7, deposited in the Nanjing Institute of Geology and Palaeontology, Chinese Academy of Sciences, impression fossil, from Longtan Formation (Lopingian), Jiangning, Jiangsu Province, China.

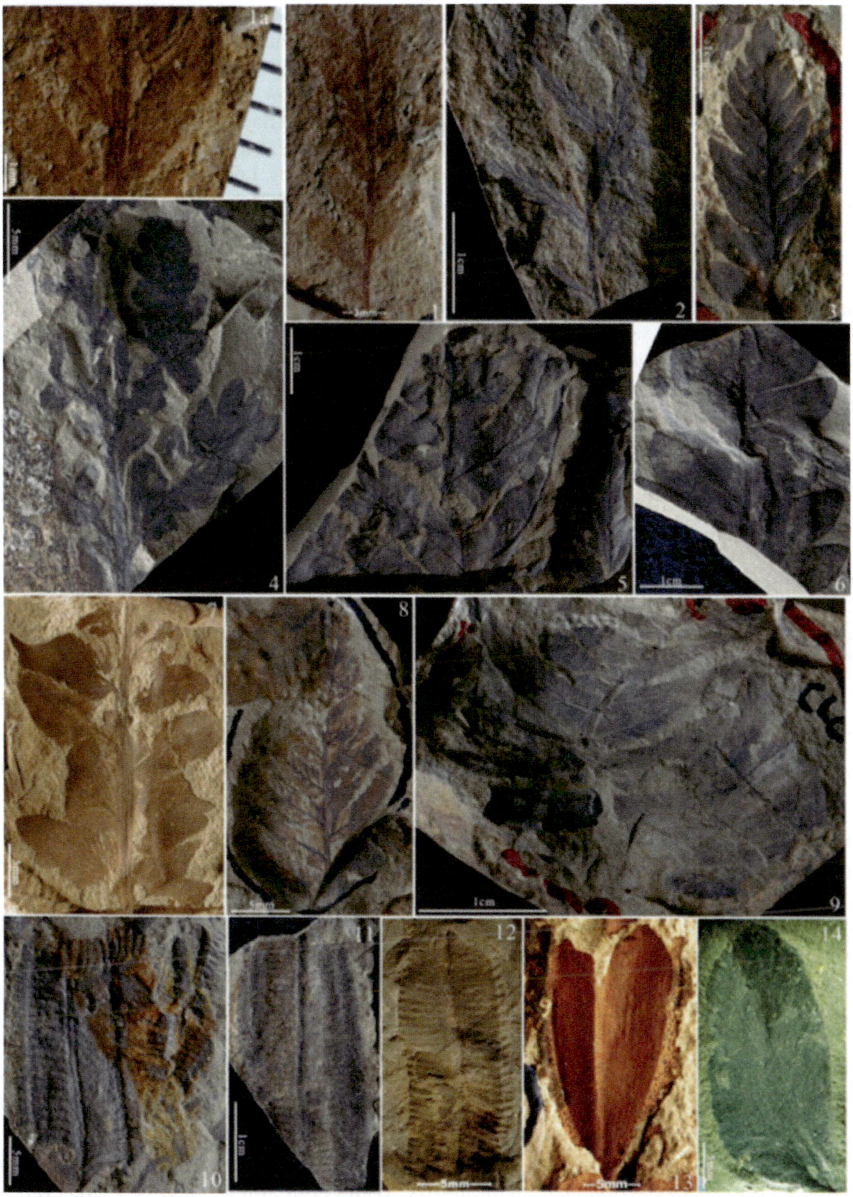
1a
1
2
3
4
5
6
8
9
10
11
12
13
14
1cm
5mm

◀**Fig. 9.6** 1, 1a *Sphenopteris* cf. *tenuis* Schenk Locality: Tucheng section, Panxian County, Guizhou Province Sample Number: GPT-10-4, 2 *Sphenopteris tenuis* Schenk Locality: Chinahe section, Xuanwei City, Yunnan Province Sample Number: YXC-(-1)-44, 3 *Cladophlebis* sp. Locality: Kele section, Hezhang County, Guizhou Province Sample Number: CLC-4-5, 4 *Cladophlebis manchurica* Kaw Locality: Chinahe section, Xuanwei City, Yunnan Province Sample Number: CNH-0-17, 5 *Cladophlebis nystroemii* Halle Locality: Chinahe section, Xuanwei City, Yunnan Province Sample Number: YXC-P3*x*-10, 6, 7 *Cladophlebis permica* Lee et Wang Locality: 6 Chinahe section, Xuanwei City, Yunnan Province Sample Number: YXC-(-1)-169, 7 Chahe section, Weining County, Guizhou Province Sample Number: GWC-18-12a, 8, 9 *Cladophlebis fuyuanensis* Zhang Locality: Kele section, Hezhang County, Guizhou Province Sample Number: CLC-35-37, CLC-35-17, 10-12 *Guizhoua gregalis* Zhao Locality: 10, 11 Chinahe section, Xuanwei City, Yunnan Province Sample Number: YXC-P3*x*-1, CNH-0-10B 12 Chahe section, Weining County, Guizhou Province Sample Number: GWC-58-17a, b, 13 *Tingia gerardii* Stockmans et Mathieu Locality: Chahe section, Weining County, Guizhou Province Sample Number: GWC-3-1, 14 *Linopteris* sp. Locality: Chahe section, Weining County, Guizhou Province Sample Number: GWC-58-29

Diagnosis (emended): The rachis thick. The pinnules bilateral symmetry, not uneven in length. The midrib distinct and slightly decurrent; the lateral veins bifurcating once near the midrib, and the upper branches bifurcating once again.

Material examined: GWC58-7b.

Occurrence: Chahe section, Weining County, Guizhou; Late Permian Xuanwei Formation. Chinahe section, Xuanwei County, Yunnan; Late Permian Xuanwei Formation.

Description: The ultimate pinnae is linear to lanceolate, the apex is obtuse. The pinnules are confluent at the base, ovate, oblong to lingulate, rather small, attaining a length of 2.5–3.5 mm and a breadth of 1.5–2 mm. with a rounded apex. The midrib is distinct, slightly decurrent. The lateral veins give off at an angle of 45° from the midrib, bifurcating once near the midrib. The branches of the lateral veins retain almost the same thickness before and after bifurcating.

Discussion and Comparison: The specimens are the same as *P. lativenosa* in the pinnule's shape, but the venations of the present specimens are denser than *P. lativenosa.*

Pecopteris* (*Asterotheca*) *guizhouensis (Figs. 9.5-2, 9.5-3, 9.5-4 and 9.5-5)

1980 *Pecopteris (Asterotheca) guizhouensis*, Zhang et al., p. 78, Pl.VIII, Figs. 2–3; Pl.X, Figs. 2–5, 2a

2011 *Pecopteris (Asterotheca) guizhouensis*, Wang et al., pp. 254–255, Pl.II, Fig. 9

Holotype: *Pecopteris (Asterotheca) guizhouensis* Zhang, p. 78, Pl.VIII, Figs. 2–3; Pl.X, Figs. 2–5, 2a, deposited in the Nanjing Institute of Geology and Palaeontology, Chinese Academy of Sciences, impression fossil, from Xuanwei Formation (Lopingian), Xuanwei, Yunnan Province, China.

Diagnosis (emended): Frond gigantic, at least bipinnate; ultimate pinnae linear, their sides parallel and contracting slowly to the apex. Pinnules relatively large, 5 mm in length, 2 mm in width, alternate and touching closely each other. The midrib distinct and thick, extending strightly and separating near the apex. The fertile frond with same shape of the sterile one; the sporangia synangium belonging to *Asterotheca*-type, bearing in the reverse side of the fertile frond, slightly touching each other, generally 5–6 couples with 4 sporangia respectively.

Material examined: YXC-(-1)-98A, YXC-(-1)-175, GWZ-38-10, GWC-58-3.

Occurrence: Chinahe section, Xuanwei County, Yunnan; Late Permian Xuanwei Formation. Chahe and Zhejue sections, Weining County, Guizhou; Late Permian Xuanwei Formation.

Description: The pinna is linear. The pinnules are closely set, forming an angle of about 70–80° with the rachis, rectangular, straight or slightly falcate, with a broadly rounded apex. The margin thickens. The midrib is straight, separating near the apex. The lateral veins are simple and straight, not difurcating or once, arising at a 60–70 degree angle. The fertile frond seems to be of the same shape of the sterile one, margin of the pinnules indistinct. The sporangia are ovate, slightly separating each other, generally 5–6 couples, forming rounded synangia of 5–6 sporangia. The synangia are situated on the lateral veins about midway between the midrib and the margin and form two rows, one on each side of the midrib.

Discussion and Comparison: The samples of this species are gigantic relatively, with a broadly rounded apex, which differs from *P (Asterotheca). hemitelioides.*

Pecopteris* (*Asterotheca*) cf. *hemitelioides (Brongniort, 1974) (Fig. 9.5-1)

1974 *Pecopteris (Asterotheca) hemitelioides* Brongniart, Paleozoic plants of China, p. 87; Pl.55, Figs. 9–11; Pl.56, Figs. 1–2

2007 *Pecopteris* cf. *hemitelioides* Brongniart, Liu et Yao, p. 200, Pl.1, Fig. 10

Holotype: *Pecopteris (Asterotheca) hemitelioides* Brongniart, 1834, deposited in Muséum national d'histoire naturelle (France)., impression fossil, from Loire France

Diagnosis (emended): Multiple pinnate frond, Pinnules ligulate-oblong, the length about three times of the width, attaching in the ultimate rachises vertically; pinnules sides parallel, with a obtuse apex. Midrib thick and straight; the lateral veins not difurcating. The sporangia synangium belonging to *Asterotheca*-type.

Material examined: YXC-(-1)-44.

Occurrence: Chinahe section, Xuanwei County, Yunnan; Late Permian Xuanwei Formation.

Description: Frond fragment is at least bipinnate, the ultimate rachises are 2 mm wide, with fine longitudinal lines. Pinnules are alternate, forming a wide-angle with the rachis, ligulate oblong in shape, about 30–35 mm long and 7–10 mm wide; pinnules sides are parallel, with a obtuse apex; they are entirely attached in the ultimate rachises and slightly decurrent at the base. The midrib is 1 mm wide; thick and straight, separating near the apex; slightly decurrent. The lateral veins are thinner than the midrib, extending outward near the midrib and not bifurcating, arising at a 50–60° angle.

Discussion and Comparison: Due to the lack of the fertile frond, this specimen is identified according to the morphology of the sterile pinnules. This specimen is typical of the ligulate-oblong pinnules and the simple lateral veins, which is seem to the *Pecopteris (Asterotheca) hemitelioides* Brongniart (Gu and Zhi 1974).

Pecopteris lativenosa (Halle 1927) (Figs. 9.5-11 and 9.5-12)

1927 *Pecopteris lativenosa*, Halle, pp. 86–87, Pl.25, Figs. 1–7

1974 *Pecopteris lativenosa*, Paleozoic plants of China, p. 94, Pl.62, Figs. 7–11

1989 *Pecopteris lativenosa*, Sze, p. 36, Pl.44, Figs. 1. 1a

2010 *Pecopteris lativenosa*, Sun et al., p. 346, Pl.204, Figs. 10, 11

Holotype: *Pecopteris lativenosa* Halle, pp. 86–87, Pl.25, Figs. 1–7, deposited in impression fossil, from Shanxi Province, China.

Diagnosis (emended): The pinnules bilateral symmetry, not uneven in length, with a slight contraction in the upper side and slightly decurrent in the lower side. The midrib distinct and decurrent; the lateral veins bifurcating once near the midrib, and the upper branches bifurcating once again; the first lateral vein at the proximal base arising from the decurrent part of the midrib, bifurcating more than 3 times, with a thicker apex near the margin.

Material examined: YXC-(-1)-15, GWC-58-20.

Occurrence: Chahe section, Weining County, Guizhou; Late Permian Xuanwei Formation.

Description: The figured specimen shows the very distinctive venation of this species. The rachis is stout. The pinnules are ovate in shape, slightly overlapping each other, with the broadly rounded apex and very slightly auriculate base. The midrib is distinct, slightly decurrent at the base, arising at a narrow angle, bending slightly forwards, divising near the apex. The lateral veins are thick and dense, arching, bifurcating close to the midrib, the distal branch occasionally dividing once.

Discussion and Comparison: Like the typical specimens of Central Shanxi (Halle 1927), the most prominent feature of the specimen is the venation. The midrib is strong, decurrent at the base, bending slightly forwards, persisting almost to the apex. The branches of secondary veins retain almost the same thickness the whole way to the margin. These characteristic are completely in accordance with which Halle pointed out. It is likely that the present specimen belongs to this species.

Pecopteris longifolioides (Figs. 9.5-16 and 9.5-17)

1980 *Pecopteris longifolioides* Zhang, p. 79, Pl. VII, Figs. 3–5, 3a, 5a

Holotype: *Pecopteris (Asterotheca) guizhouensis* Zhang, p. 79, Pl. VII, Figs. 3–5, 3a, 5a, deposited in the Nanjing Institute of Geology and Palaeontology, Chinese Academy of Sciences, impression fossil, from Xuanwei Formation (Lopingian), Xuanwei, Yunnan Province, China.

Diagnosis (emended). Frond gigantic, Ultimate pinnae lanceolate, 12 cm in length and 5 cm in width, with a slow contraction at the base; pinnule apices tapering toward the apex; the rachises surface with tumors. Pinnules ligulate-oblong, or slightly bending as the falcate shape; with bluntly rounded apex and slightly separated base; margin entire or undulate; uneven in length. The midrib distinct, not decurrent at the base. The lateral veins relatively dense and irregularity, difurcating several times.

Material examined: GWC-18-2, CLC-29-1-1.

Occurrence: Chahe Section, Weining County, Guizhou; Late Permian Xuanwei Formation. Kele Section, Hezhang County, Guizhou; Late Permian Xuanwei Formation.

Description: The ultimate rachis is strong, about 2 mm. broad. The pinnules are alternate, oblong or slightly bending as the falcate shape, with bluntly rounded apex and slightly separated base, attached by the whole of their base, 2.5 times longer than broad. The midrib is distinct, not decurrent at the base, almost extending to the apex of the pinnule. The lateral veins are very dense, giving off at narrow angle, bifurcating into two branchs near the midrib, of which the posterior branch bends outwards, dividing again; the anterior branch extends to a short distance onwards and dividing 2–3 times as co-axial type. All of the branches make up of a fascicle.

Discussion and Comparison: Compared to *P. longifolia,* the present specimens are distinguished by the large pinnae with elongating pinnules, complicated lateral veins. The ultimate pinnae of *P. longifolia* are short and its lateral veins are correspondingly simple.

Pecopteris marginata (Gu and Zhi 1974) (Fig. 9.5-18)

1974 *Pecopteris marginata* Gu and Zhi, Paleozoic plants of China, p. 91, Pl.56, Figs. 10–13

Holotype: *Pecopteris marginata* Gu and Zhi, p. 91, Pl.56, Figs. 10–12, deposited in the Nanjing Institute of Geology and Palaeontology, Chinese Academy of Sciences; impression fossil; from Xuanwei Formation (Lopingian), Panxian, Guizhou Province, China.

Diagnosis (emended). Frond fragment at least bipinnate; ultimate pinnae linear. Pinnules ligule or subfalcate with a thickened margin, attaching in the ultimate rachises vertically. Midrib distinct and thick, not decurrent at the base usually. The lateral veins difurcating twice regularly, giving off at wide angle.

Material examined: GWC-58-14a.

Occurrence: Chahe and Zhejue sections, Weining County, Guizhou; Late Permian Xuanwei Formation.

Description: This specimen is fragments of an ultimate pinna. The rachis of the ultimate pinna is comparatively broad in proportion to the size of the pinnules. The ultimate pinna is linear, with the apex and base not preserved. The pinnules are ligule or subfalcate, the length is 2.5 times longer than the breadth, with an obtuse apex and thickened margin; the base slightly enlarging. The midrib is very distinct, arising at a wide angle or right angle from the rachis. The lateral veins are fine and dense, but indistinct, dividing twice. The margin of the pinnule shows a narrow zone.

Discussion and Comparison: This specimens is identical in all respects with those described and figured by Gu and Zhi (1974, p. 91, Pl.56, Figs. 10–13) as *P. marginata* from Xuanwei Formation of Panxian, Guizhou.

Pecopteris (Asterotheca) orientalis ((Schenk) Potonie, 1883) (Figs. 9.5-6, 9.5-7, 9.5-8 and 9.5-9)

1883 *Callipteridium orientale*, Schenk, p. 212, 217; Pl.30, Figs. 5–7; Pl.45, Figs. 2–12

1893 *Pecopteris orientalis*, Potonie, p. 90

1927 *Pecopteris (Asterotheca) orientalis*, Halle, p. 73, Pl.16, Figs. 1–7; Pl.17, Figs. 1–5

1974 *Pecopteris (Asterotheca) orientalis*, Paleozoic plants of China, p. 93, Pl.62, Figs. 1–4

1987 *Pecopteris orientalis*, Zhao et al., p. 80, Pl.12, Figs. 1–3, 3a; Pl.13, Figs. 4,4a

1992 *Pecopteris (Asterotheca) orientalis*, Sun, p. 34, Pl.10, Figs. 9–10

Holotype: Pecopteris (Asterotheca) orientalis (Schenk) Potonie,

Diagnosis (emended): The pinnules bilateral symmetry, not uneven in length, with a slight contraction in the upper side and slightly decurrent in the lower side. The midrib distinct and decurrent; the lateral veins bifurcating once near the midrib, and the upper branches bifurcating once again; the first lateral vein at the proximal base arising from the decurrent part of the midrib; the lateral veins bending occasionally, bifurcating once or not.

Material examined: GWC-18-59, GWC-18-61, GWC-58-31, YXC-(-24)-1A.

Occurrence: Chahe section, Weining County, Guizhou; Latest Permian and Earliest Triassic Xuanwei Formation. Chinahe section, Xuanwei County, Yunnan; Late Permian Xuanwei Formation.

Description: The rachis is broad. The ultimate pinnae are linear, opposite, with a rounded or obtuse apex. The pinnules are small, oblong to ovate, their sides more or less parallel; apex broadly rounded; they are closely set, slightly decurrent and slightly confluent, with a slight contraction of the proximal margin at the base; upper side often slightly convex. The midrib is not much thicker than the lateral veins, decurrent at the base, bending slightly forward in the upper half of the pinnule. The lateral veins are straight and arching, bifurcating once generally close to the midvein. The first lateral vein at the proximal base arises from the decurrent part of the midrib.

Discussion and Comparison: According to the shape of the ultimate pinnae and pinnules, size and veination of the pinnules, these specimens belonged to this species.

Pecopteris sahnii (Hsu 1952) (Figs. 9.5-14 and 9.5-15)

1952 *Pecopteris sahnii*, Hsu, p. 250, Pl.3, Figs. 30, 33

1974 *Pecopteris sahnii*, Paleozoic plants of China, pp. 89–90, Pl.58, Figs. 14–16, Pl.59, Figs. 1–3

Holotype: *Pecopteris sahnii* Hsu, p. 250, Pl.3, Figs. 30, 33

Diagnosis (emended): Frond at least bipinnate. Ultimate pinnae linear, tapering toward the apex. Pinnules contiguous, thick, lingulate or nearly rectangular, with a rounded apex and a slightly decurrent base. Midvein thick, slightly decurrent or not at the base; lateral veins not bifurcate or occasionally bifurcate once; basal acroscopic lateral vein departing at an acute angle, then arching and bending inward, and the the first lateral vein at the proximal base arising from the decurrent part of the midrib.

Material examined: CLC-29-11-1, GWC-58-28.

Occurrence: Chahe section, Weining County, Guizhou; Late Permian Xuanwei Formation. Kele Section, Hezhang County, Guizhou; Late Permian Xuanwei Formation.

Description: The specimen is small and fragmentary, but well characterized by the veins. The ultimate rachis is straight. The pinnules are alternate or subopposite, touching each other, forming an angle of about 70–80° or a right angle with the rachis, ligule or nearly rectangular in shape, with an obtuse apex. The midrib is distinct but slender, slightly decurrent, attaining the apex. The lateral veins are straight, the same width as the midrib, forming an angle of about 40° with the midrib, not dividing or bifurcating once close to the midrib.

Discussion and Comparison: Compared to *P. hemitelioides* whose pinnules are narrow and long, the length being 4 times longer as the breadth, the midveins not decurrent, the lateral veins not bifurcating, the present specimen is different with it. The characters of the shape and veination of the present specimens is accordant with *Pecopteris sahnii*, so we assigin them to this species.

Pecopteris taiyuanensis (Halle 1927) (Fig. 9.5-13)

1927 *Pecopteris taiyuanensis*, Halle, p. 84, Pl. 21, Figs. 1–4

1974 *Pecopteris taiyuanensis*, Paleozoic plants of China, p. 93, Pl. 62, Figs. 5–6

1992 *Pecopteris taiyuanensis*, Sun, p. 35, Pl. 10, Fig. 11

Holotype: *Pecopteris taiyuanensis,* Halle, p. 84, Pl. 21, Figs. 1–4

Diagnosis (emended). Frond at least bipinnate; rachises thick and smooth; ultimate pinnae lanceolate. Pinnules bilateral symmetry, not uneven in length, with a with a rounded apex and a slightly decurrent base. Midrib distinct and decurren, persisting straightly and divising near the apex; the lateral veins thick and dense, curved, bifurcating once near the midrib, and the upper branches bifurcating once again occasionally; the first lateral vein at the proximal base arising from the decurrent part of the midrib, bifurcating twice or three times.

Material examined: YXC-(-6)-4.

Occurrence: Chinahe section, Xuanwei County, Yunnan; Late Permian Xuanwei Formation.

Description: The specimens partly show bipinnata pinnae, the ultimate pinnae are lanceolate. The pinnules are lingulate or ovate in shape, about 4–6 mm in length and 2–3 mm in width; pinnules are attached by the whole of their base, with a rounded apex and a slightly decurrent base. The midrib is distinct, more or less thicker than the lateral veins; slightly decurrent at the base, divising near the apex. The lateral veins are curved, bifurcating once near the midrib, and the upper branches bifurcating once again. The first lateral vein at the proximal base arises from the decurrent part of the midrib, arching strongly.

Discussion and Comparison: Compared to *P. anderssonii* whose pinnules are wide and long, the the midveins not distinct, the lateral veins thin and bifurcating more than three times, the present specimen is different with it. The characters of the shape and veination of the present specimens is accordant with *Pecopteris taiyuanensis*, so we assigin them to this species.

***?Pecopteris* sp.** (Figs. 10.3-11, 10.3-13 and 10.3-14)

Material examined: YXC-(-25)-6.

Occurrence: Lower Triassic Kayitou Formation in Western Guizhou and Eastern Yunnan, South China.

Description: Preserved as single pinule. Relatively thin midrib and lateral veins branch at acute angle. Broken impression fossil, without distinct branching pattern.

Discussion and Comparison: Only based on the pinnule's shape and vein system, those are probably preserved Pecoteridales.

Genus ***Rajahia*** Kon'no, 1970

Type species:

Diagnosis (emended): The sacs are arranged in a single line into linear sacs, which are attached to the back of kohlrabi slices, are wrinkled and extremely small, and are tightly packed on both sides of side veins or branches.

Discussion: Occures in the late Permian. *Rajahia* in different areas has different characteristics, such as *R.guizhouensis'* axis width is small, kohane leaves are nearly

Fig. 9.7 1 *Rajahia guizhouensis* Zhang Locality: Chahe section, Weining County, Guizhou Province Sample Number: YWC-69-01

perpendicular to the axis, and there are more capsular piles; *R.changduensis*' kohane-shaped leaves are close at the base but not united, and its near-top part is sterile and does not have a capsular pile.

Rajahia guizhouensis (Figs. 9.7-1, 9.8-1)

1974 *Danaeites saraepontanus* Stur, Paleozoic plants of China, pp. 98–99, Pl.67, Figs. 9–12

1980 *Rajahia guizhouensis* Zhang, p. 80, Pl.XI, Figs. 2–4, 2a

◀**Fig. 9.8** 1 *Rajahia guizhouensis* Zhang Locality: Chinahe section, Xuanwei City, Yunnan Province Sample Number: YXC-(-6)-5B, 2 *Fascipteris sinensis* (Stockm. et Math.) Gu et Zhi, Locality: Chinahe section, Xuanwei City, Yunnan Province Sample Number: YXC-(-1)-2, 3 *Fascipteris hallei* Kaw Locality: Chinahe section, Xuanwei City, Yunnan Province Sample Number: CNH-0-9, 4, 5 *Fascipteris densata* Gu et Zhi Locality: Chahe section, Weining County, Guizhou Province Sample Number: GWC-18-47, GWC-18-42, 6, 7, 8 *Fascipteris stena* Gu et Zhi Locality: 6 Kele section, Hezhang County, Guizhou Province Sample Number: CLC-3-4, 7 Chahe section, Weining County, Guizhou Province Sample Number: GWC-4-6a 8 Chinahe section, Xuanwei City, Yunnan Province Sample Number: YXC-(-1)-71A, 9, 10 *Neuropteridium guizhouensis* Zhang Locality: 9 Chinahe section, Xuanwei City, Yunnan Province Sample Number: YXC-(-11)-1A, 10 Chahe section, Weining County, Guizhou Province Sample Number: GWC-18-69, 11, 12 *Protoblechnum (Compsopteris) contracta* Gu et Zhi Locality: 11 Chahe section, Weining County, Guizhou Province Sample Number: GWC-58-4, 12 Zhejue section, Weining County, Guizhou Province Sample Number: GWZ-38-4, 13 *Compsopteris imparis* Gu et Zhi Locality: Chahe section, Weining County, Guizhou Province Sample Number: GWC-3-15, 14, 15 *Compsopteris punctinervis* Mo Locality: 14 Chahe section, Weining County, Guizhou Province Sample Number: GWC-28-1, 15 Chinahe section, Xuanwei City, Yunnan Province Sample Number: YXC-(-2)-3, 16, 17 *Compsopteris wongii* (Halle) Zal. Locality: 16 Kele section, Hezhang County, Guizhou Province Sample Number: CLC-35-20, 17 Zhejue section, Weining County, Guizhou Province Sample Number: GWZ-38-13

Holotype: *Rajahia guizhouensis*, Zhao Xiuhu et al. 1980. p. 80, pl. 11, Figs. 2–4, 2a, deposited in the Nanjing Institute of Geology and Palaeontology, Chinese Academy of Sciences, a impression fossil, from Xuanwei Formation (Changhsingian), Guizhou and Yunnan provinces, China.

Diagnosis (emended): Odd pinnate compound leaves, the last pinnae is linear-lanceolate, the tip is tapered, the last pinnae is thick and the small pinnae is sub-opposite, irregular, nearly perpendicular to the ventral edge of the pinnae, tightly packed or slightly separated, and the base is disconnected. The fully developed plumule has a long tongue shape, which can reach 13 mm in length and 5 mm in width, the base is slightly contracted, and the top is blunt; Midrib is thick and straight, extending to near the top; Lateral veins are sparse and generally do not diverge. There are 16–20 pairs of linear capsular piles on both sides of midvein on the back of reproductive plumule; The saccule pile consists of up to 24 saccules, which are oval to long oval, arranged in a row along the lateral veins and cling to each other.

Material examined: GWC-69–58, YXC-(-6)-5B.

Occurrence: Chahe section, Weining County, Guizhou; Earliest Triassic Xuanwei Formation.

Description: The present specimens show that the frond isat least odd-bipinnate, probably tripinnate. Rachis of the penultimate pinna is very strong, attaining a breadth of 5 mm on the impression, with longitudinal striation. The ultimate rachis is slender, about 1 mm broad. Ultimate pinnae closely set, alternate, linear to linear-lanceonate, gradually tapering to the apex, forming an angle about 60–80° with the rachis. Pinnules are closely set or slightly separate, forming an angle of 80–90° with the rachis, sub-alternate, oblong, 8–10 mm long and 3–4 mm wide, with slightly contracted base and obtuse apex. The midrib is strong and straight, persisting almost to the apex. The lateral veins are sparse, simple.

Discussion and Comparison: Compared with *Rajahia minor* (Zhao 1980), *Rajahia guizhouensis*'midrib has more pairs of sacs on both sides, while the *R. minor* has wider feather axis.

Genus ***Cladophlebis*** (Brongniart 1849)

Type species: *Cladophlebis permica* Lee and Wang (1956).

Diagnosis (emended): Multiple pinnate compound leaves, small pinnate, mainly sickle-shaped, asymmetrical on both sides, slightly pointed or rounded at the top, attached to the feather axis with the whole base, with whole or serrated leaves, obvious midvein, scattered up to or near the top, and bifurcated lateral veins. As a pioneer, it began in Permian, but its number is not very large.

Discussion: *Cladophlebis* is similar to *Pecopteris*, with multiple pinnate compound leaves, and the pinnae is attached to the feather axis. However, the small feathers of *Cladophlebis* are mostly curved in sickle shape, and the top is sharp and convex and large; The small feather of *Pecopteris* is mainly tongue-shaped, oval or rectangular.

Cladophlebis yunnanensis

1980 *Cladophlebis yunnanensis* Zhang, p. 82, Pl.XIII, Figs. 6, 6a

Holotype: *Cladophlebis yunnanensis* Zhang, 1980, p. 82, Pl.XIII, Figs. 6, 6a, deposited in the Nanjing Institute of Geology and Palaeontology, Chinese Academy of Sciences, a impression fossil, from Xuanwei Formation (Changhsingian), Guizhou and Yunnan provinces, China.

Diagnosis (emended): The width of the last feather can reach more than 4.5 cm, and the length is unknown; Feather axis is thin, about 1 mm wide, with a longitudinal stripe. The pinna extends out at an angle of about 60, is tightly packed, sickle-shaped, 2.4 cm long and 1.1 cm wide, with sharp tip, connected base, and split into thick teeth with sharp forward extension at its side. Midrib is thin and obvious, extending at about 60, quite straight, reaching the top; The lateral veins extend at an angle of 20–30, and they diverge after about 1 mm, and then most of them diverge twice in a coaxial way.

Occurrence: Chahe section, Weining County, Guizhou; Late Permian Xuanwei Formation.

Description: The thin rachis is about 1 mm wide with a longitudinal striation. The falcate pinnule comes off at about 60-degree angle, is about 8 mm long and 4 mm wide; the apex acute, the base is confluent, the lateral margin is wave-shaped. The slender but prominent midvein departs at a 60-degree angle, straight, almost to the apex, decurrent at the base; the lateral veins come off at a 20–30 degree angle about 1 mm, then bifurating once or twice by the sympodium.

Discussion and Comparison: Compared to *Cladophlebis yunnanensis* which has thin and acute crenate margin, each tooth only has a lateral vein, the margin of the present specimen is wave-shaped, but the shape and vein of its pinnule is similar to *C. yunnanensis.*

Genus ***Fascipteris*** (Gu and Zhi 1974)

Type species: *Fascipteris hallei* (Kawasaki) Gu and Zhi (1974).

Diagnosis (emended): Feather-like compound leaves, small pinnate linear, base often contracted, whole or wavy. The middle vein is thick, and the lateral veins are bifurcated or branched several times to form vein bundles. Each vein bundle is equivalent to a shallow lobe at the edge of the pinna, and there is no adjacent vein or interbeam vein.

Discussion: Differ from *Validopteris* (Bert) in Europe as *Fascipteris* has no bundle vein. *Fascipteris* occurs in late Permian.

Fascipteris sinensis (Stockm. et Math.) (Gu and Zhi 1974) (Fig. 9.8-2)

1957 *Validopteris sinensis* Stockm. et Math., p. 22, pl.XIII, Figs. 2–3

1974 *Fascipteris sinensis* (Stockm. et Math.) Gu and Zhi, p. 100, pl.69, Figs. 5–7

Holotype: *Fascipteris sinensis* (Stockm. et Math.) Gu and Zhi 1974, p. 100, pl.69, Figs. 5–7, deposited in the Nanjing Institute of Geology and Palaeontology, Chinese Academy of Sciences, a impression fossil, from Xuanwei Formation (Changhsingian), Guizhou and Yunnan provinces, China.

Diagnosis (emended): The feathery leaf is narrow-band to linear, 1–2 cm wide, and its edge is microwave-like or blunt-toothed. The midvein is thick, about 2 mm, and the lateral veins bifurcate 3–4 times in a narrow angle and form bundles, almost perpendicular to the midvein. The width of vein bundle is less than 3 mm, and the outer two veins are slightly higher than the midrib and parallel to the edge of the chalcedony. Each vein of the vein bundle is nearly equal in thickness, but it is obviously thickened at the base of the vein bundle.

Material examined: YXC-(-1)-2.

Occurrence: Chahe section, Weining County, Guizhou; Late Permian Xuanwei Formation; Niulanshan, Yongding County, Fujian; Early Permian Tongziyan Formation.

Description: The present specimen is very small fragment of pinnule, probably linear in shape, with slightly sinuate margin. The breadth of the pinnule is about 1 cm, length unknown. The midrib is wide, thicker than the lateral veins which forming fascicles. The lateral veins of the each fascicle give off at a right angle, and bifurcating twice or three times at an acute angles. After twolateral veins lying in the margin of each fascicle run a short distance to the midrib, they are parallelly to the margin of the pinnule.

Discussion and Comparison: The vein's type of the present specimen is similar to that of *F. stena* Gu et Zhi, but the fascicle of the latter is not perpendicularity to the midrib. This species is compared to *F. recta* Gu et Zhi and *F. chongsoensis* (Kaw.) Li, Yao and Deng, their pinnules are the same as *F. sinensis* in shape, and the lateral veins give off at a right angle from the mibrib. The difference among them is that the lateral of the latter is very slender.

Fascipteris densata (Gu and Zhi 1974) (Figs. 9.8-4 and 9.8-5)

1974 *Fascipteris densata* Gu and Zhi, Paleozoic plants of China, pp. 100–101, pl.69, Figs. 8–14

Holotype: *Fascipteris densata* Gu and Zhi (1974), pp. 100–101, pl.69, Figs. 8–14, deposited in the Nanjing Institute of Geology and Palaeontology, Chinese Academy of Sciences, a impression fossil, from Xuanwei Formation (Changhsingian), Guizhou and Yunnan provinces, China.

Diagnosis (emended): Feathery leaves lanceolate, whole or crenate, blunt at the top, and eccentric at the base. The midvein is thick, and the lateral veins are bifurcated like narrow angle and axis for many times, forming a slightly outward curved vein bundle, and sometimes the branches are bifurcated again.

Material examined: GWC-18-47, GWC-18-42.

Occurrence: Chahe section, Weining County, Guizhou; Late Permian Xuanwei Formation; Funiushan, Zhenjiang City, Jiangsu; Late Permian Longtan Formation.

Description: The pinnules are well preserved, falcate or lanceolate in shape, with 2–3 pairs of lobes at the base of each pinnule, entire at the upper part of the pinnule. The midrib is straight or slightly curving forwards. The lateral veins bifurcating once, generally at a short distance and an acute angle from the midrib, forming fascicles. The anterior and posterior branches divide 2–3 times as dichotomous branching. The fascicles obviously bend upward.

Discussion and Comparison: The winglet of *Fascipteris densata* is lanceolate or sickle-shaped, while *Fascipteris* hallei's winglet is linear, and the branch of *F.hallei*'s side vein is no longer bifurcated, reaching the edge of the winglet, but the branch of *F.densata* sometimes bifurcates again.

Fascipteris stena (Gu and Zhi 1974) (Figs. 9.8-6, 9.8-7, 9.8-8 and 10.4-12)

1974 *Fascipteris stena* Gu et Zhi, Paleozoic plants of China, p. 101, pl.69, Figs. 15–17

Holotype: *Fascipteris stena* Gu and Zhi, 1974, p. 101, pl.69, Figs. 15–17, deposited in the Nanjing Institute of Geology and Palaeontology, Chinese Academy of Sciences, a impression fossil, from Xuanwei Formation (Changhsingian), Guizhou and Yunnan provinces, China.

Diagnosis (emended): The leaves are large, the small feathers are whole, the top is blunt, slightly curved, and the midrib is slightly thin; The lateral veins extend at wide angles, and bifurcate at narrow angles for 5–6 times. The vein bundle is about 2 mm wide, long and narrow, and almost bends to one side.

Material examined: CLC-3–4, GWC-4-6a, YXC-(-1)-71A, YXC-(-25)-30.

Occurrence: Chahe section, Weining County, Guizhou; Late Permian Xuanwei Formation; Jiangning City, Jiangsu; Late Permian Longtan Formation; Upper Permian Xuanwei Formation and Early Triassic Kayitou Formation in Western Guizhou and Eastern Yunnan, South China.

Description: A single pinnule well preserved, oblong in shape, entire or slightly sinuate, with an obtuse apex. The midrib is slightly bending forwards. The lateral veins dividing once, generally at a short distance and an acute angle from the midrib, forming fascicles. The anterior and posterior branches divide 3–4 times as the dichotomous branching. The fascicles obviously bend, about 2 mm broad.

Discussion and Comparison: Compared to *F. hallei*, the difference is that the bifurcating times of the present specimens is less than *F. hallei*, the fascicle breadth is narrower than *F. hallei*.

Division **Progymnospermophyta**

Order **Noeggerathiales** (Nemejc 1931)

Family **Tingiales** (Halle 1925)

Genus ***Tingia*** (Halle 1925)

Type species: *Tingia carbonica* (Schenk 1883; Halle 1925)

Diagnosis (emended): Branches are feathery and dorsolateral, with leaves arranged in four rows, with two rows on top and two rows on the bottom. The front two rows of leaves are large (large leaves), which are wide wedge-shaped, inverted-oval, long oval or linear, and the base is extended, which is semi-stalk-shaped, twisted after extending, lying in the same plane with the axis, with the whole side edge, and the top often irregularly divided into teeth. The lower two rows of leaves are narrow and small (lobules), which are located on the back of branches, and often cling to branches and point to the front of branches. On the same branch, the middle leaves are longer and larger, the two ends are shorter and smaller, the veins are parallel, the base branches are more, and they go up to the top of the leaves roughly parallel and extend into the teeth.

Discussion: Compared to *Plagiozamites,* the veins on both sides of *Tingia* are parallel, and the leaves are whole and arranged in four rows. *Tingia* mainly occurs from late Carboniferous to late Permian.

Tingia gerardii (Stockmans and Mathieu 1939) (Fig. 9.6-13)

1939 *Tingia gerardii* Stockmans et Mathieu, Mus. Roy. Hist. Nat. Belgique; pl.24, Figs. 1–2

Holotype: *Tingia gerardii* Stockmans and Mathieu (1939), Mus. Roy. Hist. Nat. Belgique; pl.24, Figs. 1–2.

Diagnosis (emended): Obovate, with narrow lobes, round top, and little difference between large and small leaves.

Material examined: GWC-3–1.

Occurrence: Chahe section, Weining County, Guizhou; Late Permian Xuanwei Formation.

Description: The large leaves are obovate in shape, 18 mm long and 6 mm, the base decurrent, the lateral margin entire, the tip blunt and round. The veins parallel to the lateral margins, thin and dense, forked at the base. Small leaves are unknown.

Discussion and Comparison: Compared with *Tingia carbonica*, *Tingia gerardii* has smaller total width of branches, smaller axial width, inverted ovate leaves and nearly equal length and width of large and small leaves.

Tingia crassinervis (Halle 1927) (Fig. 9.6-14)

1925 *Tingia crassinervis* Halle, Geol. Surv. China Bull., 7: 6–8; pl. II, Figs. 2–3

1927 *Tingia crassinervis* Halle, Pal. Sinica, ser. A, 2(1): p. 234, pl. 61

Holotype: *Tingia crassinervis*, Halle (1925), p. 6, pl. 2.

Diagnosis (emended): Feathery leaf is oval or linear, with thick veins and similar vein density. The shape of large leaves changes greatly, with rounded or truncated tips and wide lobes. The leaflets at the base of branches are very wide and short, with rounded or truncated tips and no lobes.

Material examined: GWC-3-1.

Occurrence: Chahe section, Weining County, Guizhou; Late Permian Xuanwei Formation.

Description: The axis attains a breadth on the impression of 3 mm. The leaves oblong-oblanceolate, 30 mm long and 5–7 mm wide, apex of these leaves broadly blunt, round or truncate; the veins are strong.

Discussion and Comparison: To this species I have referred a few fragmentary specimens from the locality Chahe of Guizhou. The leaves are broader and oblong-oblanceolate in shape, the veins are coarser. The diagnosis of this species was given by Halle (1927, p. 234).

Division **Pteridospermophyta**

Order **Peltaspermales** (Nemejc 1968)

Family **Peltaspermaceae** (Thomas 1933)

Genus ***Peltaspermum*** (Harris 1937)

Type species: *Peltaspermum rotula* Harris, 1937. Rhaetian and Lower Liassic; East Greenland.

Diagnosis (emended): Plants with bipinnate, rarely pinnate or tripinnate fronds. Primary rachis robust, basal part naked. Pinnae suboppositely or alternately attached to the primary rachis. Rachides may be covered with blister-like swellings. Intercalary pinnules inserted in two lateral rows on the adaxial side of the primary rachis. Pinnules lanceolate, alethopteroid entire-margined to pinnately lobed. Intercalary pinnules resemble the other pinnules. Pinnule apex rounded or acute. Venation pinnate. Ovuliferous organs consist of elliptical to radially symmetrical, umbrella-shaped ovule-bearing discs. Discs with radiating ribs and lobed margin. Ovules attached to the lower surface of the discs in a ring around the insertion of the central stalk.

Discussion: *Peltaspermum* has been redefined by Poort and Kerp (1990) as a natural genus including sterile foliage and ovuliferous organs for the species *P. martinsii* (Galtier and Broutin 2008). They combined the sterile part *Callipteris martinsii* with the fertile part *Peltaspermum*, and these authors use the *Peltaspermum* as a natural genus. More recently Naugolnykh and Kerp (1996) combined some peltate discs, reported by Gomankov and Meyen (1986) from the Kungurian of the Fore-Urals, with sterile pinnae described by Zalessky as *Callipteris retensoria*, and regrouped into a natural taxo *Peltaspermum retensorium* (Zalessky). Nevertheless, in most cases the reproductive organ *Peltaspermum* hardly can be connected with the sterile part of plant, while based on the present sterile pinnule and fertile organ occurring in the same beds, I prefer to use the generic name *Peltaspermum* in sense of Harris (1937).

Peltaspermum lobulatum (Wang and Wang 1989) (Fig. 11.2–22)

1989 *Peltaspermum lobulatum* Wang et Wang, Shanxi Geology, Vol. 4, p. 34, Pl.V, Figs. 1, 4

Holotype: *Peltaspermum lobulatum* Wang and Wang (1989).

Diagnosis (emended): Small to medium-sized bipinnate, rarely pinnate or tripinnate fronds. Fronds overtopping in the apical portion, in some cases with double apical (pseudo-) dichotomies in larger, (partially) tripinnate fronds. Pinnae subopposite to alternate. Primary rachis with intercalary pinnules. Pinnules small, semi-circular, linguiform to elongate linguiform, broadly attached, decurrent to basally fused in the basal and apical frond portions and pinna apices, entire-margined, crenulate to pinnatifid. Pinnules thick and fleshy with rounded apices and a pinnate venation that is usually hidden by the thick cuticle. Leaves amphistomatic, stomata irregularly scattered; stomatal complexes sunken, monocyclic with 4–8 (usually 5–6) subsidiary cells; papillae may be present. Ovuliferous organs consisting of peltate discs with a central stalk and marginal lobes; upper surface with radiating ribs.

Material examined: GPT-22-1, GPT-22-5, GPT-23-6.

Occurrence: Tucheng section, Panxian County, Guizhou; Early Triassic Layitou Formation.

Description: Disc-like sporophyll, about 1 cm in diameter, with a rounded lobelet in margin consists of lobelets 8–10 in number. Lobules are sphenoidal in shape, with rounded apex, radially given off from a concave center, gradually becoming convex. Lateral margins of rib-lobules uniting side by side.

Discussion and Comparison: The main characteristic of the present specimens is with rounded lobelet, small in size. This species is similar to P. multicostatum Zhang et Shen named by Zhang et Shen (1987) from Nanshan Section, Gansu Province. They have a rounded lobelet in margin, but not forming a circular zone making up of the rounded lobelets to P. multicostatum. Compared to P. lobulatum named by Wang and Wang (1989), their features are completely same except that the present specimens are smaller than the latter.

***?Peltaspermum* sp.** (Figs. 11.2-21, 11.1-23)

Material examined: GPT-22-6.

Occurrence: Tucheng section, Lower Triassic Kayitou Formation in Panxian County, Guizhou, South China.

Description: Fertile sporophyll consists of radially symmetrical peltate discs, about 6 mm in diameter, with a sinuate margin. The surface of peltate disc slightly depressed in central part. The restorable figure shows that the sporophyll has about 12 lobelets in number. Each lobule is spade-type in shape, radiating to the centre part, apex concave. Lateral margins of rib-lobules slightly uniting side by side.

Discussion and Comparison: As shown by the present specimen, the structure of peltate disc is unique. Based on fragments preserved, we temporarily assign this specimen as to genus? *Peltaspermum.*

Order **Peltaspermales**

Family **Peltaspermaceae**

Genus ***Germaropteris*** (Kustatscher et al. 2014)

Type species: *Germaropteris martinsii* (Germar in Kurtze 1839) Kustatscher, Kerp et Van Konijnenburg-van Cittert 2014.

Diagnosis (emended): Small to medium-sized bipinnate, rarely pinnate or tripinnate fronds. Fronds overtopping in the apical portion, in some cases with double apical (pseudo-)dichotomies in larger, (partially) tripinnate fronds. Pinnae subopposite to alternate. Primary rachis with intercalary pinnules. Pinnules small, semi-circular, linguiform to elongate linguiform, broadly attached, decurrent to basally fused in the basal and apical frond portions and pinna apices, entire-margined, crenulate to pinnatifid. Pinnules thick and fleshy with rounded apices and a pinnate venation that is usually hidden by the thick cuticle. Leaves amphistomatic, stomata irregularly scattered; stomatal complexes sunken, monocyclic with 4–8 (usually 5–6) subsidiary cells; papillae may be present. Ovuliferous organs consisting of peltate discs with a central stalk and marginal lobes; upper surface with radiating ribs.

Discussion: This genus is named after Ernst Friedrich Germar (1786–1853), which is monotypic but the type species has been long understood. Recently several species of *Alethopteris, Callipteris, Lepidopteris, Peltaspermum* are revised to *Germaropteris* (Kustarscher et al. 2014). Foliage and ovuliferous organs are described by Germar at first and Göppert (1850) respectively as different species. The species name was firstly assigned to *Callipteris martinsii* (Germar in Kurtze). After longtime arguing, then the ovuliferous organ *Peltaspermum rotula* and pollen organ *Antevsia zeilleri* were reported and correlated with *Lepidopteris martinsii* foliage (Poort and Kerp 1990). Whereas, different types of foliage seemed to share same type of ovuliferous organ i.e., Lepidopteris and Scytophyllum Bornemann (1856). Based on the morphology analysis and cuticle study, Kustarscher established a new genus to refer to the first and well-studied peltasperms with both foliage and ovuliferous.

Germaropteris martinsii (Germar in Kurtze 1839) Kustatscher, Kerp et Van Konijnenburg-van Cittert (2014) (Figs. 11.2-1, 11.2-2, 11.2-3 and 11.2-4)

Foliage-selected synonymy

1839 *Alethopteris martinsii* Germar in Kurtze, pp. 34, 35, 38, pl. III, 2

1840 *Alethopteris martinsii* Germar, pp. 35–36

1846 *Alethopteris martinsii* Althaus, pl. 1, 3

1846 *Caulerpites crenulatus* Althaus, pl. I, 4

1848 Pecopteris (Alethopteris) martinsii Geinitz, p. 20

1862 *Alethopteris goepperti* Geinitz, p. 142, pl. XXV1, 7, 8

1906 Callipteris martinsii Zeiller, p. 71

1907 *Callipteris martinsi* Gothan pp. 1–4, Figs. 1, 2

1921 *Callipteris martinsi* Gothan and Nagalhard, pp. 451–453, pl. 6, 5, 6, pl. 7, 13

1928 *Callipteris martinsi* Weigelt, pp. 457, 458, pl. II, 14, 15, 26; pl. VII, 1–14, pl. 3, 7, 14, (22?)

1958 *Callipteris martinsi* Stoneley, pp. 313–315, Figs. 5, 6, pl. 37, 2, 5

1960 *Lepidopteris martinsii Townrow,* pp. 345–347, figs. lL, 2 J, 3G-K,4A, B, 5 J, 6D

1962 *Callipteris martinsi* Schweitzer, pp. 339–340, Fig. 4, pl. 1, 3, 4, pl. 3, I

1990 *Peltaspermum martinsii* Poort and Kerp (per partem), pp. 203, 206–208, 210, 214, 216, pl. I-V, pl. VI, 1 (per partem)

2012 *Lepidopteris martinsii,* Kustatscher et al., p. 4, pl. I, 7

Ovuliferous organs—selected synonymy

1850 *UIlmannia bronni* Göppert, p. 188, pl. 20, (?21, 22), 24–26.

1864–65 *Ullmannia bronni* Göppert, pp. 226, 227, pl. 45, 21 23, 25.

1884 *Strobilites bronnii* Solms-Laubach, pp. 19–24, pl. II, 2–9, 16–19.

1944 *Strobilites bronnii* Florin, p. 447, pl. 169–170, 5–9, 13–2.

1958 *Strobilites bronni* Stoneley, p. 329, pl. 40, 7–11.

1990 *Peltaspermum martinsii* Poort and Kerp, pp. 207, 216–218, pl. VI, 1 (parte), 2–7, pl. VII, VIII.

2012 *Peltaspermum* sp. Kustatscher et al., p. 4, pl. I, 3.

Basionym: *Alethopteris martinsii* (Germar in Kurtze 1839), Commentatiode petrefactis quae schisto bituminoso mansfeldensi repriuntur, pp. 34–35.

Holotype: *Germaropteris martinsii* Germar in Kurtze, 1839, Kustatscher, Kerp et. Van Konijnenburg-van Cittert 2014.

Diagnosis (emended): Small to medium-sized bipinnate, rarely pinnate or tripinnate fronds. Fronds overtopping in the apical portion, in some cases with double apical (pseudo-)dichotomies in larger, (partially) tripinnate fronds. Pinnae subopposite to alternate. Primary rachis with intercalary pinnules. Pinnules small, semi-circular, linguiform to elongate linguiform, broadly attached, decurrent to basally fused in the basal and apical frond portions and pinna apices, entire-margined, crenulate to pinnatifid. Pinnules thick and fleshy with rounded apices and a pinnate venation that is usually hidden by the thick cuticle. Leaves amphistomatic, stomata irregularly scattered; stomatal complexes sunken, monocyclic with 4–8 (usually 5–6) subsidiary cells; papillae may be present. Ovuliferous organs consisting of peltate discs with a central stalk and marginal lobes; upper surface with radiating ribs.

Material examined: Registered specimens: YXM (A)-46-6; GPT-22-3, GPT-22-4, GPT-23-3, GPT-23-5, etc.

Occurrence: Mide section, Xuanwei City, Yunnan; Tucheng section, Panxian County, Guizhou; Early Triassic Kayitou Formation.

Description: These broken specimens shown in Fig. 14, possibly represent the terminal part of a pinna. Rachis is very thick. Pinnules alternate to subopposite, open to more or less right angle to the rachis, ovate. Leaf veins indistinct, lateral veins forking or sometimes simple, some of the proximal basal veins being apparently derived from the rachis.

Discussion and Comparison: All of the pinnae from Early Triassic Kayitou Formation, Panxian County, Guizhou, are incomplete, however the pinnae are slender and the sterile pinnules are somewhat similar to those in specimens of *Peltaspermum martinsii* illustrated by Poort and Kerp (1990) and Visscher et al. (2001).

Division **Pteridophyta** (Schimper 1869)

Order **Osmundales**

Family **Osmundaceae**

Genus ***Neuropteridium*** Schimper and Schenk, 1879

Type species: *Neuropteridium voltzii* (Brongniart 1828a) Schimper, 1879.Grès bigarre (Buntsandstein, early Anisian); Vosges (France).

Diagnosis (emended): *Neuropteridium voltzii* has longer (up to 7 cm) but narrower (4–6 mm) pinnules than *N. elegans*, resulting in a much larger length/width ratio. The pinnules are attached by almost their entire base, the apex is roundly acute. The venation is neuropterid with a clear midrib extending along about two-thirds of the pinnule. Secondary veins are numerous and fork two–three times.

Discussion: Brongniart's holotype (Brongniart 1828b) from the Vosges is a frond fragment over 20 cm long with a rachis c. 5 mm wide and pinnules that arise more or less perpendicularly. The pinnules are 4–5 cm long and 8–10 mm wide with a constricted base and a pointed apex. Schimper and Mougeot figured a frond fragment c. 25 cm long that includes the apical region. The width of the rachis decreases from 8 mm basally to 3 mm apically, and the apical pinnules arise at a smaller angle and also diminish in size. This material originates from the Upper Buntsandstein of the Vosges. Blanckenhorn also described and figured material from the surroundings of Floisdorf (Berg) near Kommern that consists of long frond fragments (up to 1 m long according to Blanckenhorn) with a rachis varying from 1 cm in width at the base to 2 mm at the apex. The pinnules are usually 4–5 cm long and 8–12 mm wide. The venation is in all cases the same as in our material. Blanckenhorn also described and figured one broad pinnule fragment as *N. voltzii* var. *latifolium*. This fragment is too small to be sure of its identification. As discussed above, Blanckenhorn created the species *N. bergense* from the same locality, which is characterized by shorter fronds (30–45 cm long), a rachis 5 mm wide, and pinnules 5–6 mm with an obtusely acute apex. Earlier he stated that the species were difficult to distinguish and we believe that *N. bergense* is actually a smaller variety of *N. voltzii*. We have both forms in our collection as well.

Brongniart (1828b) described and figured the species *Sphenopteris palmetta* based on a specimen from the same beds in the Vosges as *N. voltzii*, stating that the pinnules resemble those of *N. voltzii* that were dissected during preservation, but that he thought these incisions were so regular that it was probably a different species. Goeppert (1836) transferred the species to his new genus *Asplenites*. Schimper (1869) considered that *Sphenopteris palmetta* was based on a poorly preserved *Neuropteris intermedia* frond. Fliche (1910) decided that the specimen would be better placed in the genus *Acrostichites* and that it was probably a poorly preserved specimen of.

A. *densifolius* Fontaine. However, we agree with Schimper (1869) that it is quite possible that the specimen is poorly preserved specimen of *N. voltzii*.

Neuropteridium guizhouense (Figs. 9.8-9 and 9.8-10)

1980 *Neuropteridium guizhouense* Zhang, p. 82, Pl.XIV, Figs. 3, 3a

Holotype: *Neuropteridium guizhouense* Zhang, deposited in the Nanjing Institute of Geology and Palaeontology, Chinese Academy of Sciences, from Late Permian Xuanwei Formation, Guizhou Province, China.

Diagnosis (emended): The last pinna may be up to 8 cm long and 4 cm wide, slowly contracting forward, with a single apical barb, with a thick Rachis 3 mm wide. The pinnules are attached to the ventral surface and are ovate or lingual in different locations. Middle vein thick, base not extended; Lateral veins straight, extremely fine 'membranous with sparse fine markings.

Material examined: GWC-18-69.

Occurrence: Late Permian Xuanwei Formation, Chahe section, Weining County, Guizhou.

Description: The shape of the pinna unknown, probably linear, attaining a breadth of at least 2.5 cm. Rachis is very thick, with a maximum breath of 4 mm. Pinnules closely arrange, forming an angle of about 80° with the rachis, oblong, with broadly rounded apex, strongly contracted and cordate at the base, more or less auriculate. The midvein is distinct, no decurrent, disappearing near the apex. The lateral veins are straight but dense, dividing 2–3 times.

Discussion and Comparison: The main characters of the present specimen are its thick and no decurrent midvein, straight, dense and slender lateral veins. *Neuropteridium coreanum* Koiwai is similar to the present species but they differ in that *N. coreanum* has smaller angle (40–50°) between pinnule and rachis, sinuate margin.

Division **Pteridospermatophyta**

Order **Peltaspermales**

Genus ***Compsopteris*** (Zalessky)

Type species: *Compsopteris adzvensis* Zalessky (1934).

Diagnosis (emended): Fronds pinnate, petiolate, with prominent rachis. Pinnules oblong, straight or slightly curving, with entire margins and rounded apex; pinnules small and attached about perpendicular to rachis in the basal frond portion, becoming longer and arising at increasingly acute angles towards the apex; lowermost pinnules with a constricted base; towards the apex, pinnules becoming more broadly attached to decurrent or auriculate. Frond apex forked to form a single, basally confuent pair of terminal pinnules. Venation alethopteroid; midrib strong, reaching close to tip; secondary veins arising at acute angles, simple or forking up to four times. Wing-like decurrent basiscopic pinnule portions with veins arising directly from rachis. Fronds amphistomatic; stomata haplocheilic, cyclocytic.

Discussion: The recent emendation of the genus lacked important information on cuticle features and included the following problematic terms that we have omitted here. The frond architecture was described as 'imparipinnate' whereas all sufficiently complete specimens instead show a forking apex with a terminal pinnule pair (Halle 1927). In addition, the terms 'sphenopteroid' and 'pecopteroid' were used to describe a constricted versus broadly attached pinnule base; these terms are, however, conventionally also used to describe venation and margin features, and are thus potentially misleading.

Compsopteris imparis (Gu and Zhi 1974) (Fig. 9.8-13)

1974 *Compsopteris imparis* Gu and Zhi, Paleozoic plants of China, p. 115, Pl.82, Figs. 7–8

Holotype: *Compsopteris imparis* Gu and Zhi (1974), deposited in the Nanjing Institute of Geology and Palaeontology, Chinese Academy of Sciences,

Diagnosis (emended): Fronds pinnate, petiolate, with prominent rachis. Pinnules oblong, straight or slightly curving, with entire margins and rounded apex; pinnules small and attached about perpendicular to rachis in the basal frond portion, becoming longer and arising at increasingly acute angles towards the apex; lowermost pinnules with a constricted base; towards the apex, pinnules becoming more broadly attached to decurrent or auriculate. Frond apex forked to form a single, basally confuent pair of terminal pinnules. Venation alethopteroid; midrib strong, reaching close to tip; secondary veins arising at acute angles, simple or forking up to four times. Wing-like decurrent basiscopic pinnule portions with veins arising directly from rachis. Fronds amphistomatic; stomata haplocheilic, cyclocytic.

Material examined: GWC-3-7a, b, GWC-3-15.

Occurrence: Chahe section, Weining County, Yunnan; Late Permian Xuanwei Formation.

Description: The rachis is about 4 mm broad. The pinnule is oblong-ovate or lanceolate, alternate, attaining a breadth of 2.0 cm and a length of more than 3.5 cm, with oblique, contracted base, the apex not preserved. The margin is entire or slightly sinuate. The midvein is strong, about 1.5 mm. broad. The lateral veins arise at angles of about 30–40° from the midvein, arching upwards and dividing once near the midrib; each branch then bends outwards and bifurcating once, bending slightly forward near the margin and reaching the margin at wide angles. The veins at the margin number 30–35 per centimeter.

Discussion and Comparison: The main characters of this species are its petiole and obliquely cordiform base, thin and dense veins, which different from *C. contracta* in having a rare vein, the veins at the margin number 30–35 per centimeter.

Compsopteris punctinervis (Figs. 9.8-14, 9.8-15)

1980 *Compsopteris punctinervis* Mo, p. 82, pl. XVIII, Figs. 1–5, 1a

Holotype: *Compsopteris punctinervis* Mo, deposited in the Nanjing Institute of Geology and Palaeontology, Chinese Academy of Sciences,

Diagnosis (emended): Fronds pinnate, petiolate, with prominent rachis. Pinnules oblong, straight or slightly curving, with entire margins and rounded apex; pinnules small and attached about perpendicular to rachis in the basal frond portion, becoming longer and arising at increasingly acute angles towards the apex; lowermost pinnules with a constricted base; towards the apex, pinnules becoming more broadly attached to decurrent or auriculate. Frond apex forked to form a single, basally confuent pair of terminal pinnules. Venation alethopteroid; midrib strong, reaching close to tip;

secondary veins arising at acute angles, simple or forking up to four times. Wing-like decurrent basiscopic pinnule portions with veins arising directly from rachis. Fronds amphistomatic; stomata haplocheilic, cyclocytic.

Material examined: GWC-28-1, GWC-69-18a, b.

Occurrence: Chahe section, Weining County, Guizhou; Late Permian and Early Triassic Xuanwei Formation.

Description: The pinnules are linear or lanceolate, the length and the breadth are unknown, with an entire margin. The apex not preserved, the base possibly contracted. The midvein disappears before reaching the apex (Pl.21, Fig. 3). The lateral veins are thin and dense, giving off at a 35-degree angle from the midvein as the fascicular-type, dividing 2–3 times, each branch is parallel and extends to the margin. The veins at the margin number 34–38 per centimeter.

Discussion and Comparison: The present specimens are distinguished by varying in shape and size, fascicular and dense veins which differ from *C. wongii* having rare vein, 15–20 per centimeter. Compared with the holotype species, the punctatus is not found on the surface of the present specimens.

Compsopteris wongii (Halle) (Zal 1974) (Figs. 9.8-16, 9.8-17)

1927 *Protoblechnum wongii* Halle, Halle, pp. 135–138, Pls.35, 36, pl.64, Fig. 12

1974 *Compsopteris wongii* (Halle) Zal., Paleozoic plants of China, p. 115, Pl.82, Figs. 1–3; Pl.130, Fig. 6

Holotype: *Compsopteris wongii* (Halle) Zal, 1974.

Diagnosis (emended): Fronds large (exceeding 40 cm in length), ovate in outline, petiolate, pinnate. Frond apex forking into two equally sized pinnules. Rachis prominent, channeled on upper side, keeled on lower. Pinnules closely spaced, arising at angles of 50–60° from the rachis in median portion of frond, in upper portion more oblique, in lower portion more extended, at the base commonly perpendicular. Pinnules broadly attached, basally decurrent toalmost confuent with or slightly overlapping the subjacent pinnule margin. Decurrent wings wide, in the upper frond portion fat, in lower portion curving towards upper side of rachis in an auriculate manner, becoming gradually more indistinct towards frond base, lacking in lowermost pinnules. Pinnules linear, entire-margined, gradually tapering towards obtuse or subacute apex. Venation alethopteroid; midrib strong, reaching close to tip; secondary veins arising at narrow angle, arching, reaching margin at angles of 50–60° in median pinnule portion, usually forking once or twice, rarely simple. Veins in auricles directly arising from rachis, almost perpendicularly, bifurcating once. Fronds amphistomatic. Rachis, midrib, and pinnule lamina stomatiferous. Epidermal cells of rachis, midrib, and costal felds polygonal-elongate, epidermal cells of intercostal felds polygonal-isodiametric. Haplocheilic stomata longitudinally oriented on rachis and midrib, randomly oriented and distributed on lamina; monocyclic to incompletely tricyclic. Large multicellular trichomes randomly distributed on rachis.

Material examined: GWZ-38-13; YXM(A)-21-3; GPT-10-6.

Occurrence: Chahe section, Weining County, Guizhou; Tucheng section, Panxian County, Guizhou; Mide section, Xuanwei City, Yunnan; Late Permian Xuanwei Formation.

Description: The specimen is very fragment, but its veins are clear. The midrib is strong. The lateral veins departing at an angle of 50°, arching, dichotomizing once or twice. Numbers of veins near the margin are 18–20 per centimeter at the margin.

Discussion and Comparison: *Compsopteris wongii* belongs to a group of foliage genera from the Late Palaeozoic, such as *Glenopteris, Nanshanopteris, Supaia*, and *Megalopteris* Dawson 1871. *Glenopteris* from the Lower Permian of Euramerica has often been proposed as a close relative of *Compsopteris wongii* (e.g., Halle 1927; White 1929; Krings et al. 2005). The type species *Glenopteris splendens* Sellards 1900 shares many similarities in macromorpholgy and especially epidermal anatomy with C. wongii; the main differences concern the architecture of the frond tip and the simple, indistinct venation in the former. Furthermore, cuticles of *C. wongii* do not show the peg-like projections in the anticlinal walls that are a typical feature of *G. splendens* (Krings et al. 2005), although anticlinal walls are thickened. Differences dfrom the second glenopteroid species *Nanshanopteris nervosa* Wan and Wang (2015) from China are mainly the same as from *G. splendens*, with the addition that epidermal cells of the former bear papillae (Wan and Wang 2015). However, papillae are not necessarily of high taxonomic relevance; they may instead refect paleoecological adaptation (Poort and Kerp 1990). Both of these glenopteroids occured in seasonal Permian environments: *G. splendens* from the Artinskian of Euramerica in a coastal environment periodically affected by seasonality and soil and ground water salinity (Krings et al. 2005), and *N. nervosa* from the Changhsingian of China on a well-drained lacustrine shore under a seasonally wet climate (Wan and Wang 2015). *C. wongii* from the Guadalupian of China clearly lived in a tropical, everwet environment close to a meandering river system (Liu and Yao 2007). It is therefore not surprising that epidermal features such as papillae, peg-like projections, or other supposedly xerophytic adaptations are absent.

Another interesting genus is *Supaia*, initially described from the Permian of Euramerica (White 1929). It differs from *Compsopteris* in its bipartite, imparipinnate frond architecture. Wang (1997) recorded *Supaia* fronds from the Capitanian Tianlongsi Formation (another name for the Upper Shihhotse Formation) of Shanxi and proposed it might represent a descendent of *Compsopteris wongii* that is morphologically adapted to a semi-arid environment. Pinnules of *S. contracta* Wang (1997) are amphistomatic, with monocyclic to dicyclic and slightly sunken stomata somewhat comparable to those of *C. wongii*. These similarities in the epidermal anatomy to *C. wongii* suggest a close relationship, with the main differences being the architecture, the contracted leaf bases, and the papillae on the cells of lower leaf sides of *S. contractum.*

Megalopteris from the Pennsylvanian of Euramerica is superfcially similar, but most species were described based on small fragments, yielding limited information on frond architecture and morphology that is insufcient for detailed comparison. Moreover, those three species for which cuticles have been described difer from *Compsopteris wongii* in that they are indicative of hypostomatic leaves (Florin 1933).

Genus ***Protoblechum*** (Lesquereux 1877)

Type species: *Protoblechum holdeni* (Lesquereux 1877).

Diagnosis (emended): Odd-numbered pinnate compound leaves, rachis ca. 3 mm wide, pinnules lanceolate, lateral entire, occasionally undulate, 4–5 cm long, ca. 1.5 cm wide, pinnules base oblique, apex acuminate. The width of mid vein is more than 1 mm, reaching to the tip of the pinnacle; Lateral veins fine, bifurcation 2–3 Times; The first bifurcation is close to the mid vein, then the second bifurcation bends outwards and divides again, and the first bifurcation extends a short distance toward the front of the pinnacle, then sharply bends outwards and divides again. At margin, about 50 veins per centimeter.

Description: *Protoblechum* was established by Lesquereux in 1897 based on fossil materials in Ohio, USA. Its model species is *P. holdeni* (Andrews) Lesquereux. This type species has been identified as a *Lethopteris holdeni* by Andrew S. The relationship of *Protoblechum* to *Glenopteris* Sellars, *Supaia* White and *Compsopteris* Zalcsssky is a question that paleobotanists often discuss and still have to be resolved. The root of the contradiction lies in how to treat the bifurcation character of feather shaft. The author agrees with the view of He et al. (1996) that: "Whether or not the feather shaft bifurcates has some significance in morphological taxonomy. The specimen of *Protoblechum wongii,* originally belonging to the genus *Pteropteris*, should be assigned to genus *Supaia,* while all others belong to the genus *Compsopteris.* Because none of the species in *Compsopteis* have been analyzed cuticles, it is not possible to compare with the plantain fern with cuticles characteristics in the former Soviet Union, temporarily placed in genus *Protoblechum.*" The feathered axes found in the Mojiang area have not been bifurcated, nor have they been analyzed by cuticle, so they are tentatively classified into the genus *Protoblechum.*

Protoblechnum contracta (Gu and Zhi 1974) (Figs. 9.8-11, 9.8-12)

1974 *Compsopteris contracta* Gu et Zhi, Paleozoic plants of China, p. 115, Pl.82, Figs. 4–6; Pl.83, Figs. 1–2, 3?

1999 *Protoblechnum contracta* Gu et Zhi, Chinese Science Bulletin, p. 382, Fig. 1

Holotype: *Protoblechnum contracta* Gu and Zhi (1974), deposited in the Nanjing Institute of Geology and Palaeontology, Chinese Academy of Sciences.

Diagnosis (emended): Fronds pinnate, petiolate, with prominent rachis. Pinnules oblong, straight or slightly curving, with entire margins and rounded apex; pinnules small and attached about perpendicular to rachis in the basal frond portion, becoming longer and arising at increasingly acute angles towards the apex; lowermost pinnules with a constricted base; towards the apex, pinnules becoming more broadly attached to decurrent or auriculate. Frond apex forked to form a single, basally confuent pair of terminal pinnules. Venation Alethopteroid; midrib strong, reaching close to tip; secondary veins arising at acute angles, simple or forking up to four times. Wing-like decurrent basiscopic pinnule portions with veins arising directly from rachis. Fronds amphistomatic; stomata haplocheilic, cyclocytic.

Material examined: GWC-3-20, GWC-3-21, GWC-58-4, GWC-58-6; GWZ-38-4 and so on.

Occurrence: Chahe and Zhejue sections, Weining County, Guizhou; Late Permian Xuanwei Formation and Early Triassic Kayitou Formation.

Description: The pinnule, with a petiole and obliquely cordiform base, is linear to lanceolate. The margin is entire or slightly sinuate. The midvein is distinct, about 1 mm broad, longitudinally striated. The lateral veins are dense, bifurcating twice or thrice. The first branching is near the midrib; the posterior branch bends downwards, dividing once. The anterior branch bends upwards for a short distance, dividing once or twice. All branches parallelly extend to the margin. The veins at the margin number 40–56 per centimeter.

Discussion and Comparison: The main characters of this species are its petiole and obliquely cordiform base, thin and dense veins, which different from *C. impairs* in having a rare vein, only 30–35 per centimeter.

Genus ***Linopteris*** Presl. 1838

Type species: *Linopteris gilkersonensis* Presl. 1838.

Diagnosis (emended): Frond pinnate or pinnatifid; pinnule oblong, straight, sessile; veins reticulate; areolm hexagonal, elongate in the centre, and somewhat parallel to the margins of the pinnule, afterwards oblique and then free near the margin; no midrib.

Discussion: This remarkable genus was established by Gutbier for those fossil ferns possessing the general habit of *Neuropteris*, but differing from it in having a somewhat radiate reticulate venation and no distinct midrib, in which latter character it differs from both *Lonchopteris* and *Woodwardites*. Four species are described,—the type, *Dictyopteris* Brongniartii, Gutbier (Verst. der Zwickauer Schwartz-Kohle, p. 63, pl. 2. f. 7, 9, 10), from the coal-schists of Saxony; *D. muensteri* is very frequent in the coal-measures of Southern Russia; *D. obliqua* Bunbury (Quart. Journ. Geol. Soe. iii. p. 427), from the coal-measures of Nova Scotia; and M. Brongniart mentions a fourth species from the eastern part of Egypt. Presl subsequently established a recent genus, *Dictyopteris*, for certain species of *Polypodium* having a reticulate venation and no free veins, as *P. attenuatum* Brown, and changed the name of the fossil genus to *Linopteris*; but the right of priority appears to belong to Gutbier. Mr. J. Smith, whose valuable researches and works on the recent ferns are well known, has substituted the name *Dictymia* for the living ferns assigned to *Dictyopteris* by Presl.

***Linopteris* sp.** Gutb (Fig. 9.6–14)

Diagnosis (emended): Frond pinnate or pinnatifid; pinnulm oblong, straight, sessile; veins reticulate; areolm hexagonal, elongate in the centre, and somewhat parallel to the margins of the pinnule, afterwards oblique and then free near the margin; no midrib.

Material examined: GWC-4-3a, b; GWC-58-29.

Occurrence: Chahe Section, Weining County, Guizhou; Late Permian Xuanwei Formation.

Description: The pinnule is oblong, 2.8 cm. Long and 1.3 cm broad, with oblique, contracted base and obtuse apex. The margin is entire or slightly sinuate. The midrib is no more than the half length of the pinnule. The lateral veins are fine and dense, repeatedly bifurcating, connected the network. The mesh is short, polygonal.

Discussion and Comparison: It is difficult to classify into known species due to incomplete preservation.

Order **Gigantopteridales**

Genus ***Gigantonoclea*** (Koidzumi 1936)

Type species: *Gigantonoclea lagrelii* (Halle) Koidzumi (1936).

Diagnosis (Wang 1999 emended): Frond pinnate compound, moderate to large in size. Pinnae mostly opposite or rarely subopposite-alternate sometimes coherent in terminal ones. Margins crenate or serrate, rarely undulate or even entire; upper margin decurrent at base; lower margin more or less constricted. Cordate bases andrarely modified pinnae triangular with entire base in attachment. Foliar vasculature of three orders of main vein: rachis of frond, misveins of pinnae, and secondary vein. All main veins delineate intercostal fields with two orders of anastomosing vein: tertiary veins and veinlets forming areolae. One or two dark-dots in each areola.

Discussion: Although the above diagnosis mainly follows the original by Koidzumi (1936), it has been emended to make it agree with current concepts of leaf architecture, and with the designation of venation based on the original description of *G. lagrelii* by Halle (1927). Halle's specimens distinctively have: pinnately compound fronds with more or less coherent terminal pinnae; pinnae mostly opposite with crenate or undulate margins; foliar vasculature consisting three orders of main veins (rachis, midveins and secondary veins) and two orders of anastomosing veins: tertiary veins and veinlets; and dark dots within each areola or mesh.

In order to avoid confusion, it is essential to define precisely the terms used to describe the various orders of veins. Following Foster and Hickey, main veins delimit the intercostal field(s) of the lamina and extend outward up to its margins. There are three orders of main vein: a rachis, reaching to the top of a frond; a midvein extending to the pinna apex; and a secondary vein extending to the pinna margins. Anastomosing veins form a network within an intercostal field, consisting of the tertiary veins,

Fig. 9.9 1, 1a *Gigantonoclea guizhouensis* Gu et Zhi Locality: Mide section, Xuanwei City, Yunnan Province Sample Number: YXM-2-4, 2, 3 *Gigantonoclea lagrelii* (Halle) Koidz. Locality: Chinahe section, Xuanwei City, Yunnan Province Sample Number: YXC-(-4)-1, CNH-0-15, 4, 4a *Gigantonoclea plumosa* Mo Locality: Tucheng section, Panxian County, Guizhou Province Sample Number: GPT-5-2a, 5 *Gigantonoclea rosulata* Gu et Zhi Locality: Chinahe section, Xuanwei City, Yunnan Province Sample Number: YXC-P3*x*-2, 6, 6a *Gigantonoclea hallei* (Asama) Gu et Zhi Locality: Chahe section, Weining County, Guizhou Province Sample Number: GWC-29-1 7, 8, 9 *Gigantopteris nicotianaefolia* Schenk Locality: 7 Mide Section, Xuanwei City, Yunnan Province Sample Number: YXM (B)-12-1a, 8, 9 Chinahe section, Xuanwei City, Yunnan Province Sample Number: YXC-(-1)-84, YXC-(-1)-199, 10-13 *Gigantopteris dictyophylloides* Gu et Zhi Locality: 10, 11 Tucheng section, Panxian County, Guizhou Province Sample Number: GPT-5-11, GPT-5-12, 12, 12a Chahe section, Weining County, Guizhou Province Sample Number: GWC-69-13, 13 Chinahe section, Xuanwei City, Yunnan Province Sample Number: YXC-(-24)-2 ▶

and veinlets which include branches of the teriary vein, and veins developing into accessory meshes along both sides of the main vein. At its point of attachment, the rachis of a frond is much thicker than the midvein, which in turn is thicker than the secondary vein of a pinna, and which in turn must be rather thicker than a proceeding tertiary vein at its starting point. Finally, the tertiary veins must be thicker than the veinlets. All veinlets are of the same venation order, no matter what the number or grade of branching. Some veinlets may vary slightly in thickness but the variation is local and not strong enough to justify a stable and distinctive order.

The evolutionary series in leaf coherence proposed by Asama (1959) was reputedly reflected in changes in the venation, in particular in the increase in branching grade of the anastomosing vein. For example, a simple grade of vein branching occurs within a pinnatified lobe of a *Gigantonoclea lagrelii* pinna, a double grade of vein branching within a *G. hallei* pinna with coalesced pinnules, and a triple grade of vein branching within a 'leaf' of *G. taiyuanensis* consisting of coalesced pinnae. However, according to Hickey's (1979) rule, all these branching grades belong to the same venation order, consisting of a thicker tertiary vein and many thinner veinlets. A similar change occurs in the plumose-reticulate venation of the extant dicotyledon Quiina acutangula Ducke, where increasing grades of anastomosing veins within an intercostal field is related only to leaf ontogeny and not to phylogeny. Also, Trivett and Pigg (1996) indicated that the reticulate venation is not an adequate indicator of phylogeny.

Gigantonoclea lagrelii ((Halle) (Koidzumi 1989) (Figs. 9.9-2, 9.9-3)

1927 *Gigantopteris lagrelii*, Halle, pp. 170–172, Pl.46, Figs. 2–11, 12?

1934 *Gigantopteris lagrelii*, Kawasaki, Pl.LXVII, Figs. 189

1935 *Gigantopteris lagrelii*, Sze, p.159, Pl.1, Fig. 6

1936 *Gigantonoclea lagrelii* (Halle) Koidzumi, p. 138

1959 *Gigantonoclea lagrelii*, Asama, p. 57, Pl.5, Figs. 7–11

1
1a
2
3
4
4a
5
6
6a
7
8
9
10
11
12
12a
13

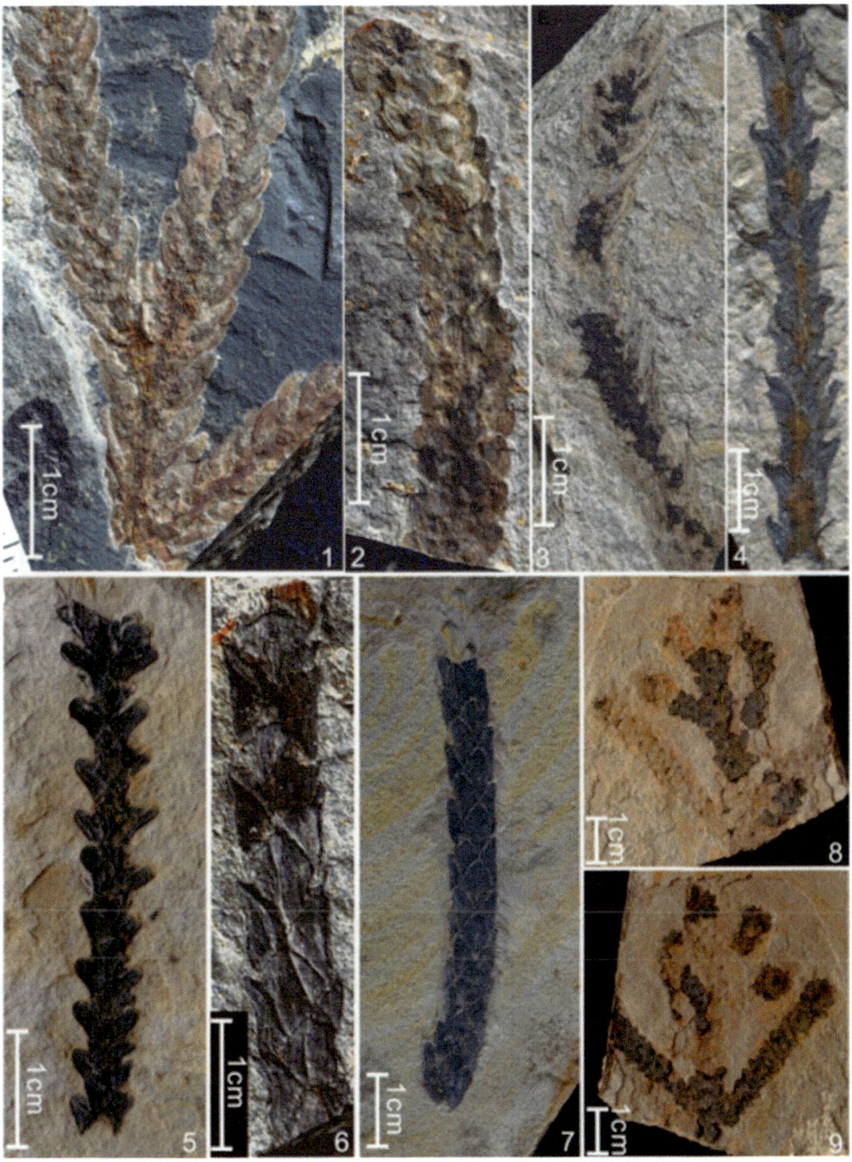

Fig. 9.10 1, 2 *Szecladia multinervia* Yao, Liu, Rothwell and Mapes Locality: Xinmin section, Anshun City, Guizhou Province Sample Number: GAXM11013, GAXM11016, 3 *Pseudoullmannia frumentarioides* He, Liang and Shen Locality: Duanshan-A section, Huishui County, Guizhou Province Sample Number: GHDSA3-1002, 4 *Anshuncladus* sp.3 Locality: Xinmin section, Anshun City, Guizhou Province Sample Number: GAXM11008, 5 *Anshuncladus* sp.3 Locality: Duanshan-A section, Huishui County, Guizhou Province Sample Number: GHDSA3-33, 6 *Anshuncladus* sp.2 Locality: Xinmin section, Anshun City, Guizhou Province Sample Number: GAXM11007, 7 *Anshuncladus* sp.2 Locality: Duanshan-B section, Huishui County, Guizhou Province Sample Number: GHDSB2-26, 8, 9 *Anshuncladus* sp.1 Locality: Duanshan-A section, Huishui County, Guizhou Province Sample Number: GHDSA3-458, GHDSA

1974 *Gigantonoclea lagrelii* (Halle) Koidz., Paleozoic plants of China, p. 126, Pl.95, Figs. 5–7

1980 Gigantonoclea lagrelii (Halle) Koidzumi; Zhang, pl. 1, Figs. 1–2

1989 Gigantonoclea lagrelii (Halle) Koidzumi; Si, p. 61, pl. 71, Figs. 3–5

Lectotype (Wang, 1999): Halle's plate 46, Fig. 6 from the 18th bed at East Hill, Taiyuan. Halle did not designate a type specimen. Subsequently, Si (1989, p. 203) proposed two of the original syntypes (Halle 1927, pl. 46, Figs. 6–7) as 'comparable types'.

Diagnosis (emended): At least one pinnate compound leaf; Ultimate rachis about 2 mm in width; Pinnae oblong, nearly 6 cm in length, 2.5 cm in width, crenated or sinuate margin, occasionally entire, tapered near apex, usually a little constricted at the base with a decurrent lower side but in rare cases are strongly constricted. More or less vaulted laminae on the adaxial surface, rolled downwards along the dentate-crenate margins. Midvein straight, prominent on the adaxial surface, grooved on adaxial surface, and extends from the rachis at 60–70°, maintains a thickness of about 1 mm for more or less equal intervals and disappears near the marginal teeth. Fine tertiary veins, rather prominent at the base and anasomosing. Veinlets form elongate areolae in the longitudinal direction, parallel to the tertiary areolae, black dots situate in the middle of the areolae. With accessory mesh at both side of the midvein and primary lateral vein; occasionally with blind vein, Laminae surfaces bear wrinkled relief.

Material examined: GWC-3-8a, b

Occurrence: Chahe section, Weining County, Guizhou; Late Permian Xuanwei Formation.

Description: Ultimate rachis about 2 mm broad. Frond oblong, the length is more than 2 cm, the broadest is 0.8 cm. Base of frond abruptly constracted, with deflected cordate outline, apex unknown. Margin of frond is entire or slightly sinuate. Midrib distinct but straight. First veins arise at an angle of 80° from the midrib, breaking off before reaching the margin; veinlets freely branching and anastomosing to form a network of elonglated, deltoid or more rarely polygonal meshes. On the specimens there are some black dots (possibly glandular traces or resin bodies) irregularly scattered on the surface of the pinnae. Accessory mesh is unclear and blind vein visible.

Discussion and Comparison: The present specimen is similar to *G. tenuinervis* Yang in having wavy pinna margin, and elongate ovate pinna shape, but they differ in that the latter had thin lamina, acute apex, thin veins. The meshes of the present specimen are similar to *G. cathaysiana* Yang, but it is differs from the latter in having entire margin. The main characters of *G. lagrelii* are that the pinnules have the first lateral veins, wavy margins. The present specimen is similar to this species.

Gigantonoclea guizhouensis (Gu and Zhi 1974) (Figs. 9.9-1, 1a)

1974 *Gigantonoclea guizhouensis* Gu and Zhi, Pl.96, Figs. 7–10

1980 *Gigantonoclea guizhouensis*, Tian, Tian et al., p. 27, text-Fig. 20

2006 *Pinnagigantonoclea guizhouensis* (Gu et Zhi) Yang, p.309, Pl.61, Figs. 4, 4a

Holotype: *Gigantonoclea guizhouensis*, Gu and Zhi (1974). Changhsingian (Upper Permian); Panxian county, Guizhou Province, southwestern China.

Diagnosis: At least one pinnate compound leaf. Pinnae oblong, nearly 10 cm in length, 4 cm in width; shallow serrate at the edge, sometimes entire margin. Midwein 1 mm in width; with two lateral veins, veinlets from elongate-polygonal areolae. No blind veins, vague accessory mesh, suture vein loom, situate downwards.

Material examined: GWC-18, GWC-21-6, GWC-29-3, GWC-GWZ-36-3; GWZ-42, GWZ-4; YXM (A)-2-4, YXM(A)-3, YXM (A)-4; YXM(B)-0-2; GPT-4-4, GPT-5-1, GPT-10-4, GPT-14-1; GPT-15-1, etc.

Occurrence: All of four sections; Late Permian Xuanwei Formation.

Description: The specimens are fragmentary, but to good advantage show the fine venations. The pinnate compound leaf, thin; the margin sinuate toothed. Midvein is strong, probably reaching to the apex, about 1 mm wide. First veins distinct, forming an angle of generally 50° with the midvein, straight or curving slightly forwards, running out into the margin. Secondary veins forming an angle of about 80° with the first veins, giving off fine lateral veins. The first veins are stronger than the secondary ones. The fine lateral veins (veinlets) originating at acute angles and dichotomizing to form elongate-polynogal meshes. No blind veins giving off from the fine veins. Suture veins are unclear.

Discussion and Comparison: The present specimens are the same as the holotype of *G. guizhouensis* named by Gu and Zhi in shape and veins. This species is similar to *G. hallei* in leaflet shape, but the latter have shorter polynogal meshes. Yang et al. assigned the species to genus *Pinnagigantonoclea* based on pinnate compound leaf. This species is abundant in upper Late Permian, and is widely distributed in western Guizhou and eastern Yunnan. It represents one of the dominant species during the late Cathaysian flora in South China.

Gigantonoclea hallei ((Asama) Gu and Zhi 1974) (Figs. 9.9-6, 6a)

1927 *Gigantopteris nicitianaefolia*, Halle, p. 162, Pls.43–44, 45

1959 *Bicoemplectopteris hallei* Asama, p. 57, pl. 2, Figs. 6–7 [refigured from Halle 1927, pl. 45, Figs. 3, 5]; pl.5, Figs. 1–6 [*non* pls 6–7]

1959 *Tricoemplectopteris taiyuanensis* Asama, p. 59, pl. 3, Fig. 4 [refigured from Halle 1927, pls 43–44, Fig. 9]

1974 *Gigantonoclea hallei* (Asama) 'Gu and Zhi', p. 127, pl. 99, Figs. 1–3 [refigured from Halle 1927, pls. 43–44, Fig. 1; pl. 45 Figs. 2, 5]; text-Figs. 103–3, 105

1974 *Gigantonoclea hallei* (Asama) Gu et Zhi, Paleozoic plants of China, p.127, Pl.99, Figs. 1–3

Lectotype: Halle' specimen (1927, pls 43–44, Fig. 1) from the 21st plant-bearing Bed of the 'Upper Shihhotze Series' at the East section of Taiyuan.

Diagnosis: Large in size, probably dichasial bifurcate in the upper part, odd pinnate compound leaf. Pinnae oblong, up to 14 cm in length, 5 cm in width, sharp apex, constricted base; deep serrated margin; several un-divided pinnae coalesce into apical pinnae, wide rhombus. Midvein nearly 1 mm in length, reach to the terminal of pinnule, with two lateral veins. Primary lateral vein reach margin tooth; secondary lateral vein form polygonal areolae. Accessory meshes long. Occasionally blind vein.

Material examined: GWC-29-1, GWC-29-3; GWZ-4-7; GPT-5-6.

Occurrence: Chahe and Zhejue sections, Weining County, Guizhou; Tucheng section, Panxian County, Guizhou; Late Permian Xuanwei Formation.

Description: The specimen shown by the following figure is part of frond. Pinna margin with sinuate-toothed. The midvein is about 1 mm broad, straight. The first lateral veins distinct, forming an angle of 70° with the midvein, straight or curving slightly forwards. Secondary veins are slender more than the first one, forming an angle of 80° with the first veins, giving off the veinlets. The veinlets freely branching, and anastomosing to form a network of irregularly polygonal mesh. The base veinlets from the adjacent secondary veins join each other to form a series of narrow and elongated rectangular accessory meshes extending along the first veins. No sutural and blind veins are seen.

Discussion and Comparison: This species is characterized by wide-form pinnae with large dentate-serrate margins, and a distinctive venation consisting of three orders of main veins (rachis, midvein and secondary veins) and two orders of anastomosing veins (tertiary veins and veinlets) which show two or three gradate branchings. Areolae are polygonal or a little elongate, and accessory areolae are developed along both the sides of the main veins. *G. lagrelii* has the same orders of venation, but differs markedly from *G. hallei* in having narrow-form pinnae, an anastomosing vein with only one grade of branching, and one or two clear black dots within each areola, but the dot is not visible or rather faint in the later areolae. *Gigantonoclea taiyuanensis* (Asama) 'Gu and Zhi' has been published based only on one specimen (Halle 1927, pls 43–44, Fig. 9). However, a careful examination of it shows a complete transition of venation between this specimen and the more typical *G. hallei*. For example, the specimen in Halle's plate 45, Figs. 3–5 shows the tertiary vein having the trace of 'the quaternary veinlet' which is that seen in the type of *G. taiyuanensis*. The best preserved specimens of Halle's (1927) *Gigantonoclea hallei* (e.g. pls 43–44, Fig. 1) could be inferred to represent the upper part of a rather large frond, whereas other fragmentary specimens could represent pinna fragments also from large fronds. What is certain is that the species must have had large enough fronds in order to yield such huge lamina segments as the '*G. taiyuanensis*' specimen.

Gigantonoclea plumosa (Figs. 9.9-4, 4a)

1980 *Gigantonoclea plumosa* Mo, p. 83, Pl.XVI, Figs. 1–3, 1a, 2a)

Holotype: *Gigantonoclea plumosa* Mo, 1980, pl. XVI, Figs. 1–3, 1a, 2a, deposited in the Nanjing Institute of Geology and Palaeontology, Chinese Academy of Sciences, from Xuanwei Formation (Changhsingian), western Guizhou and eastern Yunnan area, China.

Diagnosis: Pinnae linear, 4 cm in width, margin straight or slightly sinuate. Midwein nearly 1.5 mm in width. Primary lateral vein thick, extend from the midvein at about 70° then go straight, slightly curve upwards near the margin, and gradually disperse till disappear. Secondary vein thin, extend from the primary vein at 30°, straight or slightly curve, fork 1–3 times, forking vein parallel to each other and connect to narrow lanceolate mesh during extending, forking vein of secondary vein extend slightly downwards to the middle of adjacent primary vein, connect with each other to loom suture line.

Material examined: YXM (A)-21-2; GPT-5-2, GPT-5-4, etc.

Occurrence: Mide section, Xuanwei City, Yunnan; Tucheng section, Panxian County, Guizhou; Late Permian Xuanwei Formation.

Description: The lateral margin is slightly sinuate. The midvein is about 1 mm wide. The primary lateral veins are strong and thick, depart at a 70-degree angle from the midvein and straightly extend nearly to the margin, then slightly curve upward and gradually disperse. The secondary lateral veins are fine, depart at an 80-degree angle from the primary lateral veins, straight or slightly curved, fork 3–4 times; the forking veins almost are parallel to extend ahead, each other connect to form narrow but long, lanceolate-shaped meshes. Forking veins of the secondary lateral veins extend to the middle of two adjacent primary lateral veins, each other connects to form the obscure seam.

Discussion and Comparison: This species is different from *G. guizhouensis* in that the latter's mesh being short. This species is different from *Cathaysiopteris whitei* by the secondary lateral veins of the latter forking 1–2 times, not forming the network, having a prominent seam veins.

Genus ***Gigantopteris*** Schenk ex Potonié, 1902 emend

Type species: *Gigantopteris nicotianaefolia* Schenk ex Potonié, 1902, emend.

Diagnosis: Megaphylls with pinnate venation. Lamina displays at least four orders of venation. Penultimate order veins form large reticulate polygonal meshes within which finer ultimate order veinlets anastomose to form meshes and blind endings. Areoles imperfect to incompletely closed, randomly oriented, of irregular shape.

Discussion: The generic diagnosis is emended to highlight the complex nature of the higher order venation, particularly the occurrence of meshes within meshes, otherwise called *Zuzammengesetzte Maschenaderung* (Li and Yao 1983), that act to differentiate this genus from other genera of gigantopterid, particularly *Gigantonoclea*. The diagnosis is also emended such that it is not based on the unproven status of the leaf/leaflet as compound or simple.

Gigantopteris dictyophylloides (Gu and Zhi 1974) (Figs. 9.9-10, -11, 9.9-12 and 9.9-13)

1927 *Gigantopteris dictyophylloides* Gu and Zhi, Paleozoic plants of China, pp. 130–131, Pl.102, Figs. 1–6

1974 *Gigantopteris dictyophylloides* Gu and Zhi, pp. 130–131, pl. 102, Figs. 1–6; text-Figs. 11. 2, 103.4

1989 Gigantopteris dictyophylloides Guo et al., p. 166

1993 *Gigantopteris dictyophylloides* Guo et al., pp. 94–95, pl. 2, Figs. 1–15

2004 *Gigantopteris dictyophylloides* Guo et al.; Glasspool et al., pp. 1356–1357, Pl. 2, Fig. 5

Holotype: *Gigantopteris dictyophylloides* Gu and Zhi (1974), pl. 102, Figs. 1–2, deposited in the Nanjing Institute of Geology and Palaeontology, Chinese Academy of Sciences, from Xuanwei Formation (Changhsingian), western Guizhou and eastern Yunnan area, China.

Diagnosis: Widely crenate megaphylls with pinnate venation. Tertiary veins anastomose, branches may be of higher or same order. Anastomosed branches form a reticulum of irregular polygonal meshes. Within these meshes veinlets anastomose into small meshes forming well-developed orientated, typically polygonal areoles, with frequent blind endings. No accessory meshes occur along the higher order veins.

Material examined: GPT-5-7, GWC-58-23, GWC-32-3; GPT-4-5, GPT-5–3, GPT-10-5; YXM (B)-12-7, GWC-69-13, etc.

Occurrence: Late Permian and Early Triassic Xuanwei Formation.

Description: Material referable to this species consists of many fragments which show the distinct venation and the shape of mesh. The midvein is strong and flat, reaching to the apex. The primary veins are distinct, giving off at an angle of 60–65° from the midvein, disappearing without reaching to the margin. Secondary veins are narrower more than the primary veins, giving out at about a 50° angle from the primary veins. The tertiary veins give out at a wide-angle and irregularly branching and anastomosing with each other, forming big and polygonal meshes. The fine veins in the big meshes connect each other forming regular but small polygonal meshes. There constantly are the blind veins in the small meshes.

Discussion and Comparison: The main characters of the present specimens are the primary veins shorten, disappearing before extending to the margin, and the shapes of small mesh regular, having blind veins in small meshes. The diagnosis has been emended to provide a standardised terminology (sensu Hickey 1979) and to emphasize less strongly the order of the finer veins. These veinlets are ultimately reduced in width. However, just above the point of anastomosis, they are frequently of the same width as below it, and as such may be considered of comparable order. The species is distinguished from *G. nicotianaefolia* in having a crenate margin, in the regular distribution of the veinlets forming the areoles, in the frequent occurrence of blind endings and, in the absence of even poorly defined accessory meshes.

Gigantopteris* (*Megalopteris*) *nicotianaefolia (Figs. 9.9-7, 9.9-9)

1883 *Megalopteris nicotianaefolia* Schenk, pp. 238–239, pl. 32, Fig. 6

1902 *Gigantopteris nicotianaefolia* Schenk ex Potonie´, in Engler and Prantl, vol. 1, part 4, p. 513

1930 *Gigantopteris nicotianaefolia* Schenk; Chu, p. 6, pl. 4, Fig. 1

1933 *Gigantopteris nicotianaefolia* Schenk; Gothan and Sze, pp. 23–26, pl. 4, Fig. 1a

1940 *Gigantopteris nicotianaefolia* Schenk; Shimakura, pp. 11–13, Pl.1, Figs. 1–2, 5, Pl. 7, Figs. 1–2

1974 *Gigantopteris nicotianaefolia* Schenk, emend. Gu and Zhi, p. 130, pl. 100, Figs. 2–4; pl. 101, Fig. 1; text-figs 103 (5), 108

1983 *Gigantopteris nicotianaefolia* Schenk; Yao, pp. 1–8, pl. 1, Figs. 1–6; pl. 2, Figs. 1–5; pl. 3, Figs. 1–5; text-Fig. 3b

2004 *Gigantopteris nicotianaefolia* Schenk; Glasspool et al., pp. 1357–1358, Pl. 2, Figs. 1–4

Basionym: *Megalopteris nicotianaefolia* Schenk, A., 1883. Steinkohlenpflanzen. In v. Richthofen, China IV, pp. 238–239, pl. 31, Fig. 3c; pl. 32, Figs. 6, 8; pl. 33, Figs. 1–2; pl. 35, Fig. 6.

Lectotype: *Gigantopteris nicotianaefolia* (Schenk 1883, pl. 32, Fig. 6; this paper, Pl. 1, Fig. 1; Pl. 2, Fig. 1; Pl. 3, Figs. 1–6).

Diagnosis: Obovate-elliptic-oblong, entire margined, sometimes slightly revolute megaphylls with pinnate eucamptodromous venation. Lamina apex roundly acute. Primary vein stout-massive. Tertiary veins anastomose, branches may be of higher or same order. Anastomosed branches form reticulum of irregular polygonal meshes. Within these meshes veinlets anastomose into small meshes forming imperfect to incompletely closed, randomly oriented areoles, with occasional blind endings. Poorly developed accessory meshes may occur along secondary veins.

Material examined: GWC-48-1, GWC-58-48a, b, GWC-69-32; YXM-12-1a, b; etc.

Occurrence: Chahe section, Weining County, Guizhou; Mide section, Xuanwei City, Yunnan; Late Permian Xuanwei Formation.

Description: The specimens shown in the following figures possibly represent the middle- upper parts of the frond. Lager simple leaf, oblongate or ovate, attaining a length of more than 12 cm and a breadth of more than 6 cm, with entire or slightly sinuate margin and obtuse apex. The midvein is wide but flat, reaching to the apex.

The primary veins are distinct, giving off at an angle of 70–80° from the midvein, reaching to the margin. Secondary veins are narrower more than the primary veins, giving out at about a 50 degree angle from the primary veins. The tertiary veins give out at a wide-angle and irregularly branching and anastomosing with each other, forming big rhombic or polygonal meshes. The fine veins in the big meshes connect each other forming irregularly small polygonal meshes.

Discussion and Comparison: The major characters of this species are its simple leaf, fine veins forming the irregularly small meshes and occasional blind veins, which can be used to distinguish it from *G. dictyophylloides*.

Order **Gymnospermae**

Family **Coniferopsida**

Genus ***Anshuncladus*** gen. nov.

Type species: *Anshuncladus xinminensis* sp. nov. Late Permian Changhsingian Talung Formation in South China.

Diagnosis: Typical of monopodial branching, spiral to opposite-decussate arrangement of leaf shoots and scale-like leaves. Leave oval, triangular or rhombic with varying outlines. Apices beaked, bluntly rounded or bluntly pointed. Leaves helically arrayed, slightly overlapped, strongly decurrent with broad base, detached in upper part, concave ventrally, conves dorsally, with distinct median keel. Leaf margin sometimes wrinkled. Leaf amphistomatous. Epidermal cell polygonal and isodiametric, with broad, straight anticlinal wall and smooth periclinal wall. Papillae absent, radial line sometimes appear on surface of epidermal cell, of which the corners are rounded. Stomata evenly distribute in discontinuous row. Stomatal complexes often circular, oval and occasionally filled with plugs. Stomatal arrangement cyclitic with two kidney-shaped guard cells, sunken and always badly preserved. Number of subsidiary cells commonly four or five, rarely three and six. Cuticles on subsidiary cell surface thicker than other parts, forming distinct ring.

Discussion: *Anshuncladus* is a morphotype established after the macro morphology and micro character of the leaf and shoot fossils. The scientific name originate in the name of fossil locality and the preserved shoots: Anshun (County) + Cladus (Shoot). It could be distinguished from the same layer preserved *Pseudoullmannia* for the scale-like leavies. The later has long lanceolate-shape leaves; it is differ from *Szecladia* according to the vein structure, *Anshuncladus* only has one midvein, while *Szecladia* has 7–8 veins. The subsidiary cell number is the main difference between *Anshuncladus* and *Ullmannia*, the later of which is wildly distributed in our country and elsewhere like Europe and America. According to the leaf outline, whether leaf is amphistomatous or not, the shape of epidermal cell, with papillae or not, the amount of subsidiary cell and stomata in row or not, *Anshuncladus* could be distinguished from other similar conifer plants of Permian and Triassic. The comparison between *Anshuncladus* and other coeval or similarly-aged fossil conifers is preliminarily displayed in Table 9.3. Recently only one shoot with female cone has been found. Together with the leaf outline and epidermal structure, *Anshuncladus*

Table 9.3 Comparison between *Anshuncladus* ar.d similar genus

	Genus	*Anshuncladus*	*Ullmannia*	*Pseudoullmannia*	*Szecladia*	*Pagiophyllum*	*Brachyphyllum*	*Pseudovoltzia*	*Voltzia*	*Ortiseia*	*Quadrocladus*	*Majonica*
	Reference	This paper	Taylor, 1982; Wang et al., 1986; Yang et al., 1994	He et al., 1996	Yao et al., 2000	Kendall, 1948; Adams, 1951; Liu and Yao, 2013	Kendall, 1947; Kendall, 1949; Raab et al., 1986; Ham et al., 2003	Arnold, 1947; Clement-Westerhof, 1984; Schweitzer, 1986	Zhou, 1979; Diez, 1996	Clement-Westerhof, 1984	Kerp, 1996; Meyen, 1997	Clement-Westerhof, 1987
Leaf morphology	Outline	oval, triangular, rhombic; apex beaked, bluntly rounded, bluntly pointed, inturned; spiral arrangement	oval, lanceolate; apex bluntly pointed, bandy; spiral arrangement	linear, long lanceolate; apex pointed; spiral arrangement	oval; apex obtuse, inturned;spiral or imbricate arrangement	wide triangular; strongly decurrent; apex acute, inturned; loosely spiral arrangement	wide triangular; strongly decurrent; apex short and bluntly pointed; margin folds inward; spiral arrangement	narrow triangular; lanceolate; apex acute or obtuse; spiral arrangement (fake lateral arrangement)	lanceolate, triangular, falcate in lateral view; apex pointed or bluntly pointed; loosely spiral arrangement	oval, triangular, lanceolate; apex acute or bluntly pointed; tapering slightly at the base; spiral arrangement	linear, shovel-shaped, sub-triangular; apex acute or bluntly pointed; spiral arrangement	narrow sub-triangular, oval; apex acute or bluntly pointed; decurrent; spiral arrangement
	Branching	irregularly	irregularly	irregularly	irregularly	irregularly,at an angle of 40°	pinnately or alternately	–	irregularly	pinnately	irregularly	–
	Vein	mid-ridge	mid-vein,fine longitudinal stripes	fine longitudinal stripes	7-8 veins	mid-ridge, fine longitudinal stripes	mid-ridge, serveral longitudinal stripes	–	mid-vein	–	–	–
	Trichome bases, papillae	none	none	–	–	–	one papillae on each subsidiary cell	papillae	–	papillae	papillae occasionally	papillae
	Epidermal cells	polygonal, isodiametric, straight anticlinal walls	isodiametric, retangle, straight anticlinal walls	retangle, straight anticlinal walls	–	retangle, wide and straight anticlinal walls	retangle and short; straight anticlinal walls	polygonal	–	isodiametric, straight anticlinal walls	–	small, isodiametric, thin anticlinal walls
Stomata	Stomata	amphistomatic	amphistomatic	–	–	amphistomatic	amphistomatic	–	–	–	amphistomatic	–
	Distribution	evenly, in discontinuous rows, not in bands	–	–	–	in bands, wide as one stoma	gernerally in rows	–	–	in rows	irregularly	irregularly
	Subsidiary cells	commonly 4-5; rarely 3 and 6	8-10	–	–	3-6	4-7	7-8, much smaller that epidermal cell	–	4-8	3-7, usually 4-5, very similar to epidermal cell	–
	Stomatal porus	occasionally filled with plugs	–	–	–	–	–	–	–	–	–	–
	Taxonomic position	Voltziaceae?	Voltziaceae	Lepidodendrales-Lepidodendron?	Podocarpaceae ?			Voltziaceae	Voltziaceae	Walchiaceae		Majonicaceae

could be included in coniferopsida. However, due to the lack of seed-scale complex, in-situ pollen grain, fossil wood anatomical structure and the vague of the cone, it is hard to classify *Anshuncladus* to any recent family.

Occurrence: Marine facies Xinmin, Duanshan and Kejiao sections, Talung Formation of Changhsingian in western Guizhou and eastern Yunnan area.

***Anshuncladus* sp.1 sp. nov.** (Figs. 9.10-8, 9.10-9)

2019 *Anshuncladus* sp.1 Li and Yu, p. 35, pl. 3, fig A-I; pl. 4, fig. F–H; pl. 6, fig A-E; pl. 7, pl. 8

Holotype: *Anshuncladus* sp.1 Li and Yu, 2019, GHDSA3-58, pl. 3, Fig. E, deposited in China University of Geosciences (Wuhan) State key laboratory of Biogeology and Environmental Geology, an impression fossil with cuticle preserved, from Dalong Formation (Changhsingian), Duanshan County, Guizhou Province, South China.

Paratype. *Anshuncladus* sp.1 Li and Yu, 2019, GHDSA3-61, pl. 3, Fig. E, deposited in China University of Geosciences (Wuhan) State key laboratory of Biogeology and Environmental Geology, an impression fossil with cuticle preserved, from Dalong Formation (Changhsingian), Duanshan County, Guizhou Province, South China.

Diagnosis: More than tertiary branch, monopodial branching from leaf axil at 30°–50°. Single preserved longest terminal shoot is more than 12 cm long, Every branch accordingly shorten. Shoot loosely helically arranged, subalternate. Leaf scale like, spirally arrayed and imbricate arranged; Leaf distal portion away from axis, medium Degree of Freedom (DOF); Leaf base decurrent, rhombic or oval, middle and lower portion widest and taper till the apex; with a distinct mid-ridge. Leave apex blunt. Leaf margin curves occasionally.

Occurrence: Upper Permian, western Guizhou and eastern Yunnan, South China.

Description: The specimen is at least tertiary branch, and is monopodial branching from the leaf axil at 30°–50°. The single preserved longest terminal shoot is more than 12 cm long. Every branch accordingly shorten. Longer leaf is about 6–8.5 mm long, approximately 3.5–5.5 mm broad, shorter leaf is about 2–5 mm long, roughly 2.5–3.5 mm broad. The shoot is loosely helically arranged, and is subalternate. Leaf is scale like, spirally arrayed and imbricate arranged. The distal portion of leaf is away from the axis and the Degree of Freedom (DOF) is medium. Leaf base is decurrent. Leaves exposed in the front is usually rhombic or oval, the middle and lower portion is the widest and taper till the apex. Leave apex is blunt. Leaf margin curves occasionally. Leaf has a distinct mid-ridge.

Fertile shoot is up to 6 cm long, and could be divided into three potion from the bottom to the top. The base is shoot bearing sterile leaves and is 2.5 cm in length, 1.2 cm in width. The middle portion is shoot with reduced sterile leaves, and is 2 cm in length and 0.3 cm in width. The reduced top bears a female cone. It is turbinate, slightly nutant, 15.cm in length, and the widest part is the top, is 1 cm broad. Seed scale is oval, and spiral arrayed.

Discussion and Comparison: This kind of specimen is the largest amounts and the branching structure is the best preserved, also is the only one bearing female cone. However, it's hard to establish a new species without more anatomic data. The difference between Anshuncladus sp. 1, A. sp. 2 and A. sp.3 are listed in the key tab. 4.

***Anshuncladus* sp.2 sp. nov.** (Figs. 9.10–6, 9.10–7)

2019 *Anshuncladus* sp.2 Li and Yu, p. 35–36, pl. 5, Fig. A-G

Holotype: *Anshuncladus* sp.2 Li and Yu, 2019, GAXM11007, pl. 5, Fig. B, deposited in China University of Geosciences (Wuhan) State key laboratory of Biogeology and Environmental Geology, an impression fossil with cuticle preserved, from Dalong Formation (Changhsingian), Anshun County, Guizhou Province, South China.

Diagnosis: Width of shoots and branches even. Leaf scale like, spirally arrayed and closely detached to axis; upper portion slightly off axis, DOF extremely low; partly regularly arranged, subalternate; leaf thick, rhomic, oval to triangular, with blunt apex; leaf base decurrent and encase. Leaf margin occasionally curved, with a distinct mid-ridge. Concave ventrally, convex dorsally near the mid-ridge.

Occurrence: Upper Permian, western Guizhou and eastern Yunnan, South China.

Description: The width of shoots and branches are even. Leaf is scale like, spirally arrayed and closely detached to axis. The upper portion is slightly off axis, and the DOF is extremely low. Part of the specimen are arranged more regularly, and is subalternate. Leaf is thick, rhomic, oval to triangular, with blunt apex. The base of leaf is decurrent and encase. Leaf margin occasionally curved. Leaf has a distinct mid-ridge. The later leaf indicates the adaxial surface is concave and the abaxial surface is convex near the mid-ridge.

Discussion and Comparison: This specimen resembles *Anshuncladus* sp.1, so is the epidermal structure. The main difference is the detached degree of leaf to the axis. Due to the closely detachment to the axis, the arrangement of leaf transfer from irregular spiral to more regular. In some specimen, there are apparent subalternate arrangement of leaves, which is important to the conifers' evolution. So we take this as the key identification of different species.

***Anshuncladus* sp.3** (Figs. 9.10-4, 9.10-5)

2019 *Anshuncladus* sp.3 Li and Yu, p. 36, pl. 4, Fig. A–E

Holotype: *Anshuncladus* sp.3 Li and Yu, 2019, GAXM11008, pl. 4, Fig. A, deposited in China University of Geosciences (Wuhan) State key laboratory of Biogeology and Environmental Geology, an impression fossil with cuticle preserved, from Dalong Formation (Changhsingian), Anshun County, Guizhou Province, South China.

Diagnosis: With fine shoots. Leaf small, scale-like, loosely spirally arrayed, spacing 0.1–1.0 cm unequally, subalternate, detached with the axis at wide angle (60°–90°),

Table 9.4 Identification key of *Anshuncladus* species

I.Leaves densely helically arranged
1. Leaves' distal portion away from the axis, medium Degree of Freedom (DOF)
Anshuncladus sp.1
2. Leaves' distal portion closely detached to the axis, extremely low DOF
Anshuncladus sp.2
I. Leaves loosely arranged, subalternate, distal portion away from the axis, high DOF
Anshuncladus sp.3

DOF is high. Leaf pex is blunt or bluntly round, usually recurved in the direction of adaxial portion. Leaf base intensely decurrent. Mid-ridge unclear.

Occurrence: Upper Permian, western Guizhou and eastern Yunnan, South China.

Description: The specimen has fine shoots. Leaf is small, scale-like, loosely spirally arrayed, and spacing 0.1–1.0 cm unequally. The leaf is subalternate, detached with the axis at wide angle (60°–90°), and the DOF is high. The leaf pex is blunt or bluntly round, usually recurved in the direction of adaxial portion. The leaf base intensely decurrent. Due to the bad preservation of leaves in the front of the middle axis, so the mid-ridge is not clear.

Discussion and Comparison: This species is distinct from other two sspecies due to the thin shoots, small leaves, low amounts of epidermal stomata which is scattered and the structure is hard to recognize. It is suspected to be the young shoots of the other two species or belongs to totally different genus (Table 9.4).

Genus ***Szecladia*** (Yao et al. 2000)

Type species: *Szecladia multinervia* (Yao et al. 2000). Changhsingian (Upper Permian); Guizhou and Guangxi provinces, South China.

Diagnosis: Vegetative conifer shoots with irregular branching, helical phyllotaxis, and small imbricating leaves. Leaves roughly ovate, broadly attached, with several veins. Stems eus.

Discussion: The generic name *Szecladia* consists of the surname Sze and cladus (Greek klados 5 branch), and is proposed in honor of the late Professor H. C. Sze, distinguished palaeobotanist and founder of paleobotany in China, who first described fossil conifers of this kind from the Guangxi Autonomous Region (Sze 1940).

Szecladia multinervia (Yao, Liu, Rothwell and Mapes, 2000) (Figs. 9.10-1, 9.10-2)

1940 *Ullmannia* aff. *bronnii* Goeppert (? sp. nov.), Sze, p. 156, taf. I, Figd. 1–8

1974 *Ullmannia* cf. *bronnii* Goeppert, Gu and Zhi, p. 154, pl. 121, Fig. 6; pl. 122, Figs. 1, 3, 4

1980 *Ullmannia* cf. *bronnii* Goeppert, Zhao et al., pl. XVII, Figs. 6, 6a

1985 *Ullmannia bronnii* Goeppert, Feng and Zhu, p. 203, pl. I, Figs. 1–3

1996 *Pseudoullmannia bronnioides* He, Liang and Shen, p. 169, part (non holotype, pl. 69, Fig. 6)

2000 *Szecladia multinervia* Yao, Liu Rothwell and Mapes, p. 525, Figs. 3–5

Holotype: *Szecladia multinervia* Yao et al. (2000), Figs. 3.1, 3.2, 4.1, 4.3, 4.4, 4.7, 5.1–5.7, deposited in the Nanjing Institute of Geology and Palaeontology, Chinese Academy of Sciences, from Tulong Formation, Upper Permian, Guangxi Province, South China.

Diagnosis: Ultimate branches generally 3–8 mm wide with leaves up to 6 mm long and 3 mm wide, with inwardly bending tip; leaves commonly contain seven to eight parallel veins with uniseriate bundle sheaths. Secondary xylem tracheids commonly 11–22 mm in diameter with uniseriate circular bordered pits; rays uniseriate, one to three cells high with large, simple cross field pits. Ground tissues of pith, cortex and leaves contain nests of sclereids. Well developed periderm interior to leaf bases of penultimate branches.

Occurrence: Duanshan, Xinmin sections of Changhsingian in western Guizhou and eastern Yunnan area, South China. Dongluo Section of Changhsingian in Guangxi.

Description: The largest specimens of *Szecladia multinervia* show two orders of irregularly branched stems with imbricating, helically-arranged leaves attached throughout leafy stems range 3.3–8 mm in diameter. Leaves are up to 6 mm long and 3 mm wide, and are loosely appressed to the stem with a tip that bends slightly inward. In surface views the leaves are roughly ovate, broadly attached, widest in the midregion, and taper to a bluntly rounded apex. Each leaf displays several prominent longitudinal striations that represent the positions of approximately seven to eight veins. In cross sections the leaves are elliptical near the base, becoming wider and more blade like distally.

Discussion and Comparison: *Szecladia multinervia* displays a distinctive suite of conifer characters that clearly distinguish it from previously described genera. Among Paleozoic seed plants, irregularly branched shoots with uniformly small helically arranged leaves like those of *Szecladia* are known only for conifers. Also characteristic of conifers, are the endarch eustele with leaf traces that initially diverge as a single bundle and the dense wood with uniseriate circular bordered pits of *Szecladia.* Distinct longitudinal striations of the leaves are shared by *Szecladia*, species of *Ullmannia* (Goeppert 1850), and Pseudoullmannia (He et al. 1996). However, in *Ullmannia* the striations represent stomatal furrows on the surface of leaves with a single midvein, whereas in *Szecladia* they are formed by several parallel veins. The structural nature of the striations in *Pseudoullmannia* is not known. Among Paleozoic conifers with small leaves, only *Szecladia* displays several leaf traces. Although leaf size and shape in *Szecladia* and *Ullmannia* are similar, as are the inwardly curved, bluntly rounded apices, multiveined leaves clearly distinguish *Szecladia* from species of *Ullmannia*. The leaves of both *Pseudoullmannia frumentalioides* and the holotype of *P. bronnioides* are relatively longer and narrower than those of *Szecladia* and *Ullmannia*, and are terminated by a relatively straight, acutely pointed apex (He et al. 1996). Multiveined leaves are found in several Paleozoic and Triassic coniferophyte taxa worldwide. These include Paleozoic Cordaitean leaves from Euramerica, Cathaysia and Angara that are usually referred to *Cordaites* (Rothwell 1988; Halle 1927; Meyen 1987), coniferophyte leaves from Angara that are assignable to *Rufloria* and *Lepeophyllum* (Meyen 1978, 1987), and others of unknown affinity (e.g., *Pelourdea* Halle 1927; *Phylladoderma*, Meyen 1988). The Triassic conifers *Aethophyllum* from France (Grauvogel-Stamm 1978) and *Heidiphyllum* from Gondwana deposits (Anderson and Anderson 1985) also have several parallel veins, but those genera comprise species with leaves that are all much larger, mesomorphic, strap-shaped forms.

The growth form and reproductive structures of *Szecladia multinervia* remain unknown. General features of the stelar architecture and leaf-trace divergence in *Szecladia* are similar to those of the Pennsylvanian *Walchian* conifers from North America (Rothwell et al. 1997) as well as to those found in more recent conifer families. The relatively widely spaced, uniseriat circular bordered pits of the small, secondary tracheids, and the simple cross field pitting of *Szecladia* are suggestive of the Podocarpaceae, but similar characters also occur in the wood of some pinaceous species. However, in the Pinaceae there are usually prominent resin canals that do not occur in *Szecladia*. Multiveined leaves are present in both Szecladia and many species of the Podocarpaceae, including the Triassic *Notophytum krauselii* from Antarctica (Axsmith et al. 1998) and the Upper Triassic Stalagma samara from Hunan, China (Zhou 1983). This feature further suggests podocarpaceous affinities for *Szecladia*. However, multiveined strap-shaped leaves are characteristic of many other Mesozoic conifers whose familial affinities are uncertain (Axsmith et al. 1998). Although we suspect that *Szecladia* may be allied with the Podocarpaceae, in the absence of reproductive organs, familial affinities of the genus cannot be determined with confidence.

Genus ***Pseudoullmania*** (He et al. 1996)

Type species: *Pseudoullmania frumentalioides* He et al. (1996). Late Permian of South China.

Diagnosis: See that of *Pseudoullmania frumentalioides.*

Discussion: The same specimen from South China were referred to *Ullmania* Goeppert. He studied the shape and epidermal structure and insisted those specimen were greatly different from those of West Europe. However there is no evidence of their higher taxanomy, and He et al. (1996) suspected it may not belong to conifers for the reason that the rectangular contour of cell walls more resemble young shoot of lepidodendroid plant.

Pseudoullmannia frumentarioides (He et al. 1996) (Fig. 9.10-3)

1977 *Ullmannia furmentaria* Goeppert, Feng et al., p. 671, pl. 251, Figs. 1–2

1979 *Ullmannia furmentaria* (Schlotheim) Goeppert, Yang and Chen, p. 137, pl. 46, Figs. 2–4; pl. 47, Figs. 1–3

1996 *Pseudoullmannia frumentarioides* He, Liang and Shen, p. 87, pl. 69, Figs. 1–5; pl. 70, Fig. 2; pl. 96, Figs. 1, 2

Holotype: *Pseudoullmannia frumentarioides* He et al. (1996), pl. 69, Figs. 1–5; pl. 70, Fig. 2; pl. 96. Deposited in the Nanjing Institute of Geology and Palaeontology, Chinese Academy of Sciences, from Lower Laoshan Submember and Wangpanli Member of Loping Formation in South China.

Diagnosis: Shoot irregularly arranged, with spirally and densely arranged leaves. Leaves lanceolate in shape, apex acute, about 8–15 mm in length and about 2 mm in width at the base, base decurrent, leaf surface striated. Midrib invisible. Lower epidermal cells comparatively small, about 10 μm in length and 2 μm in width, rectangular in shape, lateral walls straight and regularly arranged, no stomata recognized. Upper cuticle unknown.

Occurrence: Duanshan, Xinmin sections of Changhsingian in western Guizhou and eastern Yunnan area, South China. Dongluo Section of Changhsingian in Guangxi.

Description: The penultimate shoots of our specimens are 6–10 mm in width, both ultimate and penultimate shoots bear spirally arranged leaves. Leaves are lanceolate, about 13–10 mm in length and 2 mm in width at the broadest lower part. Leaf apex is acute and the base is slightly contracted. There are not midvein found, and several parrallel veins can be seen on the better preserved dorsal surface.

Discussion and Comparison: The present specimen are completely identical with those of *Ullmania frumentaria* described by Fong, Yang and Chao et al. While our specimen has linear to leaves, obtuse leaf apex and strongly convex and geniculated lateral sides. Compared with *Ullmania frumentaria*, our specimen has no midvein. The epidermal cells are isodiametric polygonal and stomata are regularly longitudinally oriented. According to all the difference listed, our specimens are different from *Ullmania frumentaria.*

Danaeites saraepontanus Stur, Paleozoic plants of China.

References

Anderson JM, Anderson HM (1985) Palaeoflora of Southern Africa: prodromus of South African megafloras, devonian to lower cretaceous. Balkema, Rotterdam, p 567

Asama K (1959) Systematic study of so-called *Gigantopteris*. Sci Rep Tohoku Univ 2nd Ser (Geol) 31:1–72

Axsmith BJ, Taylor TN, Taylor EL (1998) A new fossil conifer from the Triassic of North America: implications for models of ovulate cone scale evolution. Int J Plant Sci 159:358–366

Brongniart A (1822) Sur la Classification et la Distribution des Végétaux Fossiles en Général, et Sur Ceux des Terrains de Sédiment Supérieur en Particulier. Imprimerie De A. Belin, Paris, pp 1–91 (also in Mémoires du Muséum d'Histoire Naturelle, 8:203–240, 297–348)

Brongniart A (1828a) Essai d'une flora du grès bigarre. Annales de Science Naturelles, 1st Ser 15:435–460

Brongniart A (1828b) Prodrome D'une Histoire des végétaux fossiles. Dictionnaire des Science Naturelles 57:1–223

Feng Z, Wei HB, Guo Y, He XY, Sui Q, Zhou Y, Liu HY, Gou XD, Lu Y (2020) From rainforest to herbland: New insights into land plant responses to the End-Permian Mass Extinction. Earth-Sci Rev 204:103153

Fliche P (1910) Flore fossile du Trias en Lorraine et Franche-Comté (avec des considérations finales par M. R. Zeiller. Bulletin De La Société Des Sciences De Nancy III, XI 1:222–286

Florin R (1933) Studien uber die Cycadales des Mesozoikums. Kungliga Svenska Vetenskapsakademiens Handlingar 12:1–134

Galtier J, Broutin J (2008) Floras from red beds of the Permian Basin of Lodève (Southern France). J Iberian Geol 34(1):57–72

Gomankov AV, Meyen SV (1979) On representatives of the family Peltaspermaceae from the Permian deposits of the Russian platform. Paleontol Zh 2:124–138

Gomankov AV, Meyen SV (1986) *Tataria* flora (composition and distribution in the Late Permian of Eurasia). Akademii Nauk SSSR, Ordena Trudovogo Krasnogo Znameni Geologicheskii Institut, 401:1–174 (in Russian)

Göppert HR (1850) Monographie der fossilen Coniferen. Hollandsche Maatschappÿ Wetensch. Natuurk. Verh. 6:1–286

Grauvogel-Stamm L (1978) La flore du Grès à Voltzia (Buntsandstein supérieur) des Vosges du Nord (France). Morphologie, anatomie, interprétations phylogénique et paléogéographique. Sci Géol Bull 50:1–225

Gu Z, Zhi L (1974) Palaeozoic plants from China. Nanjing Institute of Geology and Palaeontology, Institute of Botany, Academia Sinica. Science press, Beijing, pp 1–277

Halle TG (1925) *Tingia*, a new genus of fossil plant from the Permian of China (preliminary note). Bull Geol Soc China 7:1–12

Halle TG (1927) Fossil plants from southwestern China. Palaeontologia Sinica Ser A 1(2):1–26

Harris TM (1937) The fossil flora of Scoresby sound, East Greenland. Part. 5: Stratigraphic relations of the plant beds. Medd Grønland 112(2):1–114

He XL, Liang DS, Sheng SZ (1996) The Permian Floras from Jiangxi, China. China University of Minings Press, Xuzhou, pp 1–201

Kon'no E, Asama K (1950) On the genus *Lobatannularia* Kawasaki 1927 from Permian Beds in South Manchuria and Shansi, China. Short Pap. IGPC 1:18–31

Krings M, DiMichele WA, Kerp H, Taylor TN (2005) Frond morphology and epidermal anatomy of Glenopteris splendens (Pteridospermopsida, Peltaspermales) from the Permian of Kansas, USA. N M Mus Nat Hist Sci Bull 30:171

Kustatscher E, Bauer K, Butzmann R, Fischer TC, Meller B, van Konijnenburg-van Cittert JHA, Kerp H, (2014) Sphenophytes, pteridosperms and possible cycads from the Wuchiapingian (Lopingian, Permian) of Bletterbach (Dolomites, Northern Italy). Rev Palaeobot Palynol 208:65–82

Lesquereux L (1877) A species of fungus recently discovered in the shales of the Darlington coal bed (lower productive coal measures, Alleghany River series) at Cannelton, in Beaver County, Pennsylvania. Proc Am Philos Soc 17:173–175

Li XX, Yao ZQ (1983) Current studies of gigantopterids. Palaeontologia Cathayana 1:319–326

Liu LJ, Yao ZQ (2007) Plant megafossils from the Permian Changhsingian marine deposits of Fusui, Guangxi, china. Acta Palaeontologica Sinica 46(2):195–212

Meyen SV (1987) Fundamentals of palaeobotany. Chapman and Hall, London

Naugolnykh SV, Kerp H (1996) Aspects of Permian palaeobotany and palynology. XIV. On the oldest known peltasperms with radially symmetrical ovuliferous discs from the Kungurian (uppermost Lower Permian) of the Fore-Urals (Russia). Rev Palaeobot Palynol 91(1–4):35–62

Poort RJ, Kerp JHF (1990) Aspects of Permian palaeobotany and palynology. XI. On the recognition of true Peltasperms in the Upper Permian of western and central Europe and a reclassification of species formerly included in *Peltaspermum* Harris. Rev Palaeobotany Palynol 63(3–4):197–225

Potonie H (1902) Ueber die fossilen Filicales im Allgemeinen etc. Emgler & Frantl: Die naturlichen Pflanzenfamilien. Teil 1, Abteil. 4, 513

Potonié R (1956) Synopsis der Gattungen der Sporae dispersae I. Teil: Sporites. Amt für Bodenforschung, Hannover

Rothwell GW (1988) Cordaitales. In: Beck CB (ed) Origin and evolution of gymnosperms, pp 273–297. Columbia University Press, New York

Rothwell GW, Mapes G, Mapes RH (1997) Late Paleozoic conifers of North America: structure, diversity and occurrences. Rev Paleobot Palynol 95:95–113

Schenk A (1883) Pflanzen aus der Steinkohlen formation. In: Richthofen F von. China, 4. Berlin, pp 238–239

Schimper WP (1869) Traité de Paléontologie végétale ou la flore du monde primitif dans ses rapports avec les formations géologiques et la flore du monde actuel. Tome premier. Baillière J.B. et Fils, Paris, p 732

Stockmans F, Mathieu FF (1939) La Flore paleozoique du Bassin houiller de Kaiping (China). Mus Roy Hist Nat, Belgique, pp 49–165

Trivett ML, Pigg KB (1996) A survey of reticulate venation among fossil and living land plants. In: Taylor DW, Hickey LJ (eds) Flowering plant origin, evolution and phylogeny, pp 8–31. Chapman and Hall, New York, NY

Visscher H, Kerp H, Clement-Westerhof JA, Looy CV (2001) Permian floras of the Southern Alps. Naturae Bresciana, Annali del Museo Civico di Scienze Naturali di Brescia, Monografia 25:117–123

Wan ML, Shi H, Wang DY (2011) Latest Permian Cathaysian Flora from Yangbazhai, Mojiang in Yunnan Province. Acta Palaeont Sinica 50(2):251–267

Wang ZQ (1997) Permian Supaia fronds and an associated Autunia fructification from Shanxi, China. Palaeontology 40:245–277

Wang ZQ, Wang LX (1989) Earlier early Triassic fossil plants in the Shiqianfeng Group in North China. Shanxi Geol 14(1):23–40 (in Chinese with English abstract)

White D (1929) Flora of the Hermit shale, Grand Canyon, Arizona. Bull Carnegie Inst Wash 405:1–221

Yao ZQ, Xu JT, Zhen ZG, Mo ZG (1980) Late Permian biostratigraphy and the Permian-Triassic boundary in Western Guizhou and Eastern Yunnan. In: Nanjing Institute of Geology and Palaeontology (ed), Stratigraphy and palaeontology of Late Permain coal-bearing formations in Western Guizhou and Eastern Yunnan. Science Press, Beijing, pp 1–69. (In Chinese)

Yao ZQ, Liu LJ, Rothwell WG et al (2000) Szecladia new genus, a Late Permian conifer with multiveined leaves from South China. J Paleontol 74:524–531

Zhao XG, Mo ZG, Zhang SZ, Yao ZQ (1980) Late Permian fossil floras of western Guizhou and eastern Yunnan. In: Nanjing Institute of Geology and Palaeontology, Academica Sinica, (ed.), Late Permian coal-bearing strata and palaeontological fauna in western Guizhou and eastern Yunnan. Science Press, Beijing, pp 70–122 (in Chinese)

Zhou ZY (1983) *Stalagma samara*, a new Podocarpaceous conifer with monocolpate pollen from the Upper Triassic of Hunan, China. Palaeontographica B 185:56–78

Zijlstra G, Kustatscher E, van Konijnenburg-van Cittert JHA (2007) Proposals to conserve the names *Schizoneura* against *Convallarites* and *S. paradoxa* against *C. erecta* (fossil Sphenopsida). Taxon 56:965–966

Chapter 10
Systematic Palaeontology: Part 2 Megaspores

Jianxin Yu, Wenchao Shu, Xiao Shi, and Zhen Xu

Sporites H. Potonie, 1893.

Turma **Triletes** (Reinsch, 1881) Potonie and Kremp, 1954

Subturma **Azonotriletes** Luber, 1935.

Infraturma **Laevigati** (Bennie and Kidston, 1886) Potonie, 1956.

Genus ***Trileites*** (Erdtmann, 1945, 1947) Potonie, 1956.

Trileites vulgaris Diikstra, 1955 (Fig. 10.1-1, 10.1-2 and 10.1-3).

1955 *Trileites vulgaris* Diikstra, Pl. 1, Figs. 1–4.

Description: Trilete megaspore. Equatorial outline is round-round triangular. 220–480 μm in diameter. Trilete is 5.5–10.0 μm in width, straight or mildly curved, the length is same with the spore's diameter. With no arcuate ridge. Smooth extine, the surface is cracked or covered because of preservation.

Comparison and Discussion: As a impression fossil, the height of the ray is hard to measure. Except from that, this specimen is a little bit smaller than the type species established by Fuglewicz, so is the width of the ray, but still ranging in the variation of this species.

Locality and Horizon: Beds 19,24 and 27; Kayitou Formation, Mide section, Yunnan.

J. Yu (✉) · W. Shu · Z. Xu
School of Earth Sciences, China University of Geosciences, Wuhan, P.R. China
e-mail: yujianxin@cug.edu.cn

X. Shi
College of Earth Sciences, Jilin University, Changchun, Jilin, P.R. China

J. Yu et al. (eds.), *Plants and Palynomorphs around the Permian-Triassic Boundary of South China*, New Records of the Great Dying in South China,
https://doi.org/10.1007/978-981-19-1492-8_10

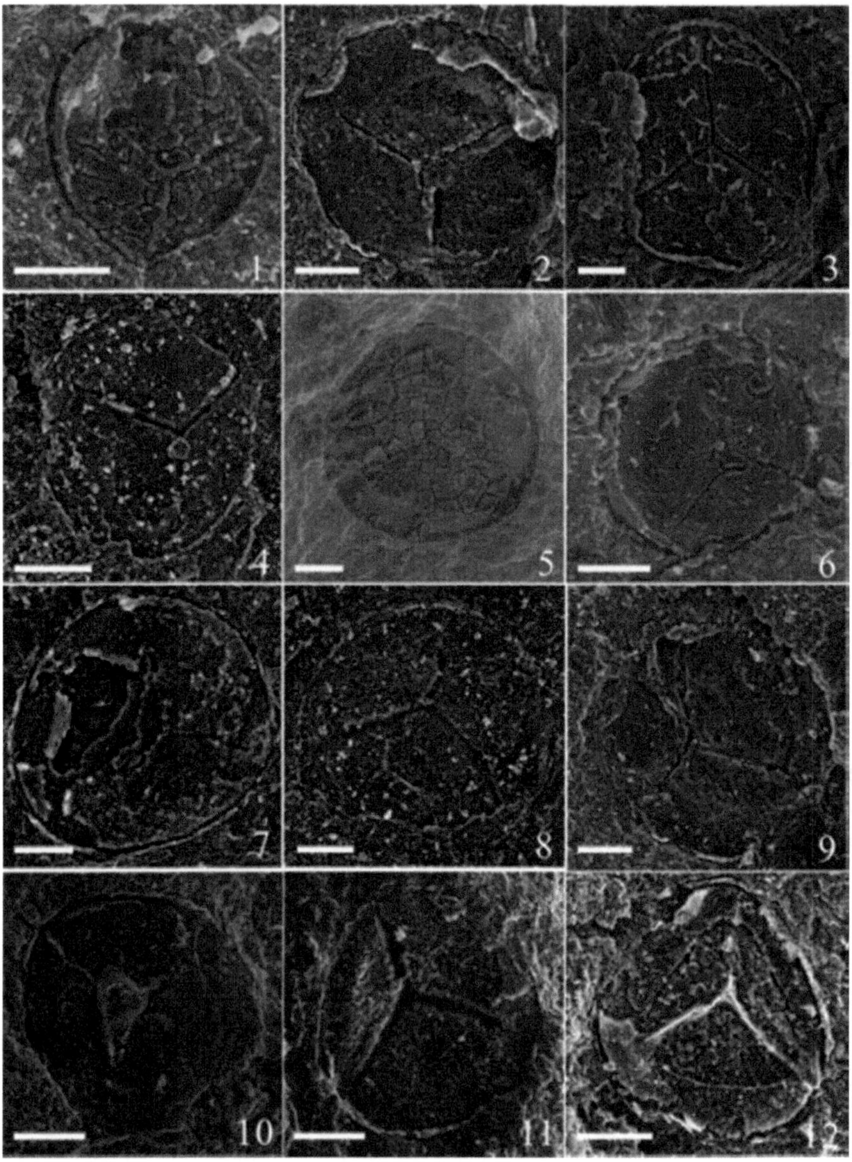

Fig. 10.1 (Scale bar is 100 μm). 1–3 *Trileites vulgaris* Fuglewicz. Locality: Xinmin section, Anshun City, Guizhou Province. Sample Number: YXM-B-M-217, 052, 010. 4–7 *Trileites levis* Fuglewicz. Locality: Xinmin section, Anshun City, Guizhou Province. Sample Number: YXM-B-M-086, 237, 242. 8–9 *Trileites* sp. Locality: Xinmin section, Anshun City, Guizhou Province. Sample Number: YXM-B-M-093, 042. 10 *Maexisporites rotundus* Fuglewicz. Locality: Xinmin section, Anshun City, Guizhou Province. Sample Number: YXM-B-M-017. 11 *Maexisporites parvus* Fuglewicz. Locality: Xinmin section, Anshun City, Guizhou Province. Sample Number: YXM-B-M-168. 12 *Maexisporites pyramidalis* Fuglewicz. Locality: Xinmin section, Anshun City, Guizhou Province. Sample Number: YXM-B-M-060

Trileites levis Fuglewicz, 1973 (Fig. 10.2-4 to 10.2-7).

1973 *Trileites levis* Fuglewicz, p. 418, Pl. 19, Fig. 8.

Description: Trilete megaspore. Equatorial outline is round. 290–590 μm in diameter. Trilete is thin and straight, the length is about 3/4 of the spore's radius. With no arcuate ridge. Smooth extine, the surface is cracked caused by extrusion or preservation..

Comparison and Discussion: Compared with the type species established by Fuglewicz,vthis specimen is relatively bigger and the ray is a little thinner and longer. This specimen is also similar to *Triletes vulgaris*, and the later is characterized by the length of trilete which directly meet the equator.

Locality and Horizon: Beds 19, 21, 25 and 27; Kayitou Formation, Mide section, Yunnan.

Trileites sp. (Fig. 10.2-8, 10.2-9).

Description: Trilete megaspore. Equatorial outline is round. 430–470 μm in diameter. Trilete is 12–13 μm in width, straight or mildly curved, the length is 3/4 of the spore's radius or longer. Arcuate ridge is not obvious. The surface is covered by dense worm-like nodules of uneven size. The nodules are connected and stretched.

Comparison and Discussion: This specimen generally belongs to *Trileites*. The most prominent character is the surface ornamentation, which is similar to the *Aneuletes acrochordonodes* reported in Jimusar, Xinjiang. As the impression is not complete, this specimen is identified as undefined species for missing character.

Locality and Horizon: Bed 21, 24; Kayitou Formation, Mide section, Yunnan

Infraturma **Apiculati** (Bennie and Kidston, 1886) Potonie, 1956.

Subinfraturma **Granulati** Dybova and Jachowicz, 1957.

Genus ***Maexisporites*** (Erdtmann, 1945, 1947) Potonie, 1956.

Maexisporites pyramidalis Fuglewicz, 1973 (Fig. 10.2-12).

Description: Trilete megaspore. Equatorial outline is round triangular, and the corner is round. 290–300 μm in diameter. Trilete is straight and convex, narrowing from the base to the top. The length is same with the spore's radius. With no arcuate ridge. The contact region is clear, separated with the outside by a distinct groove. The surface contains sparse aperture, which is probably the leftover of granular ornamentation. The surface of contact region is smooth. As impression fossil, the convex of contact surface is not distinct.

Comparison and Discussion: The character of this specimen is identical with specimens reported in Jimusar, Xinjiang and in Fuglewicz (1977). The difference is that this specimen's trilete is thinner and the outside of contact region is smoother. Compared with *Hughesisprites inflatus* Fuglewicz, the later has inflated terminus and many convex ridge and corresponding groove. Different from *Pusulosporites inflatus* Fuglewicz, the trilete of the later heighten and broaden in the spore margin.

Fig. 10.2 1–4 Aff. *Germaropteris martinsii* (Harris) Poort et Kerp. Locality: 1, 2, 3, 5-Tucheng section, Panxian County, Guizhou Province. 4-Mide section (B), Xuanwei City, Yunnan Province. Sample Number: GPT-23-2, GPT-23-3, GPT-23- 4, GPT22-4. 5-13 *Germaropteris* spp. Locality: 5, 6, 7, 11-Mide Section, Xuanwei City, Yunnan Province. 8, 9, 10-Jinzhong Section, Weining County, Guizhou Province. Sample Number: 14 Indet type 1. Locality: Mide Section, Xuanwei City, Yunnan Province. Sample Number: 15 Possible Megaspores in situ or strobile. Locality: Tucheng section, Panxian County, Guizhou Province. Sample Number: YXM (B)-25-2. 16-20 Indet type 2. Locality: Tucheng section, Panxian County, Guizhou Province. Sample Number: GPT-17-11. 21, 23 *Peltaspermum* sp. Locality: Tucheng section, Panxian County, Guizhou Province. Sample Number: GPT-22-6. 22 *Peltaspermum lobutalum* Wang et Wang. Locality: Mide section, Xuanwei City, Yunnan Province. Sample Number: GPT-22-1

This specimen is also quite similar to *Trileites crassitectatus* Fuglewicz and *T. validus* Fuglewicz because of the unconspicuous of contact region convex. The key difference is the width and height of trilete. The later two have thick and distinct trilete.

Locality and Horizon: Beds 21 and 25; Kayitou Formation, Mide section, Yunnan.

Maexisporites rotundus Fuglewicz, 1973 (Fig. 10.1-10).

1973 *Maexisporites rotundus* Fuglewicz; Fuglewicz R., p. 423, Pl. 21, Fig. 5a, b.

Description: Trilete megaspore. Equatorial outline is round. Approximately 320 μm in diameter. Trilete is short and thin, rising around the equatorial axis to a pyramid. The length is 0.7 of the spore's radius. With no arcuate ridge. The surface contains sparse aperture, which is probably the leftover of granular ornamentation.

Comparison and Discussion: Except from the pyramid in the equatorial axis, this specimen is same with the type species established by Fuglewicz in Jimusar, Xinjiang. It is worth mentioning that this specimen is quite similar to the Middle Triassic Maexisporites type dispersed megaspore reported by Wang in Shaanxi. However Wang did not identified it to species, nor described in detail.

Locality and Horizon: Beds 21 and 25; Kayitou Formation, Mide section, Yunnan.

Maexisporites parvus Fuglewicz, 1973 (Fig. 10.2-11).

1973 *Maexisporites parvus* Fuglewicz; Fuglewicz R., p. 421, Pl. 21, Fig. 3a, b.

Description: Trilete megaspore. Equatorial outline is almost round. 300 μm in diameter. Trilete is distinct, thick and straight. The length is roughly 0.8 of the spore's radius or longer. With indistinct arcuate ridge. The surface contains dense aperture, which is the impression of granular ornamentation.

Comparison and Discussion: This specimen is identical to the type species established by Fuglewicz and the specimen reported in Jimusar, Xinjiang. The difference is the granular ornamentation of this specimen is indistinct. Compared with *Maexisporites rotundus*, the radius of the former is smaller and has longer trilete.

Locality and Horizon: Bed 26; Kayitou Formation, Mide section, Yunnan Province.

Subinfraturma **Verrutriletes** Dybova and Jachowicz, 1957.

Genus ***Otynisporites*** Fuglewicz, 1977.

Otynisporites eotriassicus Fuglewicz, 1973 (Fig. 10.3-5).

1973 *Otynisporites eotriassicus* Fuglewicz; Fuglewicz R., p. 412, Pl. 30, Figs. 1, 2

Description: Trilete megaspore. Equatorial outline is round. Approximately 360 μm in diameter. Trilete develops to thick ridge, straight or mildly curved, almost reached the equator. With arcuate ridge. The surface of the spore and the ray contain small nodule, which is 6–7 μm in diameter. The terminus of the nodule indistinctly branched in some obvious region. Only some base of the short and dispersed virgate nodule is left if local region.

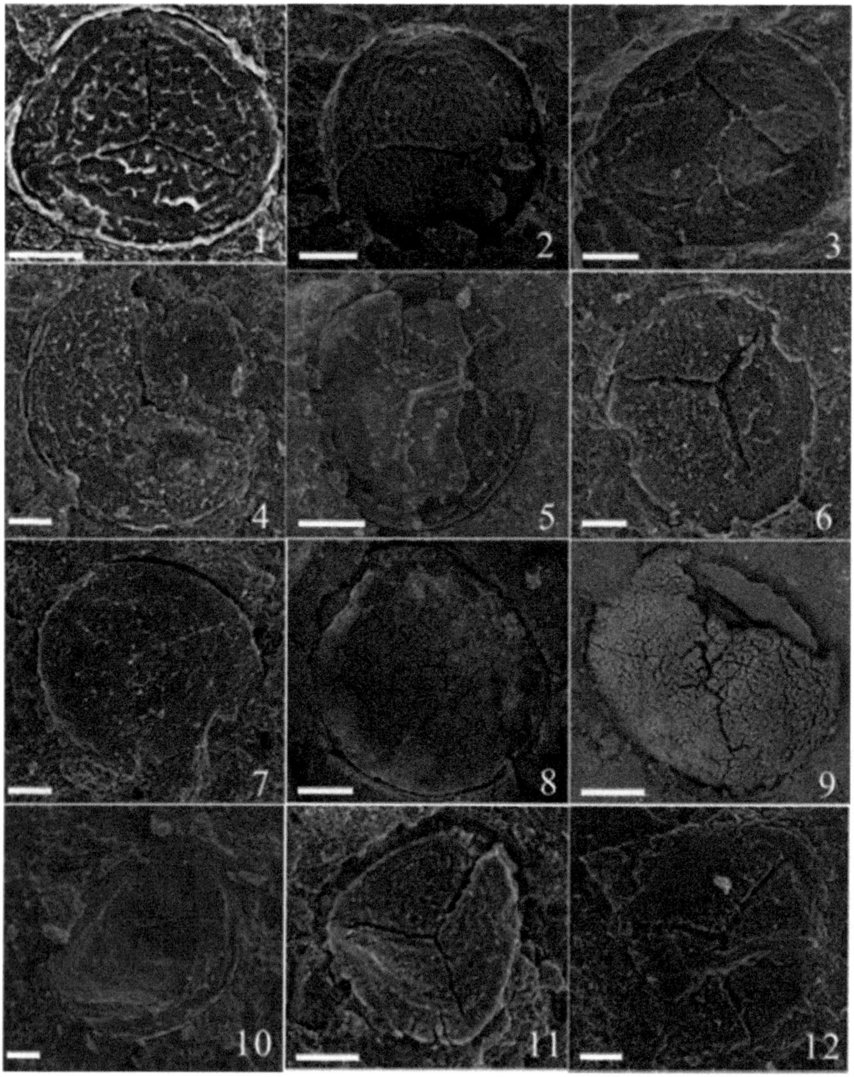

Fig. 10.3 (Scale bar is 100 μm). 1 *Zeindtisporites* sp. Locality: Xinmin section, Anshun City, Guizhou Province. Sample Number: YXM-B-M-005. 2–3 *Horstisporites sulcatus* Fuglewicz. Locality: Xinmin section, Anshun City, Guizhou Province. Sample Number: YXM-B-M-156, 165. 4 *Horstisporites* sp. Locality: Xinmin section, Anshun City, Guizhou Province. Sample Number: YXM-B-M-106. 5 *Otynisporites eotriassicus* Fuglewicz. Locality: Xinmin section, Anshun City, Guizhou Province. Sample Number: YXM-B-M-232. 6 *Hughesisporites simplex* Fuglewicz. Locality: Xinmin section, Anshun City, Guizhou Province. Sample Number: YXM-B-M-105. 7 *Banksisporites pinguis* (Harris) Dettmann. Locality: Xinmin section, Anshun City, Guizhou Province. Sample Number: YXM-B-M-091. 8–9 *Aneuletes* sp. Locality: Xinmin section, Anshun City, Guizhou Province. Sample Number: YXM-B-M-117, 123. 10 *Aneuletes rotundus* Fuglewicz. Locality: Xinmin section, Anshun City, Guizhou Province. Sample Number: YXM-B-M-103. 11 *Triangulatisporites dalongkouensis* Yang and Sun. Locality: Xinmin section, Anshun City, Guizhou Province. Sample Number: YXM-B-M-045. 12 *Triangulatisporites* sp. Locality: Xinmin section, Anshun City, Guizhou Province. Sample Number: YXM-B-M-034

Comparison and Discussion: The character of this specimen is identical with specimens reported in Jimusar, Xinjiang and in Fuglewicz (1977). The ray and arcuate ridge of this specimen character and dense of nodule is not as distinct as other specimen ever reported for the preservation condition. The character of this specimen is similar to *Otynisporites tuberculatus* Fuglewicz. The key difference is the development degree and ornamentation of arcuate ridge. The later has distinct arcuate ridge and furcate spine except from nodule.

Locality and Horizon: Bed 27; Kayitou Formation, Mide section, Yunnan Province.

Subinfraturma **Nodati** Dybova and Jachowicz, 1957.

Genus ***Hughesisporites*** Potonie, 1956.

Hughesisporites simplex Fuglewicz, 1977 (Fig. 10.3-6).

Description: Trilete megaspore. Equatorial outline is round. Approximately 520 μm in diameter. Trilete mildly curved and thick, narrowing towards the equator. The length is roughly 0.8 of the spore's radius. With delicate arcuate ridge, complete and distinct. The contact region of arcuate ridge has coarse surface and vague ornamentation. The outer surface of it is smooth.

Comparison and Discussion: Compared with the type species established by Fuglewicz (1977), this specimen is similar to it, while this specimen is shorter in radius and the radian of its arcuate ridge is greater. It could be distinguished from other species of this genus by arcuate ridge and surface ornamentation.

Locality and Horizon: Bed 24; Kayitou Formation, Mide section, Yunnan Province.

Infraturma **Murornate** Potonie and Kremp, 1954.

Genus ***Horstisporites*** Potonie, 1956.

Horstisporites sulcatus Fuglewicz, 1973 (Fig. 10.3-2, 10.3-3).

Description: Trilete megaspore. Equatorial outline is round. 470 μm in diameter. Meridian outline is round, 380–400 μm in diameter. Trilete is well developed. Arcuate ridge is developed to distinct groove. Spore surface is full of waved ridge and groove, which is connected to reticular. The reticular ridge of distal face is thicker than the proximal face.

Comparison and Discussion: The most distinct character of this species is the significant high and thick trilete, well-developed arcuate ridge, convex contact zone, and reticular like groove on the surface. Compared with the type species established by Fuglewicz (1973), this specimen has thinner trilete and arcuate ridge, indistinct contact region convex, thin reticular ridge is thin and small reticular hole. All those characters could easily distinguish this species with other genus.

Locality and Horizon: Beds 25 and 26; Kayitou Formation, Mide section, Yunnan.

Horstisporites sp. (Fig. 10.3-4).

Description: Trilete megaspore. Equatorial outline is round. 580 μm in diameter. Trilete is thin and mildly curved. The length of it is equal to the spore radius. With no arcuate ridge. Spore surface is reticular. Reticulate ridge is thin, and the size of reticulate hole is uneven.

Comparison and Discussion: The reticulation of this specimen is indistinct. The height, width of the reticulate ridge and the rediculate hole size are hard to measure. The similar species *Horstisporites heteroretiuilatus* Fuglewics is distinguished to this specimen by the smaller reticulate hole on the proximal face, bigger reticulate hole on the distal face. Because this specimen is the only one and is not able to get the proximal and distal face character, this specimen is identified as the indeterminate species.

Locality and Horizon: Bed 24; Kayitou Formation, Mide section, Yunnan Province.

Suprasubturma **Pseudosaccititriletes** Richardson, 1965.

Infraturma **Monopseudosacciti** Smith and Butterworth, 1967.

Genus ***Banksisporites*** Dettmann, 1961.

Discussion: *Banksisporites* is established by Dettmann in 1961 and is determined to the turma Triletes, subturma Azonotriletes, infraturma Laevigati. This genus is characterized by the extine is divided into the inner and outer layers (the inner layer is named as mesosporoid). Dettmann proposed the Trileites pinguis established by Harris should be attributed to *Banksisporites pinguis* and regarded it as the type species of this genus. Whereas, Fuglewicz put forward different opinions that the mesosporoid could not be the basis of establishing a new genus because there are mesosporoid of various morphology in the same genus. He reverted two species of *Banksisporites* established by Dettmann to *Trileites*. Then Scott raised objections which supported Dettmann's classification and put *Banksisporites* into different taxonomic position. He also thought lots of specimen were attributed to *Trileites* due to the lack of inner structure details and established two new species. To be concluded, there are still arguments over whether the mesosporoid could be one character of identification. The specimen described in this book has distinct mesosporoid and is different from the former *Trileites*, so we use *Banksisporites*.

Banksisporites pinguis (Harris) Dettmann, 1961 (Fig. 10.3-1, 10.3-2 and 10.3-3).

Description: Trilete megaspore. Equatorial outline is round. 470 μm in diameter. Trilete is delicate, thin and straight, reaching the equator. With mesosporoid, and is roughly 320 μm in diameter. The outer region is folded which may be resulted from contraction. The spore surface contains delicate reticulate mottles and the reticulate hole is relatively big.

Comparison and Discussion: The total characters of this specimen is identical to the species Dettamann established. The difference is that the former has longer trilete and has delicate reticulation rather than smooth.

Locality and Horizon: Bed 24; Kayitou Formation, Mide section, Yunnan Province.

Turma **Zonales** (Bennie and Kidston, 1886 non Ibr.) Potonie, 1956.

Subturma **Zonotriletes** Waltz, 1935

Infraturma **Zonati** Potonie and Kremp, 1954.

Genus ***Triangulatisporites*** (Potonie and Kremp, 1954) Karczewska, 1976.

Triangulatisporites dalongkouensis Yang and Sun, 1986 (Fig. 10.3-11).

1986 *Triangulatisporites dalongkouensis* Yang and Sun; p193, Pl.48, Fig. 4

Description: Trilete megaspore. Equatorial outline is round triangular. 280–300 μm in diameter (including zona). Trilete is straight and thin, the length is larger than the spore's radius. The widest part of zona is approximately 54 μm. There are radial groove in partial region. With no arcuate ridge. The surface of spore's noumenon has reticulate mottles. Reticulate hole is round or polygon, 9–17 μm in diameter.

Comparison and Discussion: This specimen is identical to the type species reported in Jimusar, Xinjiang. This species is quite similar to other species contains reticulate mottles. Compared with *Triangulatisporites reticulatus* Fuglewicz, the later has delicate reticulate mottles with low reticulate ridge. This specimen is distinguished from *Triangulatisporites makowskii* Fuglewicz by the thicker zona of the later. The character of this specimen resembles *Triangulatisporites triangulates* (Zerndt.) Potonie and Kremp the most. The difference is that the later has bigger reticulate hole. Compared with other similar megaspore genus with zona, like *Zonalesporites*, *Minerisporites*, the key difference is the morphology of zona.

Locality and Horizon: Beds 21 and 25; Kayitou Formation, Mide section, Yunnan.

***Triangulatisporites* sp.** (Fig. 10.3-12).

Description: Trilete megaspore. Equatorial outline is round. 500 μm in diameter. Trilete is relatively thick, the length is equal to the radius of the spore. The spore surface is covered by worm-like ornamentation, which is sightly radial near the equatorial region. The zona is indistinct.

Comparison and Discussion: The ornamentation of this specimen is similar to *Triangulatisporites jemsarensis*, while the difference of other characters are great. The *Triangulatisporites* sp. documented in Jimusar, Xinjiang resembles to this specimen. Because there is only one specimen, so it was temporally documented as indeterminata species.

Locality and Horizon: Bed 19; Kayitou Formation, Mide section, Yunnan Province.

Genus ***Zerndtisporites*** Bhardwaj, 1955.

Zerndtisporites sp. (Fig. 10.3-1).

Description: Trilete megaspore. Equatorial outline is sub-round. Approximately 310 μm in diameter (including zona). Trilete is straight and thin, broadening towards the equator and reaching the margin of the middle noumenon, not into the zona. The middle noumenon has distinct boundary with the zona and the surface contains reticulate mottles. The zona is complete and the widest part is 40 μm, with no radial stripe.

Comparison and Discussion: This genus is extracted out of the *Zonalesporites*, and the later only includes elements with trilete inserting into zona and zona more or less presenting radial stripe. This genus could also be distinguished from other genera by trilete not inserting into the zona and the character of zona. This specimen is documented as indeterminata species due to the fail in matching with similar species and the specimen is the only one.

Locality and Horizon: Bed 19; Kayitou Formation, Mide section, Yunnan Province.

Genus ***Aneuletes*** Harris, 1961.

Aneuletes rotundus Fuglewicz, 1973 (Fig. 10.3-10).

Description: Aneulete megaspore. Equatorial outline is sub-round. 560 μm in diameter. There is a concave region in the middle of the spore, distinctly separated. In the middle of this concave area there is a ridgy ridge, the widest portion is roughly 83 μm. The spore surface is smooth.

Comparison and Discussion: This specimen is identical to the description of Fuglewicz to this species. Compared to the type species, the shape of the concave zone is a little bit different, and the later is rounder.

Locality and Horizon: Bed 24; Kayitou Formation, Mide section, Yunnan Province.

Aneuletes sp. (Fig. 10.3-8, 10.3-9).

Description: Aneulete megaspore. Equatorial outline is oval-round. 290–430 μm in diameter. The ornamentation on the surface is complex. The suffused crevice develop into irregular vein, 1.9 μm in width. Every single portion separated by the crevice is full of dense and fine radial crevice on the surface. There are densely arranged paxillus in the cross section.

Comparison and Discussion: The ornamentation of this specimen is quite special and is similar to *Aneuletes acrochordonodes* Fuglewicz. Whereas, the later does not have fine radial crevice on the surface. It is still hard to know the total character of this specimen for the reason that only one side of this megaspore is detected. So it is documented as the indeterminata species.

Locality and Horizon: Beds 24 and 25; Kayitou Formation, Mide section, Yunnan.

Chapter 11
Systematic Palaeontology: Part 3 Spore and Pollen Fossils

Jianxin Yu, Hui Li, Jean Broutin, and Xiao Shi

Anteturma **Sporites** Potonié, 1893.

Turma **Triletes** (Reinsch, 1881) Potonié and Kremp, 1954.

Subturma **Azonotriletes** (Luber, 1935) Dettmann 1963.

Infraturma **Laevigati** (Bennie and Kidston 1886) Potonié and Kremp, 1954.

Genus ***Leiotriletes*** (Naumova, 1937) Potonié and Kremp, 1954.

Type species: *Leiotriletes sphaerotriangulus* (Loose) Potonié and Kremp, 1954, Upper Carboniferous, Germany.

Diagnosis (emended from Smith and Butterworth, 1967, p. 121 (translation from Potonié and Kremp, 1954, p. 120)): Microspores, trilete with smooth outline, triangular amb having straight, distinctly concave, or slightly to rather strongly convex sides. Providing the triangular shape is still obvious or the three proximal pyramidal faces are rather steep. The spores belong to *Leiotriletes*. The apex of the amb is roughly rounded or blunt. Occasionally infrapunctate to infrareticulate. Trilete rays generally greater than one-half of radius but few down to one-third of radius. Sometimes, trilete rays ended with arcuate protuberance and along with torus.

J. Yu (✉)
School of Earth Sciences, China University of Geosciences, Wuhan, P.R. China
e-mail: yujianxin@cug.edu.cn

H. Li
Jiangxi Key Laboratory for Mass Spectrometry and Instrumentation, East China University of Technology, Nanchang, P.R. China
e-mail: huili_201860188@ecut.edu.cn

J. Broutin
Faculty of Sciences and Engineering (CR2P, MNHN, CNRS), Sorbonne Université, Paris, France

X. Shi
College of Earth Sciences, Jilin University, Changchun, Jilin, P.R. China
e-mail: xiaoshi@jlu.edu.cn

J. Yu et al. (eds.), *Plants and Palynomorphs around the Permian-Triassic Boundary of South China*, New Records of the Great Dying in South China,
https://doi.org/10.1007/978-981-19-1492-8_11

Discussion: The genus *Leiotriletes* was firstly proposed by Naumova (1937) and then Potonié and Kremp (1954) emened the diagnosis of this genus but in German. After that, Smith and Butterworth (1967) gave the translation for the diagnosis of this genus from Potonié and Kremp (1954). By the way, it was mentioned that Staplin (1960, p. 14) assigned these deltoid, smooth, trilete spores of Mississippian age (Palaeozoic) to the genus *Deltoidospora* Miner, 1935 (often for post-Palaeozoic spores) but others thought the form genus *Leiotriletes* Naumova, 1937 should be conserved especially for the Palaeozoic spores or even some Mesozoic spores. (Potonié, 1960, pp. 26–27; Playford, 1962, p. 573; Smith and Butterworth, 1967, p. 121). In fact, the genus *Deltoidospora* Miner, 1935 was firstly proposed as "small unassigned deltoid or sub-deltoid spores of the type that is commonly found associated with many Mesozoic ferns, such as *Gleichenites*, *Gleicheniopsis*, *Laccopteris* and others" (Miner, 1935). So, the genus *Deltoidospora* Miner, 1935 seems to be limited to be applied for simple deltoid spores from those specific mother plants. It seems to be the genus in the half-natural and half-form system. However, the genus *Leiotriletes* is totally and widely form genus that should be more suitable for the most cases in Palaeozoic and even in Mesozoic. As it was also mentioned by Potonié (1960) in the discussion part of the genus *Leiotriletes*'s systematic palaeontology that *Leiotriletes* is more widely various forms than others and sometimes can be assigned to the other genus, e.g. *Deltoidospora*, if you can be sure. Thus, the genus *Leiotriletes* should be conserved as a useful form genus for these simple triangular trilete microspores.

Affinity: Pteridopsida (Remy and Remy, 1957; Potonié, 1960; Palyford, 1962; Smith and Butterworth, 1967; Balme 1995; Hermann et al. 2011). Some in-situ specimens with immature state were also compared as *Leiotriletes* (Balme 1995).

Leiotriletes exiguus Ouyang and Li (1980) (Figs. 11.1-5, 11.1-6, 11.2-1 to 11.2-4).

Holotype: *Leiotriletes exiguus* Ouyang and Li (1980), pl. 1, fig. 4, from the horizon FK-1 (6) of Kayitou Formation, YijintianBu12 core in Qingyun, Fuyuan, Yunnan Province, China. Preserved in the Nanjing Institute of Geology and Palaeontology, Chinese Academy of Sciences.

Diagnosis (emended from Ouyang and Li (1980), p. 125): Amb triangular, trilete, side straight to slightly concave or sometimes slightly convex, apex blunt or rounded. Size is 39 (41)-44 μm. Laesurae fine but clear, 1/3–1/2 long of the radius at the equator. Contact area triangular, slightly thickened, dark color. The exine is thin, 1 μm in usual but sometimes up to 2–2.5 μm. Smooth surface and outline. Sometimes, In proximal view, sometimes with relatively slightly thick exine near the equator.

Material examined: Over 6 grains. Some registered specimens of them: ZK4703_ZK5_033, ZK4703_ZK5_073, ZK4703_ZK10_002, ZK4703_ZK10_070, ZK4703_ZK21_017, ZK4703_ZK45_091.

Occurrence: Upper Permian Xuanwei Formation to Lower Triassic Kayitou Formation, commonly in eastern Yunnan.

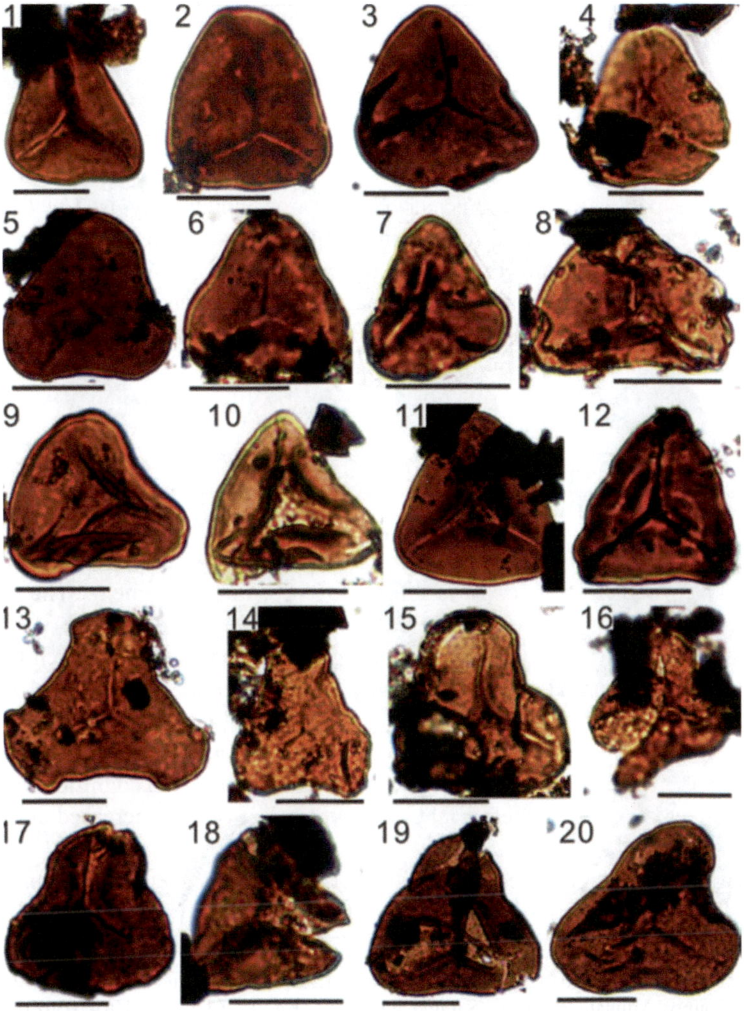

Fig. 11.1 (Scale bar = 20 μm). 1–3 *Leiotriletes directus* Balme and Hennelly. Locality: ZK4703, Fuyuan County, Yunnan Province. Sample Number: ZK4703_ZK31_349, ZK4703_ZK33_039, ZK4703_ZK39_058. 4, 17 *Leiotriletes adnatus* (Kosanke, 1950) Potonié and Kremp. Locality: ZK4703, Fuyuan County, Yunnan Province. Sample Number: ZK4703_ZK2_017, ZK4703_ZK10_038. 5–6 *Leiotriletes exiguus* Ouyang and Li. Locality: ZK4703, Fuyuan County, Yunnan Province. Sample Number: ZK4703_ZK5_068&185. 7–10, 12, 15 *Leiotriletes pulvinulus* Ouyang. Locality: ZK4703, Fuyuan County, Yunnan Province. Sample Number: ZK4703_ZK5_068&18. 11 *Leiotriletes inermis* (Waltz) Ishchenko. Locality: ZK4703, Fuyuan County, Yunnan Province. Sample Number: ZK4703_ZK33.5_192. 13–14 *Waltzispora strictura* Ouyang and Li. Locality: ZK4703, Fuyuan County, Yunnan Province. Sample Number: ZK4703_ZK2_006, ZK4703_ZK5_117. 16, 18, 19 *Leiotriletes tenuis* (Peppers, 1964) Ouyang. Locality: ZK4703, Fuyuan County, Yunnan Province. Sample Number: ZK4703_ZK5_088&178, ZK4703_ZK23_213; 20 *Leiotriletes concavus* (Kosanke, 1950) Potonié and Kremp, Locality: ZK4703, Fuyuan County, Yunnan Province. Sample Number: ZK4703_ZK33.5_062

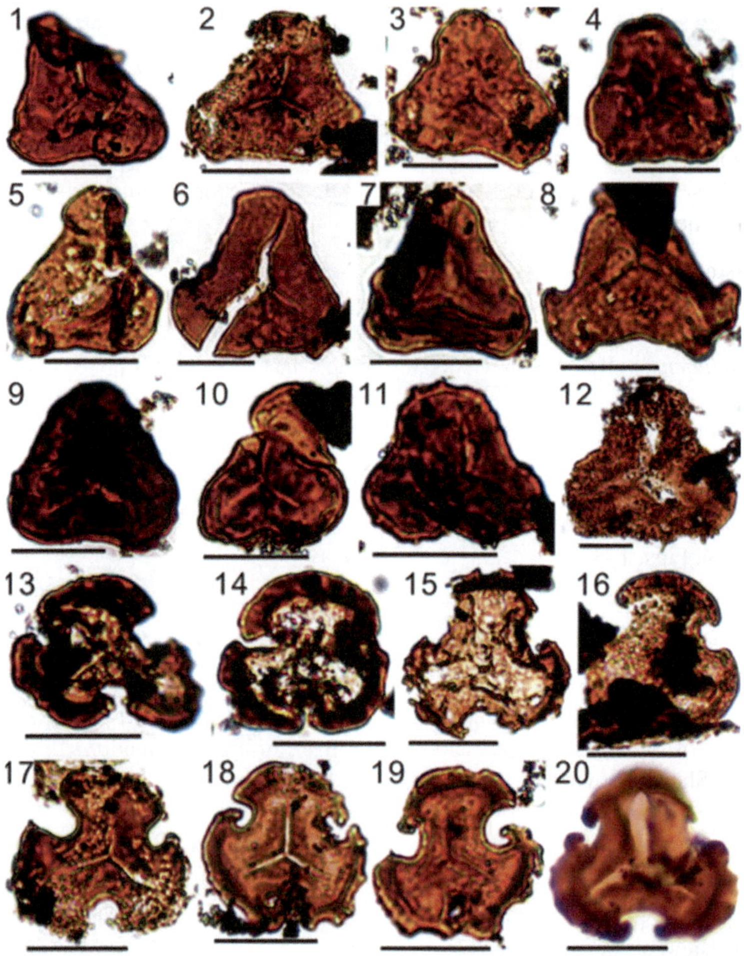

Fig. 11.2 (Scale bar = 20 μm). 1–4 *Leiotriletes exiguus* Ouyang and Li. Locality: ZK4703, Fuyuan County, Yunnan Province. Sample Number: ZK4703_ZK5_187, ZK4703_ZK14_046, ZK4703_ZK29_048&086&194. 5–6, 8 *Waltzispora strictura* Ouyang and Li. Locality: ZK4703, Fuyuan County, Yunnan Province. Sample Number: ZK4703_ZK10_013, ZK4703_ZK12_061, ZK4703_ZK25_035. 7 *Leiotriletes pulvinulus* Ouyang. Locality: ZK4703, Fuyuan County, Yunnan Province. Sample Number: ZK4703_ZK5_187. 9, 10 *Leiotriletes adnatus* (Kosanke, 1950) Potonié and Kremp. Locality: ZK4703, Fuyuan County, Yunnan Province. Sample Number: ZK4703_ZK10_062, ZK4703_ZK21_043. 11 *Lophotriletes gibbosus* (Ibrahim) Potonié et Kremp. Locality: ZK4703, Fuyuan County, Yunnan Province. Sample Number: ZK4703_ZK10_019. 12 *Neoraistrickia trilobata* Ouyang and Li. Locality: ZK4703, Fuyuan County, Yunnan Province. Sample Number: ZK4703_ZK14_038. 13–20 *Triquitrites proratus* Balme. Locality: ZK4703, Fuyuan County, Yunnan Province. Sample Number: ZK4703_ZK14_023, ZK4703_ZK5_139, ZK4703_ZK16_002, ZK4703_ZK23_002, ZK4703_ZK23_042, ZK4703_ZK25_022, ZK4703_ZK23_181, ZK4703_ZK33_394

Description: Trilete, Amb triangular, side slightly concave, its radius at the equator is about 17–20 μm long. laesurae simple, straight, short but very clear. Three commissures are straight and simple, each commissure is about 8–9 μm long. Dark color in the contact area may be due to the relatively thick exine but sometimes light color in the outline of the contact area. Contact area is triangular with relatively straight sides and blunt apex, the radius of contact area is about 11–14 μm long. And near the equator area it turn to be a little bit thick sometimes. the area ratio of contact area to outline is 0.33–0.39. With smooth surface and outline.

Discussion and Comparison: It is different from *Leiotriletes sphaerotriangulus* (Loose) Potonié and Kremp, 1954 and *Leiotriletes adnatoides* (Kos.) Potonié and Kremp, 1955 on the very short laesurae of this species. It is without torus that can be distinguished from *Leiotriletes pulvinulus* Ouyang (1986).

Leiotriletes pulvinulus Ouyang (1986) (Figs. 11.1-1 to 11.1-10, 11.1-12, 11.1-15 and 11.2-7).

1982, *Leiotriletes* sp. A, Ouyang; pp. 72, pl. 2, fig. 3.

1986, *Leiotriletes pulvinulus* Ouyang; pp. 33, pl. 1, fig. 1, 3, 4.

Holotype: *Leiotriletes pulvinulus* Ouyang (1986), pl. 1, fig. 3, YFH 121 (1), size is about 51 μm, of Xuanwei Formation, ZK-97 core in Housuoyishankou, Fuyuan, Yunnan Province, China. Preserved in the Nanjing Institute of Geology and Palaeontology, Chinese Academy of Sciences.

Diagnosis (emended from Ouyang (1986), p. 33): Amb triangular, sides straight or slightly concave or slightly convex, apex blunt or rounded. Laesurae are single ridges, 2/3–3/4 long of the radius at the equator and acuate at the end of laesurae. In the contact area, some torus are common along with laesurae, the width of torus is less than 2 μm, but sometimes they maybe blurry. Exine thin, about 1 μm thick, commonly with small folds or sometimes with irregular distributed micropitted. Smooth outline.

Material examined: Some registered specimens of them: ZK4703_ZK5_068, ZK4703_ZK5_185, ZK4703_ZK5_187, ZK4703_ZK14_046, ZK4703_ZK29_048, ZK4703_ZK29_086, ZK4703_ZK29_194.

Occurrence. Upper Permian Xuanwei Formation to Lower Triassic Kayitou Formation, commonly in eastern Yunnan.

Description: Microspore, trilete, radial, triangular outline, sides slightly concave or slightly convex or sometimes straight, apex narrow blunt or rounded. Laesurae well-developed, sometimes open, rays straight to reach the equator and sometimes with arched ridges at the end. Along with rays it bears torus and some folds across the rays. Surface smooth or sometimes few finely punctuates on surface. Outline is smooth.

Discussion and Comparison: This species is different from other species of this genus on its torus along with rays.

Leiotriletes adnatus (Kosanke, 1950) Potonié and Kremp, 1955 (Figs. 11.1-4, 11.1-17, 11.2-9 and 11.2-10).

1950 *Granulatisporites adnatus* Kosanke; p. 20, pl. 3, fig. 9.

1955 *Leiotriletes adnatus* (Kosanke, 1950) Potonié and Kremp;

Holotype: *Granulatisporites adnatus* Kosanke, 1950, pl. 3, fig. 9, with size 35*36 microns, from maceration 573 Slide 8, Coal 20 feet below the Carlinville limestone (No. 8 coal), Macoupin County, Illinois, USA, deposited at the Illinois State Geological Survey.

Diagnosis (emended after Kosanke, 1950, p. 20): Trilete radial microspores, amb roundly triangular, sides slightly concave, apex broadly rounded. The spore coat is levigate distally and proximally except for an area adjacent to the tetrad scar. This area is somewhat thicker and appears slightly granulose under high magnification. Trilete mark is distinct and extends at least three fourths of the radius at the equator. The lips are well developed and there is a definite *area contagionis*. The exine is uniformly slightly less than 2 microns thick except at the contact area.

Material examined: Some registered specimens of them: ZK4703_ZK2_017, ZK4703_ZK10_038, ZK4703_ZK10_062, ZK4703_ZK21_043.

Occurrence: Upper Permian Xuanwei Formation, commonly in eastern Yunnan in South China.

Description: Microspores, trilete radial, with triangular outline, sides slightly concave, apex rounded, Laesurae with obvious lips, the length of rays (12–13 μm) are more than 1/2 of the radius (17–21 μm) at the equator. Smooth but with few small weakly punctuates in the relatively thick contact area.

Discussion and Comparison: It differs from *Leiotriletes exiguus* on its relatively long rays and with small punctuates in contact area. It is without folds along with laesurae that can be distinguished from *Leiotriletes pulvinulus*. It differs from *Leiotriletes concavus* (Kosanke, 1950) Potonié and Kremp, 1955 on its slightly concave sides rather than strongly concave sides.

Leiotriletes directus Balme and Hennelly, 1956 (Figs. 11.1-1, 11.1-2, 11.1-3).

Holotype: *Leiotriletes directus* Balme and Hennelly, 1956, pl. 1, fig. 1–4, from seam at 688 ft, South Wallarah, No. 5 D.D.H. bore, upper part of the Newcastle Stage, N.S.W.

Diagnosis (emended after Balme and Hennelly, 1956, p. 244). Trilete radial microspore, amb triangular, sides straight or slightly convex in polar view, angles rounded. Trilete marks are well defined, rays straight or somewhat sinuous, extending almost to the periphery of the spore. Exine thin, subject to folding during compression, psilate or faintly granulate. Indistinct contact areas sometimes visible.

Material examined: Some registered specimens of them: ZK4703_ZK31_349, ZK4703_ZK33_039, ZK4703_ZK39_058.

Occurrence: Lower Triassic Kayitou Formation, commonly in eastern Yunnan.

Description: Microspore with trilete marks, radial, triangular outline, sides straight or slightly convex with narrowly or widely rounded apex. Laesurae is well developed with sometime ridge-like shapes, rays straight or sinuous and thin and it can reach near or to the equator. Smooth surface, sometimes with few weakly granulate. The contact area is weak. The length of radius at the equator is 19–24 μm and the length of laesurae is 16–21 μm.

Discussion and Comparison: It is different from *Leiotriletes sphaerotriangulus* and *Leiotriletes exiguus* on its no obvious contact area and long rays and with some foldings and finely granulates. It differs from *Leiotriletes pulvinulus* on without torus along with laesurae.

Leiotriletes concavus (Kosanke, 1950) Potonié and Kremp, 1955 (Fig. 11.1-20).

1950 *Granulatisporites concavus* Kosanke; p. 20, pl. 3, fig. 4.

1955 *Leiotriletes concavus* (Kosanke, 1950) Potonié and Kremp;

Holotype: *Granulatisporites concavus* Kosanke, 1950, pl. 3, fig. 4, with size of 55 * 58.8 microns, from maceration 318 Slide 10, "Ditney" coal bed from New Haven diamond drill core, White County, Illinois, USA, deposited at the Illinois State Geological Survey.

Diagnosis (emended after Kosanke, 1950, p. 20): Trilete radial microspore, amb subtriangular, sides strongly concave, apex rounded. The margin of the spore coat between the radii parallels the radii for a considerable distance. The spore coat is levigate and the tetrad mark extends at least three-fourths of the radius at the equator; the lips and commissure are distinct. Spore coat is less than 2 microns thick.

Material examined: Registered specimen: ZK4703_ZK33.5_062.

Occurrence: Lower Triassic Kayitou Formation, commonly in eastern Yunnan.

Description: Microspore, trilete, radial, triangular outline, sides strongly concave, apex rounded, laesurae simple but distinct, rays straight and thin, nearly reach to the equator, sides seem to be parallel to the laesurae, the contact area is blurry, some folds during compression, outline is smooth but some finely and weakly punctuates on surface.

Discussion and Comparison: This species is characterized by its strongly concave and simple laesurae.

Leiotriletes tenuis (Peppers, 1964) Ouyang (1986) (Figs. 11.1-16, 11.1-18, 11.1-19).

1964 *Granulatispontes tenuis* Peppers, pp. 20–21, pl. 2, fig. 11–12.

1986 cf. *Leiotriletes tenuis* (Peppers, 1964) Ouyang, pp. 33, pl. 2, fig. 8.

Holotype: *Granulatispontes tenuis* Peppers, 1964, Pl. 2, figure 11; maceration 1170-A, slide 10, coordinates-125.9 * 45.9, negative 6473, Fithian Cyclothem, Late Pennsylvanian, Illinois basin, USA, deposited at the Illinois State Geological Survey.

Diagnosis (after Peppers, 1964, p. 20). Trilete radial microspore, amb triangular, sides straight to slightly convex or concave, apex rounded. The spore coat is thin and usually folded during compression. The commissure is distinct, straight, extends to the spore margin, and generally is open. Lips are absent. A narrow, slightly thickened contact area is present along the aperture and extends almost to the ends of the rays. The surface of the spore coat is discernibly levigate under oil immersion objective. The exine is 1 to 1.5 μm thick.

Material examined: Registered specimens: ZK4703_ZK5_088, ZK4703_ZK5_178, ZK4703_ZK23_213.

Occurrence: Upper Permian Xuanwei Formation, commonly in eastern Yunnan in South China.

Description: Microspores, trilete, radial, triangular outline, sides slightly concave, with rounded apex. Laesurae obvious and open, can reach near the equator, with lips. And the opening of laesurae is more wider at proximal pole region than one at apex region and it is constrained to be narrow or acute at apex region. Thin exine and easy folded during compression. Relatively smooth surface but with few punctuates.

Discussion and Comparison: Its laesurae features are very special one that can be distinguished from other species of *Leiotriletes.*

Leiotriletes inermis (Waltz) Ishchenko 1952 (Fig. 11.1-11).

1938 *Azonotriletes inermis* Waltz in Luber and Waltz, p. 11, pl. 1, fig. 3, pl. 5, fig. 58 and pl. A, fig. 2

1952 *Leiotriletes inermis* (Waltz) Ishchenko, p. 9, pl. 1, fig. 2, 3.

1955 *Asterocalamotriletes inermis* (Waltz); Luber, p. 40, pl. 1, fig. 20, 21.

1955 *Leiotriletes inermis* (Waltz); Potonié and Kremp, p. 37.

1967 *Leiotriletes inermis* (Waltz) Ishchenko; Smith and Butterworth, 1967, p. 121–122, pl. 1, fig. 7, 8.

Holotype: No designated holotype by Waltz, 1938.

Diagnosis (after Smith and Butterworth, 1967): Trilete microspore, amb elongate-triangular, sides straight to slightly convex, apex rounded, laesurae simple with lips of trilete dehiscence mark slightly raised, dark colour in contact area, length of rays a little less than radius at the equator. Surface of exine smooth.

Material examined: One of the registered specimens: ZK4703_ZK33.5_192.

Occurrence: Lower Triassic Kayitou Formation, commonly in eastern Yunnan.

Description: Trilete radial microspore, slightly elongate triangular outline in proximal view, the outline is smooth, sides are slightly convex and each apex is widely rounded with a little bit thickened exine. Laesurae simple and open in the proximal pole with raised lips. And that open turns to be gradually narrow from proximal pole to the end of rays. Dark color concentrated on the proximal pole region and gradually turn to be light long with laesurae to the surroundings. Radius is about 23 micron in length, ray is about 20 micron in length and same as the radius of contact area, the area ratio of contact area to outline is about 0.45.

Discussion and Comparison: It is different from other species in *Leiotriletes* on its elongate triangular outline and slight convex sides and long open rays and dark regions.

Genus ***Waltzispora*** Staplin, 1960.

Type species: *Waltzispora* (*Azonotriletes*) *lobophora* (*lobophorus*) Waltz, 1938 in (Luber and Walz, 1938), from the Lower Carboniferous of the Moscow Basin and of the Selizharovo, Borovichi and Kizel regions, U.S.S.R.

Diagnosis (emended after Staphlin, 1960, p. 18; Playford, 1962, p. 581; Smith and Butterworth, 1967, p. 159; Ouyang and Li 1980, p. 126; Playford et al., 2010, p. 185)**:** Microspores, trilete, radial, amb subtriangular-triangular and trilobate, and various shapes at each apical extremity: reflexed or expanded, mushroom, saddle, T-shaped, shoe-last-shaped. Trilete mark distinct, with or without lip development, the exine ornament ranging from laevigate to apiculate.

Discussion: genus *Waltzispora* Staplin, 1960 is thought to be defined by Staplin (1960, p. 18). In fact the type species was firstly proposed by Waltz in (Luber and Walz, 1938) as *Azonotriletes lobophorus* Waltz, 1938.

Due to its various sculptures, ranging from smooth to apiculate. Thus, it should be considered if some species with smooth surface should be put in Infraturma Laevigati (like Playford, 1962; Ouyang and Li 1980; Ouyang 1986) while some species with apiculate sculptures should be put in Infraturma Apiculati Subinfraturma Granulati (like Sullivan, 1964; Smith and Butterworth, 1967; Playford et al. 2010) or just put in Infraturma Apiculati Subinfraturma Granulati according to the sculpture of the type species *Waltzispora lobophora*. But here we'd like to put this genus under Infraturma Laevigati because just one levigate-type species is described here.

Affinity: Unknown.

Waltzispora strictura Ouyang and Li (1980) (Figs. 11.1-13, 11.1-14, 11.2-5, 11.2-6 and 11.2-8).

Holotype: *Waltzispora strictura* Ouyang and Li (1980), pl. 1, fig. 9, size is 43 μm, from the horizon FK-1 (7) of Kayitou Formation, YijintianBu12 core in Qingyun, Fuyuan, Yunnan Province, China. Preserved in the Nanjing Institute of Geology and Palaeontology, Chinese Academy of Sciences.

Diagnosis: Amb triangular, sides strongly concave to slightly concave, apex expanded with shoe-last-shaped apical extremity and slightly pointed in the middle of apex, the junction between apex and side is contracted and angular or smooth transition. Trilete marks distinct with narrow lips, the length of ray is 1/2–2/3 of radius at the equator. Rays turn to be weak at the end. Exine is thin and about 1 μm, outline smooth but some fine punctuates on surface.

Material examined: Registered specimens: ZK4703_ZK2_006, ZK4703_ZK5_117, ZK4703_ZK10_013, ZK4703_ZK12_061, ZK4703_ZK25_035.

Occurrence: Upper Permian Xuanwei Formation, commonly in eastern Yunnan in South China.

Description: Microspores, trilete, radial, amb triangular to subtriangular, sides concave, sometime strongly concave, apex expanded with convex shape, the junction between apex and side is angular at angle of over 80 degree and then apex is contracted to sides. Laesurae clear and simple straight with thin lips, without reaching to the equator, sometimes trilete marks are short. The contact area is blurry. The outline is smooth and a few weakly granuloses or punctuates on the surface.

Discussion and Comparison: It is similar to *Waltzispora sagittata* Playford, 1962 but its apex height is short than *Waltzispora sagittate*, the apex part of *Waltzispora sagittate* is strongly pointed and the laesurae of *Waltzispora sagittate* extended almost to the equator. It is different from *Waltzispora polita* (Hoffmeister, Staplin and Malloy) Smith and Butterworth, 1967 on its simple and relatively thin laesurae rather than the broad laesurae of *Waltzispora polita*. It is different from *Waltzispora albertensis* Staplin, 1960 on its convex top part of apex rather than concave top part. It is different from *Waltzispora lobophora* (Waltz) Staplin, 1960 on its smooth outline rather than granulated outline.

Genus ***Calamospora*** Schopf, Wilson and Bentall, 1944.

Type species: *Calamospora hartungiana* Schopf, 1944 in (Schopf, Wilson and Bentall, 1944, pp. 51–52, fig. 1) from slide Y of maceration 90, deposited in the Illinois Geological Survey collections, Urbana, USA.

Diagnosis (emended after Schopf, Wilson and Bentall, 1944, pp. 49–51 and Smith and Butterworth, 1967, pp. 130–131): Microspores or some small megaspores, trilete, amb circular and very smooth, Laesurae distinct with very short, usually less than 1/2 of the radius at equator features, commissure distinct and attenuate, sometimes lips are moderately developed. Contact area distinct with color difference from other regions or unclear. Some arcuate ridges are commonly and sharp taper-point secondary folds with crescentic or narrow lenticular shape on the surface during the compression. A few minutely punctuates or slightly rugose may be seen under oil. Exine thin.

Discussion: It is characterized by its secondary folds on its thin exine. It can be also distinguished from *Leiotriletes* on its circular outline and from *Punctatisporites* on its very smooth surface and darker contact area.

Affinity: Mostly Sphenophytes, some Noeggerathiales and few megaspores of Lycophytes (Smith and Butterworth, 1967; Balme 1995).

Calamospora breviradiata Kosanke, 1950 (Figs. 11.3-1, 11.3-2).

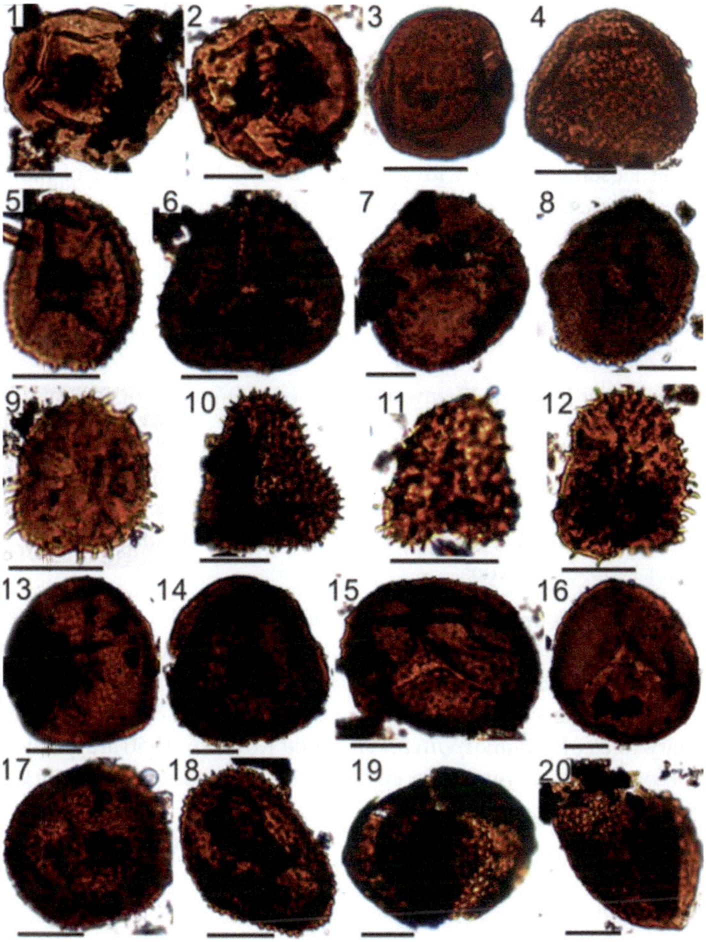

Fig. 11.3 (Scale bar = 20 μm). 1–2 *Calamospora breviradiata* Kosanke. Locality: ZK4703, Fuyuan County, Yunnan Province. Sample Number: ZK4703_ZK5_010, ZK4703_ZK23_057. 3 *Punctatisporites minutus* Kosanke. Locality: ZK4703, Fuyuan County, Yunnan Province. Sample Number: ZK4703_ZK52_152. 4 *Punctatisporites longinquus* Ouyang and Li. Locality: ZK4703, Fuyuan County, Yunnan Province. Sample Number: ZK4703_ZK52_114. 5 *Cyclogranisporites* cf. *micaceus* (Imgrund) Imgrund. Locality: ZK4703, Fuyuan County, Yunnan Province. Sample Number: ZK4703_ZK23_214. 6 *Granulatisporites mirus* Ouyang. Locality: ZK4703, Fuyuan County, Yunnan Province. Sample Number: ZK4703_ZK23_194. 7 *Cyclogranisporites aureus* (Loose) Potonié and Kremp. Locality: ZK4703, Fuyuan County, Yunnan Province. Sample Number: ZK4703_ZK14_042. 8 *Punctatisporites contractus* Ouyang and Li. Locality: Chinahe section, Xuanwei City, Yunnan Province. Sample Number: CNH13 + 1.8_031. 9 *Neoraistrickia* cf. *truncates* (Cookson, 1953) Potonié. Locality: ZK4703, Fuyuan County, Yunnan Province. Sample Number: ZK4703_ZK14_067. 10–12 *Neoraistrickia rigida* Ouyang, Locality: ZK4703, Fuyuan County, Yunnan Province. Sample Number: ZK4703_ZK34_075, ZK4703_ZK05_128, ZK4703_ZK02_049. 13–18 *Crassispora orientalis* Ouyang and Li. Locality: ZK4703, Fuyuan County, Yunnan Province. Sample Number: ZK4703_ZK23_035, ZK4703_ZK7_044, ZK4703_ZK12_034, ZK4703_ZK14_019, ZK4703_ZK23_167, ZK4703_ZK25_016. 19, 20 *Eupunctisporites chinensis* Ouyang and Li. Locality: ZK4703, Fuyuan County, Yunnan Province. Sample Number: ZK4703_ZK25_059, ZK4703_ZK14_067

Holotype: *Calamospora breviradiata* Kosanke, 1950, p. 41, pl. 9, fig. 4, size 57.7 * 65.1 microns from Maceration 579-B Slide 1, No. 2 coal bed, Bureau County, Illinois, deposited in the Illinois Geological Survey collections, Urbana, USA.

Diagnosis (after Kosanke, 1950, p. 41): Spores are radial, trilete, originally spherical, amb circular, laesurae distinct with short trilete rays, less than 1/2 of radius at the equator. The lips are distinct and elevated while the commissure is thin and attenuate. Contact area is distinct with dark color. The spore coat is levigate and is not over two microns thick and have folds generally parallel to the margin of the spore.

Material examined: Registered specimens: ZK4703_ZK5_010, ZK4703_ZK23_057.

Occurrence: Upper Permian Xuanwei Formation, commonly in eastern Yunnan.

Description: Microspores, circular outline, laesurae distinct with short rays, commissure straight with lips, contact area is circular and distinct with dark color. Levigate surface and outline, some secondary folds near the amb during compression. These folds are in narrow lenticular shapes. And few finely granulose can be seen. It radius at the equator is 55–56 microns.

Discussion and Comparison: This species was proposed by Kosanke (1950) with two distinct characters: short rays and dark contact area. But Kosanke (1950) also pointed that dark contact area is not just limited in this species. The type species is also with dark circular contact area (Schopf, Wilson and Bentall, 1944, fig. 1) but the secondary folds of *Calamospora breviradiata* Kosanke, 1950 are always near the equator and parallel to margin of spore. It can be distinguished from *Calamospora microrugosa* by its developed lips and short rays.

Genus ***Punctatisporites*** (Ibrahim, 1933) Potonié and Kremp, 1954.

Type species: *Punctatisporites punctatus* Ibrahim, 1933.

Diagnosis (emended after Kosanke, 1950, p. 18; Smith and Butterworth, 1967, pp. 124–125; Potonié and Kremp, 1954, p. 120): Trilete isospores or microspores having a circular, or near-circular, equatorial outline with a mere suggestion of triangular shape. Margin smooth, as exine devoid of ornamentation. Structure unrecognizable or only discernible owing to the presence of punctation, infrareticulation, or infragranulation (which must not be confused with a granulation causing roughness of the margin). On occasions, punctation is not more than locally visible, as, for instance, along the trilete rays. Rays generally longer than one-half the radius at the equator. Some arcuate folds or markings may be present on surface.

Discussion: This genus was first proposed by Ibrahim (1933, p. 21) as "spore with Y-marked, with finely punctate on the surface of exospore" in translation. And then Potonié and Kremp (1954) modified the diagnosis for this genus as above. But Guennel (1958) thought some species as *Punctatisporites orbicularis*, *Punctatisporites provectus* and *Punctatisporites vagus* had been carelessly transferred to

Cyclogranisporites by Potonié and Kremp (1955) because they possess finely punctate sculptures with rough outlines. In fact, in the emended diagnosis above (Potonié and Kremp, 1954) it stated that "margin or outline is smooth" but "Structure unrecognizable or only discernible owing to the presence of punctation, infrareticulation, or infragranulation (which must not be confused with a granulation causing roughness of the margin). On occasions, punctation is not more than locally visible, as, for instance, along the trilete rays". And Kosanke (1950) suggested to add arcuate markings or folds to Diagnosis: Thus, here following the emended diagnosis above by Potonié and Kremp (1954) and translated by Smith and Butterworth (1967).

Affinity: Pteridophytes (such as some Filicales: *Pecopteris*, *Scolecopteris*, *Danaeopsis*, *Todites* and so on) (Smith and Butterworth, 1967; Balme 1995; Hermann et al. 2011).

Punctatisporites minutus Kosanke, 1950 (Fig. 11.3-3).

Holotype: *Punctatisporites minutus* Kosanke, 1950, pl. 16, fig. 3 from Pennsylvanian Maceration 584 Slide 7, Woodbury (?) coal bed, Jasper County, Illinois, USA.

Diagnosis (after Kosanke, 1950): Microspore, radial, trilete, spherical shape with folded coat. Coat with minutely punctate under oil. Trilete rays distinct and lips slightly developed, commissure thin.

Material examined: Registered specimen: ZK4703_ZK52_152.

Occurrence: Upper Permian Xuanwei Formation to Lower Triassic Kayitou Formation, commonly in eastern Yunnan.

Description: This grain showed the lateral view, with circular outline and a few thin folds on surface, trilete rays on the right upper part, distinct and with thin slightly develpoed lips. Outline is nearly smooth but still some finely punctuates can be seen.

Discussion and Comparison: It can be distinguished from *Punctatisporites contractus* Ouyang and Li (1980) on its finer sculptures and thin laesurae.

Punctatisporites contractus Ouyang and Li (1980) (Fig. 11.3-8).

Holotype: *Punctatisporites contractus* Ouyang and Li, 1980, pl. 1 fig. 31 from FK-4 (8), Kayitou Formation, Fuyuan, Yunan Province, China.

Diagnosis (Ouyang and Li, 1980, p. 128; translation): Amb circular, laesurae distinct with well-developed lips, the width of laesurae is changeable ranging from 1–5 microns. The length of trilete rays is nearly 1/2 of radius at the equator. Exine thin and less than 2 microns, with finely and dense punctuates evenly distributed on it. Some folds may be developed.

Material examined: Registered specimen: CNH13 + 1.8_031.

Occurrence: Upper Permian Xuanwei Formation to Lower Triassic Kayitou Formation, commonly in eastern Yunnan.

Description: Microspore, amb circular, laesurae distinct and short with developed lips. With finely but dense punctuates on surface and some folds appeared on surface. But outline is smooth.

Discussion and Comparison: It is different from *Punctatisporites contractus* Ouyang and Li (1980) on its relatively short rays and the latter with rounded triangular outline.

Punctatisporites longinquus Ouyang and Li (1980) (Fig. 11.3-4).

Holotype: *Punctatisporites longinquus* Ouyang and Li, 1980, pl. 1, fig. 33 from FK-1 (14), Kayitou Formation, Fuyuan, Yunan Province, China.

Diagnosis (after Ouyang and Li, 1980, p. 128; translation): Amb rounded triangular (nearly circular), laesurae distinct with lips, 2–3 microns, slightly sinuous or straight, can reach to the equator sometimes with vertex. Exine thin and folded, finely and densely granuloses or slightly rough, the outline is relatively smooth.

Material examined: Registered specimen: ZK4703_ZK52_114.

Occurrence: Lower Triassic Kayitou Formation, commonly in eastern Yunnan.

Description: In proximal view, rounded triangular outline with relatively smooth outline, trilete rays distinct but thin, straight, can nearly reach to the equator but the ending of the laesurae is blurry, folds parallel to ambe and some across the rays. With finely granulose on surface.

Discussion and Comparison: It is featured by its rounded triangular outline.

Turma **Zonales** Potonié and Kremp, 1956.

Subturma **Zonotriletes** Waltz, 1935.

Infraturma **Auriculati** (Schopf) Dettmann, 1963.

Genus ***Triquitrites*** (Wilson and Coe, 1940) (Schopf, Wilson, and Bentall, 1944) Potonié and Kremp, 1954.

Type species: *Triquitrites arculatus* Wilson and Coe, 1940, p. 185, fig. 8, from the Slide Nos. 191–200. Pennsylvanian Green County Coal Mine, NW 1/4 SE 1/4s. 1, Franklin Township, Green County, Iow, USA.

Diagnosis (emended from Wilson and Coe, 1940, p. 184; Schopf, Wilson, and Bentall, 1944, pp. 46–47; Potonié and Kremp, 1954, pp. 153–154; Smith and Butterworth, 1967, pp. 201–202): Microspores, trilete, amb approximately triangular, sides slightly convex to strongly concave, apex with thickened exine, appear to be valvae or fairy small projections (some rounded projection being called auriculae), no flange in usual case at equator, but no radially plicated on valvae or auriculae.

Discussion: *Triquitrites* was proposed by Wilson and Coe (1940, p. 184) as "triradial, side concave and equatorial flange at angles". And then its diagnosis was expanded or emended (Schopf, Wilson, and Bentall, 1944, pp. 46–47; Potonié and Kremp, 1954, pp. 153–154). In addition, another similar genus *Tripartites* was proposed by Schemel (1950) as difference of "usually plicated in widest flange at corners". Especially, Sullivan and Neves (1964, p. 1088) compared the genus *Triquitrites* with other similar genus and gave the most important feature of radial plicated at apical region (auriculae) for *Tripartites* to distinguish from *Triquitrites.* And one version of diagnosis was translated from (Potonie and Kremp, 1954, pp. 153–154) in (Smith and Butterworth, 1967, pp. 201–202) but another version of diagnosis was adopted from (Schopf, Wilson, and Bentall, 1944, pp. 46–47) in (Balme, 1970). So, here the diagnosis was combined from two versions.

Affinity: Filicales: *Szea sinensis* Yao and Taylor, 1988 (Schopf, Wilson, and Bentall, 1944; Yao and Taylor, 1988; Balme 1995).

Triquitrites proratus Balme (1970) (Figs. 11.2-13 to 11.2-20).

Holotype: *Triquitrites proratus* Balme (1970), pl. 3, fig. 7, U.W.A. 57,782, Field no. K11-6D, Wargal, Salt Range, Chhidru Formation.

Diagnosis (after Balme 1970, p. 332, 334): Microspore, trilete, auriculate, amb triangular, sides concave. Trilete mark distinct, laesurae extending about three-fourths distance to the inner margin of the auriculae, sometimes bordered by slightly thickened darker exine. Exine 1–2 μm thick on proximal and distal faces, smooth or faintly maculate. Auriculae 2–5 μm wide, outer margins undulate, extremities usually markedly prolonged and incurved.

Material examined: Registered specimens: ZK4703_ZK14_023, ZK4703_ZK5_139, ZK4703_ZK16_002, ZK4703_ZK23_002, ZK4703_ZK23_042, ZK4703_ZK25_022, ZK4703_ZK23_181, ZK4703_ZK33_394.

Occurrence: Upper Permian Xuanwei Formation to Lower Triassic Kayitou Formation, commonly in eastern Yunnan.

Description: Microspore, trilete, amb approximately triangular, trilobate, sides concave, laesurae distinct with thin lips, trilete mark can reach onto apical part, slightly thickened exine on apex forming narrow auriculae or sometimes mixed with rounded projections, some granulose or levigate no plication on auriculae. In the central part, smooth to granulate surface.

Discussion and Comparison: It is different from Triquitrites arculatus Wilson and Coe, 1940 on its equatorial flange just in apical regions. It differs from *Triquitrites tribullatus* Potonié and Kremp, 1956 and *Triquitrites exceptus* Potonié and Kremp, 1956 on its strongly concave sides and relatively wide auriculae. It differs from *Triquitrites microgranifer* Ouyang, 1962 on its no flange connection between auriculate apex. It is different from *Triquitrites sinensis* Ouyang, 1962 on its sculptures without tuberculiforms.

Turma **Triletes** Reinsch 1881.

Subturma **Azonotriletes** Luber 1935.

Infraturma **Apiculati** (Bennie et Kidston 1886) Potonié et Kremp 1954.

Subinfraturma **Nodati** Dybová and Jachowicz, 1957.

Genus ***Lophotriletes*** (Naumova, 1937) Potonié et Kremp, 1954.

Type species: *Lophotriletes gibbosus* (Ibrahim) Potonié et Kremp, 1955, pl. 14, fig. 220 after Ibrahim, B61, e5, from Westphalian B, Agir Seam, Ruhr Coalfield, Germany.

Diagnosis (emended after Smith and Butterworth, 1967, p. 155; Potonié et Kremp, 1954, p. 129; translation)**:** Microspores, trilete, amb distinctly triangular rather than circular, sides generally convex, sometimes concave. Apex rounded. Trilete rays distinct, simple or flexuose, more than 1/2 radius at the equator even can reach the equator. With coni, apex of coni pointed or rounded.

Discussion: *Lophotriletes* was proposed by Naumova (1937) and emended by Potonié et Kremp (1954). The type species was defined as *Lophotriletes gibbosus* (Ibrahim) Potonié et Kremp, 1955, pl. 14, fig. 220 after Ibrahim, B61, e5. And then, Smith and Butterworth (1967) translated the diagnosis of Potonié et Kremp (1954) but some features were not defined clear, such as the trilete rays, apex and coni. Thus, here it was emended to give more details as more formal Diagnosis: It is different from *Apiculatisporis* on its triangular outline rather than circular outline.

Affinity: Filicopsida (Smith and Butterworth, 1967; Balme 1995).

Lophotriletes gibbosus (Ibrahim) Potonié et Kremp, 1955 (Fig. 11.2-11).

Holotype: *Lophotriletes gibbosus* (Ibrahim) Potonié et Kremp, 1955, pl. 14, fig. 220 after Ibrahim, B61, e5, from Westphalian B, Agir Seam, Ruhr Coalfield, Germany.

Diagnosis (after Smith and Butterworth, 1967): Microspore, amb triangular, sides concave, apex rounded, trilete rays more than 2/3 of radius at equator, tecta rather high, vertex sharp, slightly flexuose, ornament of small coni differing slightly in size, apices rounded, seldom flat. Width of coni generally equal to height, but may sometimes be greater. A negative reticulum may be appeared between the coni.

Material examined: Registered specimen: ZK4703_ZK10_019.

Occurrence: Upper Permian Xuanwei Formation, commonly in eastern Yunnan.

Description: Microspores, trilete, triangular outline with rounded apex, sides concave, laesurae distinct, slightly open, its length more than 2/3 of radius at the equator, the widest part of the commissure in the middle part and acute at the ending of the commissure. The coni is relatively evenly distributed in the proximal view and the width of coni is equal to its height. Some folds during the compression.

Discussion and Comparison: It is different from *Lophotriletes paramictus* Ouyang (1986) and *Lophotriletes mictus* Ouyang (1986) on its concave sides and regular coni. It is similar to *Lophotriletes confertus* Ouyang (1986) but the latter with very dense coni.

Turma **Triletes** (Reinsch, 1881) Potonié and Kremp, 1954.

Subturma **Azonotriletes** Luber 1935.

Infraturma **Apiculati** (Bennie and Kidston 1886) Potonié and Kremp, 1954.

Subinfraturma **Baculati** Dybová and Jachowicz, 1957.

Genus ***Neoraistrickia*** Potonié, 1956.

Type species: *Neoraistrickia truncates* Potonié, 1956, p. 34 (*Trilites turncatus* Cookson, 1953, p. 471, pl. 2, fig. 36 from pre-Tertiary Clays in the Comaum Bore, South Australia).

Diagnosis (emended after Potonié, 1956, p. 34)**:** Microspores, trilete, amb roundly triangular or subtriangular, sides biconvex or slightly concave, apex rounded, trilete rays simple, can reach to the equator, with bacula mostly on distal region and some in proximal region or smooth in proximal region. The bacula strong and straight or slightly curved, with truncated or dilated or rounded or inflated apex.

Discussion: *Neoraistrickia* was defined by Potonié (1956, p. 34) and the *Trilites turncatus* (Cookson, 1953, p. 471, pl. 2, fig. 36) was defined by Potonié (1956) as type species *Neoraistrickia truncates* (Cookson, 1953) Potonié, 1956. Dettmann (1963) thought *Cepulina* Maljavikina, 1949 and *Reticulatisporites*? Ibrahim (in Krutzsch, 1959, p. 162) are synonym of this genus. Bharadwaj and Kumar (1970) emended the diagnosis of this genus as "only distally baculate but proximally smooth". But in fact the type species *Trilites turncatus* (Cookson, 1953, p. 471, pl. 2, fig. 36) showed the proximal view with bacula sculptures. And also Dettmann (1963, p.36, pl. 5, fig. 4–5) described the proximal bacula inconspicuous and sparsely distributed on *Neoraistrickia truncates* (Cookson, 1953) Potonié, 1956. And Playford and Melo (2009) also denied that emended diagnosis of (Bharadwaj and Kumar 1970, pp. 214–215).

Affinity: Lycophytes (Dettmann, 1963).

Neoraistrickia trilobata Ouyang and Li (1980) (Fig. 11.2-12).

Holotype: *Neoraistrickia trilobata* Ouyang and Li (1980), p. 132, pl. 2, fig. 25 from FK-1 (20), Kayitou Formation, Fuyuan, Yunan Province, China.

Diagnosis (after Ouyang and Li 1980; translation): Amb subtriangular, trilobate, sides deeply concave, apex rounded, trilete distinct, sometimes with thin lips, commonly open, ray is more than 2/3 of radius at the equator with acute ending, with bacula and coni, these sculptures strong in the apex and weak in the central proximal part. The width of bacula is 1.5–4 μm, height is 3–7 μm. The apex of bacula is truncated or rounded or sometimes expanded, coni is 1–2.5 μm in height and 1–2 μm in width with pointed ending.

Material examined: Registered specimen: ZK4703_ZK14_038.

Occurrence: Upper Permian Xuanwei Formation to Lower Triassic Kayitou Formation, commonly in eastern Yunnan.

Description: Microspores, trilete, sides concave, apex rounded, trilete rays distinct with lips and open. Rays reach to the equator. In the proximal view, just some coni and few bacula can be seen but in amb more bacula can be seen. And it seems to be more bacula in the distal region. The bacula is thin and height is much larger than width with rounded apex.

Discussion and Comparison: It is different from *Neoraistrickia truncates* Potonié, 1956 on its concave sides.

Neoraistrickia rigida Ouyang (1986) (Figs. 11.3-10, 11.3-11, 11.3-12).

Holotype: *Neoraistrickia rigida* Ouyang (1986), pl. 5, fig. 26, YFH 105 (23), Upper Xuanwei Formation, Fuyuan, Yunan Province, China.

Diagnosis (after Ouyang 1986, p. 55): Amb triangular, sides concave, apex rounded, laesurae distinct and reach to the equator and sometimes with open commissures. Exine solid, with very dense and long bacula and coni, the base width is 1.5–4.0 micron sometimes up to 6.5 microns and height is 2.5–6.0 microns sometimes up to 8.0 microns, with blunt, rounded, trunctate, forked, pointed or slightly expanded apex of bacula or coni.

Material examined: Registered specimens: ZK4703_ZK34_075, ZK4703_ZK05_128, ZK4703_ZK02_049.

Occurrence. Upper Permian Xuanwei Formation to Lower Triassic Kayitou Formation, commonly in eastern Yunnan.

Description: Microspores, amb triangular, two slightly concave sides with one slightly convex side, apex rounded. Surface was covered by very dense bacula and coni, coni with pointed apex and bacula with various shaped apex, including rounded, trunctate, pointed, curved and expanded apex. Trilete rays open, covered by bacula and coni.

Discussion and Comparison: It is different from *Neoraistrickia trilobata* Ouyang and Li (1980) on its slightly concave sides and dense bacula and coni.

Neoraistrickia cf. ***truncates*** (Cookson, 1953) Potonié, 1956 (Fig. 11.3-9).

Material examined: Registered specimen: ZK4703_ZK14_067.

Occurrence: Upper Permian Xuanwei Formation to Lower Triassic Kayitou Formation, commonly in eastern Yunnan.

Description: Amb rounded triangular, two convex sides and one straight side, rounded apex, laesurae can be seen covered by bacula, the ending of bacula is trunctate or some rounded. In proximal view, bacula loose distributed.

Discussion and Comparison: It has commonly trunctate apex of bacula and laesurae existed but not very clear. But its outline is nearly circular. So, it is very similar to *Neoraistrickia truncates* (Cookson, 1953) Potonié, 1956.

Subinfraturma **Granulati** Dybová and Jachowicz, 1957.

Genus ***Granulatisporites*** (Ibrahim, 1933) Potonié and Kremp, 1954.

Type species: *Granulatisporites granulatus* Ibrahim, 1933, p. 21 from the Aegir coal bed at the top of the Westphalian B.

Diagnosis (after Smith and Butterworth, 1967)**:** Microspore, trilete, amb more or less triangular, a closely granular ornament of the exoexine. The grana are approximately circular, in lateral view, grana are flat or rounded.

Discussion: genus *Granulatisporites* Ibrahim, 1933 was emended by Schopf, Wilson and Bentall (1944), including various surfaces ranging from smooth to punctate and reticulate or apiculate. Potonié and Kremp (1954) emended it again with limitation of sculptures. Thus, here it adopted the translation of Potonié and Kremp (1954)'s diagnosis by Smith and Butterworth (1967).

Affinity: Filicales (Smith and Butterworth, 1967).

Granulatisporites mirus Ouyang (1986) (Fig. 11.3-6).

Holotype: *Granulatisporites mirus* Ouyang (1986), pl. 3, fig. 18, size 85 microns in diameter, from YFH 118 (2), Upper Xuanwei Formation, Fuyuan, Yunan Province, China.

Diagnosis (after Ouyang 1986; translation): Amb triangular, sides slightly convex or straight, laesurae is more than 3/4 of radius at the equator with relatively strong lips and open, rays slightly prominent, single and thin with slightly thickened radial region. With loose distributed grana on surface (especially in distal regions), diameter of grana is about 1 microns and its height is about 1 micron, and with also dense small grana in other regions.

Material examined: Registered specimen: ZK4703_ZK23_194.

Occurrence: Upper Permian Xuanwei Formation to Lower Triassic Kayitou Formation, commonly in eastern Yunnan.

Description: Microspores or miospores, with triangular outline in proximal view, sides slightly convex with rounded apex, laesurae distinct with well-developed lips and open. The commissures straight and can nearly reach the equator. Radial regions thickened. Surface with grana loosely distributed, in outline grana show the loose distributed pattern and the height is equal to the width.

Discussion and Comparison: It is different from *Granulatisporites granulatus* Ibrahim, 1933 on its convex sides. It is similar to *Granulatisporites adnatoides* (Potonié and Kremp, 1954) Smith and Butterworth, 1967 but the latter with more dense grana pattern.

Genus ***Cyclogranisporites*** Potonié and Kremp, 1954.

Type species: *Cyclogranisporites leopoldi* (Kremp, 1952) Potonié and Kremp, 1954, pl. 20, fig. 11.

Diagnosis (after Smith and Butterworth, 1967, p. 142)**:** Microspore, trilete, amb circular, laesurae distinct with lips, the rays can reach to the equator or short. Surface with grana in different distribution pattern.

Discussion: It is different from *Granulatisporites* on its circular shape.

Affinity: Ferns and Noeggerathiales (Smith and Butterworth, 1967).

Cyclogranisporites aureus (Loose) Potonié and Kremp, 1955 (Fig. 11.3-7).

Holotype: *Cyclogranisporites aureus* (Loose) Potonié and Kremp, 1955, pl. 13, fig. 184 from Upper Westphalian B, Bismarck Seam, Ruhr Coalfield, Germany.

Diagnosis (after Smith and Butterworth, 1967, pp. 142–143): Amb circular, laesurae distinct and simple straight, 1/2–2/3 of radius at the equator. Surface with dense grana, evenly distributed, more than 1 micron in diameter. 70–100 projects at the margin. With some narrow folds during compression.

Material examined: Registered specimen: ZK4703_ZK14_042.

Occurrence: Upper Permian Xuanwei Formation to Lower Triassic Kayitou Formation, commonly in eastern Yunnan.

Description: In proximal view, the outline is subcircular to circular, laesurae distinct and rays slightly curved during compression, contact area is not clear, rays are simple and its length is more than 1/2 of radius. Surface with dense evenly distributed grana. Near to margin some folds appeared during compression.

Discussion and Comparison: It is most different from other species on its thin exine with many folds.

Cyclogranisporites cf. ***micaceus*** (Imgrund) Imgrund, 1960 (Fig. 11.3-5).

Material examined: Registered specimen: ZK4703_ZK23_214.

Occurrence: Upper Permian Xuanwei Formation, commonly in eastern Yunnan.

Description: Subcircular outline, laesurae distinct, straight and reach to the equator. Few narrow long folds in irradial region. With dense thin grana on surface.

Discussion and Comparison: It is similar to *Cyclogranisporites* cf. *micaceus* (Imgrund) Imgrund, 1960 on narrow long folds during compression. But the size of grana is larger than the latter.

Infraturma **Murornati** Potonié and Kremp, 1954.

Genus ***Eupunctisporites*** (Bharadwaj, 1962) Ouyang and Li (1980)

Type species: *Eupunctisporites poniatiensis* Bharadwaj, 1962, pl. 1, fig. 4, from Upper Permian, Poniati Seam, Poniati Mine, East Raniganj Coalfield, India.

Diagnosis (after Ouyang and Li 1980, p. 133)**:** Miospores, circular to subcircular, trilete. Laesurae distinct, thin, low vertex and apex, sometimes with curvaturae imperfectae or curvaturae perfectae at the ending of rays, the length of rays equal to 1/2 of radius. Contact area sometimes distinct. Exine thick, distinctly punctate (pitted), mainly distributed on distal regions and equator. Surface of the exine ornamented with pits of various sizes and outline, usually small and fairly evenly spaced. Spore outline broken or smooth accordingly as the margin runs across a pit or the space between the pits respectively.

Discussion: This genus was firstly proposed by Bharadwaj (1962). And then Ouyang and Li (1980) gave some emendation on 'sometimes with curvaturae imperfectae or curvaturae perfectae at the ending of rays' and 'punctate (pitted), mainly distributed on distal regions and equator'.

Affinity: Lycophytes (Ouyang and Li 1980).

Eupunctisporites chinensis Ouyang and Li (1980) (Figs. 11.3-19, 11.3-20).

Holotype: *Eupunctisporites chinensis* Ouyang and Li (1980), pl. 2, fig. 32, size 73 micron, from FK-1(4), Kayitou Formation, Fuyuan, Yunan Province, China.

Diagnosis (after Ouyang and Li 1980, pp. 133–134): Amb subcircular to roundly triangular, subcircular in the lateral view and the distal region convex and the proximal region relatively flat in the lateral view. Laesurae distinct, with lips, the length of rays is equal to the radius, terminated by curvaturae and curvaturae perfectae for connection of laesurae, which also make contact area distinct. Exine with commonly circular punctuate and evenly mainly distributed on distal region to amb. The outline is waved. Sometimes with folds.

Material examined: Registered specimens: ZK4703_ZK25_059, ZK4703_ZK14_067.

Occurrence: Upper Permian Xuanwei Formation to Lower Triassic Kayitou Formation, commonly in eastern Yunnan.

Description: These two grains here with no perfect preservation but still with typical features of this speices. In figure 3.19, it shows the proximal region with relatively smooth surface and laesurae distinct reach to the equator and right down side broken exine show the distal region with punctuates. And In figure 3. 20 show the folded grains with laesurae distinct reach to the amb and distal region with punctuates.

Discussion and Comparison: This species can be distinguished from *Eupunctisporites poniatiensis* Bharadwaj, 1962 on its curvaturae and small punctuates.

Suprasubturma **Laminatitriletes** Smith and Butterworth, 1967.

Subturma **Zonolaminatitriletes** Smith and Butterworth, 1967.

Infraturma **Crassiti** (Bharadwaj and Venkatachala, 1961) Smith and Butterworth, 1967.

Genus ***Crassispora*** (Bharadwaj, 1957) Sullivan, 1964.

Type species: *Crassispora ovalis* Bharadwaj, 1957.

Diagnosis (emended after Sullivan, 1964, pp. 375–376)**:** Radial trilete microspore, amb circular to oval or roundly triangular. Exoexine finely to coarsely infrapunctate; crassitudinous thickening present at the equator. Distal surface ornamented with coni and occasionally spines; proximal surface without ornamentation. Intexine thin and translucent, outline rarely seen; margin conformable to equator of the spore. Apical papillae visible in intertectal areas. These are particularly well seen in over-macerated specimens. Laesurae usually indistinct but sometimes distinct with straight or sinous rays, sometimes with curvaturae imperfectae, sometimes accompanied by folds.

Discussion: This genus was proposed by Bharadwaj (1957, p. 125) and then Sullivan (1964, pp. 375–376) emended this diagnosis above. Main change is about whether the ornaments covering whole exine or just distal regions.

Affinity: Lycophytes (Smith and Butterworth, 1967).

Crassispora orientalis Ouyang and Li (1980) (Figs. 11.3-13 to 11.3-18).

1980 *Crassispora kosankei* (Potonié and Kremp) Bharadwaj, 1957; Ouyang and Li, pl. 3, fig. 5–7.

1980 *Crassispora orientalis* Ouyang and Li (1980), p. 10, pl. 2, fig. 5

Holotype: *Crassispora orientalis* Ouyang and Li (1980), p. 10, pl. 2, fig. 5, size 78 micron, from Kayitou Formation, Fuyuan, Yunan Province, China.

Diagnosis (after Ouyang 1986): Amb circular or roundly triangular, the distal region is more prominent than the proximal region at lateral view. With crassitudinous thickening at the equator but hard to be distinguished with other areas. Laesurae with developed lips and trilete mark projected, usually sinuous, sometimes with open, rays reach to the inner margin of crassitudinous thickening with curvaturae imperfectae at the ending or fused within crassitudinous thickening. Sometimes with big folds on exine, very tiny but dense grana or spongy structure. Some loosely distributed coni on surface.

Material examined: Registered specimens: ZK4703_ZK23_035, ZK4703_ZK7_044, ZK4703_ZK12_034, ZK4703_ZK14_019, ZK4703_ZK23_167, ZK4703_ZK25_016.

Occurrence: Upper Permian Xuanwei Formation to Lower Triassic Kayitou Formation, commonly in eastern Yunnan.

Description: Microspores, trilete, roundly triangular to circular, with crassitudinous thickening at the equator but narrow and hard to be separated from other areas. In proximal regions, laesurae distinct, reach to the crassitudinous thickening, with sinuous or straight rays and sometimes with open commissures. Surface with tiny grana or coni and their small size can be seen at the amb.

Discussion and Comparison: It is different from *Crassispora kosankei* (Potonié and Kremp) Bharadwaj, 1957 on its distinct laesurae.

Turma **Monoletes** Ibrahim 1933.

Subturma **Azonomonoletes** Luber 1935.

Infraturma **Laevigatimonoleti** Dybová and Jachowicz, 1957.

Genus ***Laevigatosporites*** (Ibrahim, 1933) Schopf, Wilson and Bentall, 1944.

Type species: *Laevigatosporites vulgaris* Ibrahim, 1933.

Diagnosis (after Schopf et al., 1944 and Smith and Butterworth, 1967)**:** Spores, monolete, amb oval to elongated oval and smooth, laevigate to infrapunctate, reniform shape in meridian plane. Distal margin convex, polar axis relatively short, without ornaments on exine. Monolete mark straight. Sometimes with folds during compression.

Discussion: It was proposed by Ibrahim (1933, p. 39). And then Schopf, Wilson and Bentall (1944) gave some features to emend this Diagnosis: And Smith and Butterworth (1967) translated the diagnosis given by Potonié and Kremp (1954). Here, we combined these features into diagnosis above.

Affinity: Sphenophytes and Filicales (Smith and Butterworth, 1967).

Laevigatosporites minor (Loose, 1934) Bharadwaj, 1957 (Figs. 11.4-1, 11.4-2, 11.4-3).

1934 *Laevigatosporites vulgaris minor* Loose, p. 158, pl. 7, fig. 12.

1957 Laevigatosporites minor (Loose, 1934), Bharadwaj, p. 109, pl. 29, figs. 8, 9.

Holotype: *Laevigatosporites vulgaris minor* Loose, 1934, p. 158, pl. 7, fig. 12 from Upper Westphalian B, Bismarck seam, Ruhr Coalfield, Germany.

Diagnosis (after Smith and Butterworth, 1967): Amb oval, reniform shape in lateral view. Laesurae simple, greater than 3/4 of spore diameter. Exine laevigate to faintly punctate, its size ranging from 35–64 microns.

Material examined: Registered specimens: ZK4703_ZK23_019, ZK4703_ZK02_046, ZK4703_ZK52_120.

Occurrence: Upper Permian Xuanwei Formation to Lower Triassic Kayitou Formation, commonly in eastern Yunnan.

Description: Microspore, monolete. In polar view, its shape is ellipsoidal. While in equatorial view, its shape is reniform. Laesurae simple and monolete, greater than 3/4 of spore diameter, ray is straight and slightly arching in the middle, with some folds on surface and few tiny punctuates. Their sizes ranging from 40–50 microns.

Discussion and Comparison: It is different from *Laevigatosporites minimus* and *Laevigatosporites vulgaris* on size. The size of *Laevigatosporites minimus* is the smallest one, less than 35 microns, and the size of *Laevigatosporites vulgaris* is the biggest one, more than 64 microns among these three species according to (Smith and Butterworth, 1967).

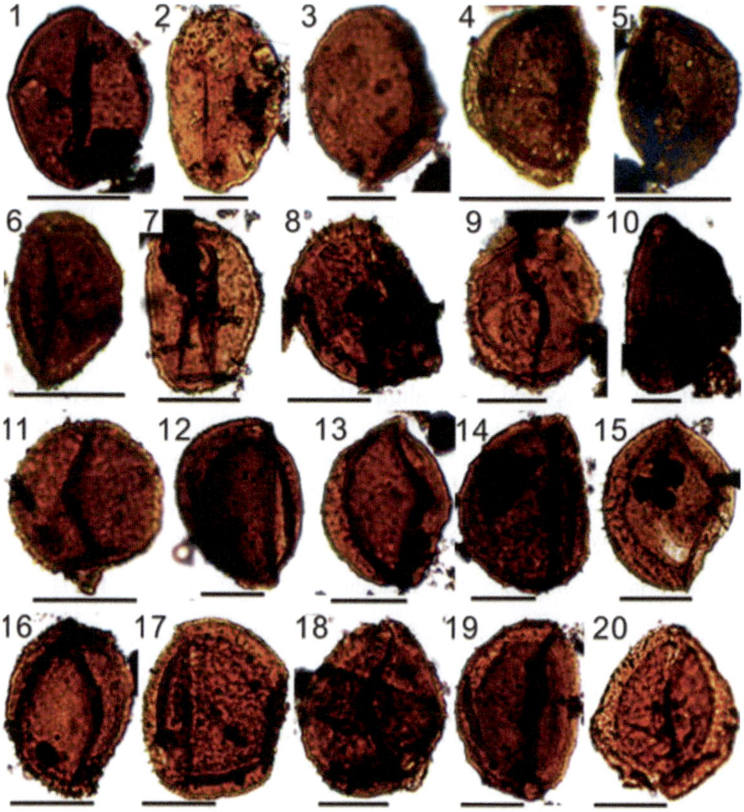

Fig. 11.4 (Scale bar = 20 μm). 1–3 *Laevigatosporites minor* (Loose, 1934) Bharadwaj. Locality: ZK4703, Fuyuan County, Yunnan Province. Sample Number: ZK4703_ZK23_019, ZK4703_ZK02_046, ZK4703_ZK52_120. 4–20 *Aratrisporites yunnanensis* Ouyang and Li. Locality: Chinahe section, Xuanwei City, ZK4703, Fuyuan County, Yunnan Province. Sample Number: CNH13 + 0.3_025, CNH13 + 0.3_038 and 039, ZK4703_ZK14_036, ZK4703_ZK23_006, 015, 016, 067, 166, 182 and 196, ZK4703_ZK25_051, ZK4703_ZK29_012, 047 and 105, ZK4703_ZK33_001 and 006

Subturma **Zonomonoletes** Naumova 1937.

Suprasubturma **Perinomonolites** Erdtman 1947.

Genus ***Aratrisporites*** (Leschik, 1955) Playford and Dettmann 1965.

1955 *Aratrisporites* Leschik, Keuperflora von Neuewelt, p. 38.

1960 *Aratrisporites* Klaus, Sporen der karnischen Stufe, p. 145.

1960 *Saturnisporites* Klaus, Sporen der karnischen Stufe, p. 142.

Type species: *Aratrisporites parvispinosus* (Leschik, 1955) Playford and Dettmann, 1965.

Diagnosis (after Playford and Dettmann, 1965, pp. 151–153)**:** Microspores monolete, bilateral. Spore wall (sclerine) two-layered, cavate; consisting of an outer structured layer (sculptine) loosely enveloping, but proximally attached to, a homogeneous inner layer. Sculptine surface finely patterned (the pattern being formed by an arrangement of structural elements) and with sculptural elevations including grana, coni, spinulae, spinae and saetae. Laesura enclosed within elevated lips which are proximal extensions of sculptine, with curvaturae imperfectae.

Discussion: Leschik (1955) put this genus into alete group but Klaus (1960) found its monolete mark and emended it. And then Playford and Dettamann (1965) included the Saturnisporites Klaus 1960 into this genus.

Affinity: Lycophytes (especially Isoetales and maybe from *Annalepis* or *Tomiostrobus*) (Harris, 1955; Playford and Dettamann, 1965; Grauvogel-Stamm and Duringer 1983; Balme 1995).

Aratrisporites yunnanensis Ouyang and Li (1980) (Figs. 11.4-4 to 11.4-20).

Holotype: *Aratrisporites yunnanensis* Ouyang and Li (1980), Pl. 4, Fig. 5, size 52*40 micron, FK-1 (5), from Kayitou Formation, Fuyuan, Yunan Province, China.

Diagnosis (after Ouyang and Li 1980, pp. 143–144; translation): Microspore, monolete, amb widely ellipsoidal to ellipsoidal. Outline in equatorial view is lenticular and the proximal region convex or flat but the distal region strongly convex. Monolete reach or nearly reach to the equator with lips, sinuous, sometimes with one nearly orthogonal but not distinct ray from the middle part of monolete laesurae, forming "false trilete" structure. Exine with two layers, intexine without ornaments and slightly thicker than exoexine. In the polar regions, two layers are separated. Sexine thin, with grana, spony structure in proximal polar. With evenly distributed coni or small spines and their apex acute or clavate apex. Ornaments mainly distributed on distal regions and equator. Monolete laesurae ended with curvaturae imperfectae.

Material examined: Registered specimens: CNH13 + 0.3_025, CNH13 + 0.3_038, CNH13 + 0.3_039, ZK4703_ZK23_016, ZK4703_ZK23_015, ZK4703_ZK23_006, ZK4703_ZK14_036, ZK4703_ZK23_166, ZK4703_ZK23_182, ZK4703_ZK23_196, ZK4703_ZK23_067, ZK4703_ZK25_051, ZK4703_ZK29_012, ZK4703_ZK33_001, ZK4703_ZK29_047, ZK4703_ZK29_105, ZK4703_ZK33_006.

Occurrence: Lower Triassic Kayitou Formation, commonly in eastern Yunnan.

Description: Microspores, monolete, cavate, widely ellipsoidal to ellipsoidal in proximal view and reniform (fig. 4.4, 4.12, 4.14) to lenticular (fig. 4.5) shape in equatorial view. Laesurae monolete with sinuous shape and always arching in the middle part, sometimes forming "false trilete" structure (fig. 4.4, 4.14, 4.18), with grana or coni on surface. And some folds can be seen on sexine.

Discussion and Comparison: It is different from *Aratrisporites fimbriatus* by the latter with hair-like fimbriae. It is different from *Aratrisporites paenulatus* by its thick exine and sinuous mid-arching laesurae.

Anteturma **Pollenites** Potonié, 1931.

Turma **Saccites** Erdtman, 1947.

Subturma **Disaccites** Cookson, 1947.

Infraturma **Disacciatrileti** Leschik, 1956.

Genus ***Alisporites*** (Daugherty 1941) Jansonius, 1971.

Type species: *Alisporites opii* Daugherty 1941, Upper Triassic, Arizona, National Monument, U.S.A.

Diagnosis (after Hart, 1965)**:** Disaccate Disacciatrileti, haploxylonoid to slightly diploxylonoid in outline and with a transverse distal sulcus running centrally across the central body. The C.B. is circular or oval with an 1-a or t-a elongation and the sacci are equal to or less than the C.B. in size and less than or about semi-circular in shape. The t-a (C.B.) is greater than or equal to the t-a (sacci). The distal zone varies from distinct to obscure.

Discussion: It contains various morphological types with alete bisaccate pollens. It is different from *Vesicaspora* on the latter with lateral bladder. It is different from *Vitreisporites* on the latter with very small size. It is similar to *Falcisporites* but without lemniscatoid transverse sulcus.

Affinity: Conifers (Balme 1995; Bomfleur, 2011).

Alisporites spp. (Fig. 11.5-1 to 11.5-15).

Material examined: Registered specimens: ZK4703_ZK45_038, ZK4703_ZK45_129, ZK4703_ZK39_005, ZK4703_ZK45_097, ZK4703_ZK39_004, ZK4703_ZK39_009, ZK4703_ZK51_048, ZK4703_ZK51_080, ZK4703_ZK34_055, ZK4703_ZK39_078, ZK4703_ZK36_019, ZK4703_ZK52_003, ZK4703_ZK51_025, ZK4703_ZK39_068, ZK4703_ZK45_105.

Occurrence: Lower Triassic Kayitou Formation, commonly in eastern Yunnan.

Description: Bisaccate pollen, alete, haploxylonoid or slightly diploxylonoid, with oval outline in polar view, body is circular to ellipsoid, sacs from semicircle to circular shape. Commonly one transverse sulcus developed in distal region (fig. 5.3). surface on body is finely punctate or few finely grana or smooth and surface on sacs is reticulate or finely punctate or finely grana.

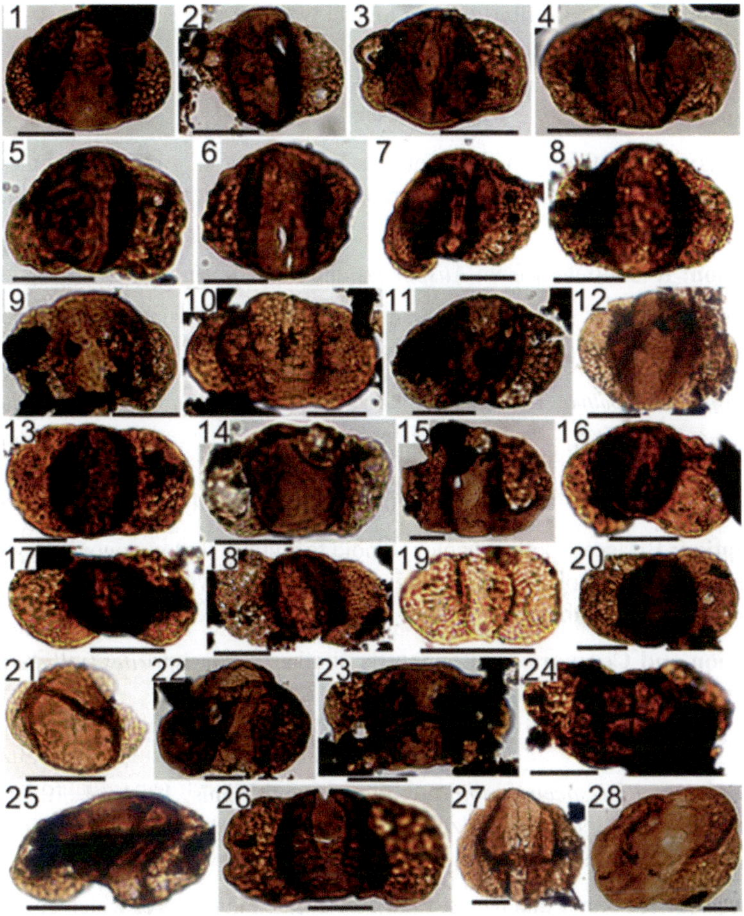

Fig. 11.5 (Scale bar = 20 μm). 1–15 *Alisporites* spp. Locality: ZK4703, Fuyuan County, Yunnan Province. Sample Number: ZK4703_ZK34_055, ZK4703_ZK36_019, ZK4703_ZK39_ 004, 005, 009, 068 and 078, ZK4703_ZK45_038, 097, 105 and 129, ZK4703_ZK51_025, 048 and 080, ZK4703_ZK52_003. 16–18, 20 *Platysaccus* spp. Locality: ZK4703, Fuyuan County, Yunnan Province. Sample Number: ZK4703_ZK39_016, ZK4703_ZK51_026, 143 and 146. 19 *Vitreisporites* cf. *pallidus* (Reissinger) Nilsson. Locality: ZK4703, Fuyuan County, Yunnan Province. Sample Number: ZK4703_ZK31_294. 21–23 *Chordasporites* spp. Locality: ZK4703, Fuyuan County, Yunnan Province. Sample Number: ZK4703_ZK31_282, ZK4703_ZK45_077, ZK4703_ZK51_045. 24 *Illinites* sp. Locality: ZK4703, Fuyuan County, Yunnan Province. Sample Number: ZK4703_ZK45_062. 25 *Lueckisporites* sp. Locality: ZK4703, Fuyuan County, Yunnan Province. Sample Number: ZK4703_ZK52_085. 26 *Striatopodocarpites* sp. Locality: ZK4703, Fuyuan County, Yunnan Province. Sample Number: ZK4703_ZK45_078. 27 *Protohaploxypinus* sp. Locality: ZK4703, Fuyuan County, Yunnan Province. Sample Number: ZK4703_ZK52_090. 28 *Vesicaspora* sp. Locality: ZK4703, Fuyuan County, Yunnan Province. Sample Number: ZK4703_ ZK31_350

Discussion: They have various morphological types here.

Genus ***Vitreisporites*** (Leschik, 1955) Jansonius, 1962.

Type species: *Vitreisporites signatus* Leschik, 1956.

Diagnosis: Bisaccate pollen, more or less haploxylonoid, small, sometimes with narrow distal furrow. Body is oval or transverse ellipsoid, sacs usually semicircle. sacs is larger than body.

Discussion: It is different from *Alisporites* and *Falcisporites* on its small size and narrow furrow.

Affinity: Caytoniales (Ouyang and Li 1980).

Vitreisporites cf. ***pallidus*** (Reissinger) Nilsson, 1958 (Fig. 11.5-19).

Material examined: Registered specimen: ZK4703_ZK31_294.

Occurrence: Lower Triassic Kayitou Formation, commonly in eastern Yunnan.

Description: Bisaccate pollen, haploxylonoid to slightly diploxylonoid, very small, outline is ellipsoid at polar view. Body is smaller than sacs with transverse oval shape. Sacs in semicircle. with narrow furrow.

Discussion and Comparison: It is mostly similar to *Vitreisporites pallidus* on its small size and narrow distal furrow.

Genus ***Platysaccus*** (Naumova, 1937) Potonié and Klaus, 1954.

Type species: *Platysaccus papilionis* Potonié and Klaus, 1954, pl. 10, fig. 12, Salzberg Hallstatt, Christina-Horizont, Schwingheim-Querschlag. Kemsalz from the Gray Core Mountains.

Diagnosis (after Potonié and Klaus, 1954)**:** Bisaccate microspores which do not have a Y mark, no proximal laesurae, no sharply marked distal area. The dumbbell-shaped overall outline at the equator. Central body is circular to oval. Sacci usually extremely over hemispherical, significantly larger than the central body. On the longest diameter the ratio of the central body to the total length is about 1: 3 because the outline of the Sacci in the equatorial section is more than semicircular. If their growth stripe is not on the longest diameter of the Sacci, but represents a shorter tendon. Sacci mostly distally approached and provided with pronounced infrared ticulum. Some folds may be seen as radial shape in the roots of sacs.

Discussion: It is different from *Alisporites* and *Falcisporites* on its much bigger sacs than body and without sulcus in distal regions. And it is different from *Vesicaspora* by it without lateral bladder.

Affinity: Seed ferns and conifers (Balme 1995; Rad and Noroozpour, 2017).

Platysaccus spp. (Fig. 11.5-17, 11.5-18, 11.5-20).

Material examined: Registered specimens: ZK4703_ZK51_026, ZK4703_ZK51_143, ZK4703_ZK51_146.

Occurrence: Lower Triassic Kayitou Formation, commonly in eastern Yunnan.

Description: Bisaccate pollen, strongly diploxylonoid, alete, without sulcus both on proximal region and distal region. It is dumbbell shape in polar view. And the body is more or less circular and sacs are more than semicircle. The size of sacs is parentally larger than the size of body. Some radial folds appeared on the roots of sacs.

Discussion: Some grains here show more than two different morphological types.

Genus ***Vesicaspora*** (Schemel 1951) Wilson and Venkatchala 1963.

Type species: *Vesicaspora wilsonii* Schemel, 1951, Mystic Coal Seam, Iowa, U.S.A.

Diagnosis (after Hart, 1965): Disaccate Disacciatrileti haploxylonoid in outline. The central body is usually obscured by the sacci exines but when seen is circular or has a t-a elongation. Running transversely across the central body there is a distal sulcus occupying the whole of the distal zone, i.e. framed by the distal roots. The outline of this sulcus may be fusiform, irregular or appear as a narrow slit. The main characteristic of the genus is the development of lateral bladders so that a completely encircling air chamber with terminal swellings appears in equatorial view.

Discussion and Comparison: This genus was firstly assigned to Monosaccites (Schemel, 1951; Wilson and Venkatchala, 1963). But Potonié (1958) started to assign it to Disaccites and then more and more studies supported this assignation (Bharadwaj, 1962; Hart, 1965).

Affinity: Ginkgophytes (Balme 1995).

Vesicaspora sp.

Material examined: Registered specimen: ZK4703_ZK31_350.

Occurrence: Lower Triassic Kayitou Formation, commonly in eastern Yunnan.

Description: Bisaccate pollen, alete. Haploxylonoid, relatively big size, central body is transverse oval and sacs are roundly crescent-shaped. Sacs are larger than body. The ends of Two sacs are separately connected by lateral bladders from extended margin of body. Some folds appeared on surface of sacs.

Discussion and Comparison: It is marked by its lateral bladders.

Genus ***Illinites*** Kosanke 1950.

Type species: *Illinites unicus* Kosanke 1950, pp. 51–52, pl. 1, fig. 3, Maceration 494 Slide 15, 10-inch coal bed exposed in Coffee Creek, Wabash County, Illinois, U.S.A.

Diagnosis (after Kosanke, 1950, p. 51; Hart, 1965): Disaccate Disaccitrileti, haploxylonoid to diploxylonoid in outline. The central body is circular or oval with a slight l-a or t-a elongation. The proximal surface of the central body has a centrally placed trilete aperture and is laevigate, punctate or finely granulate. The sutures of the trilete aperture may or may not be of equal length to one another.

Discussion: It is different from *Limitisporites* and *Vesitisporites* on its trilete mark. It is different from *Jugaesporites* on its dilete mark.

Affinity: Voltzian conifers (Grauvogel-Stamm and Grauvogel, 1973; Balme 1995).

Illinites sp.

Material examined: Registered specimen: ZK4703_ZK45_062.

Occurrence: Lower Triassic Kayitou Formation, commonly in eastern Yunnan.

Description: Bisaccate pollen, with trilete mark in proximal region of body. Haploxylonoid to slightly diploxylonoid, body is bigger than sacs. Body is circular shape and sacs are semicircular. Some folds can be seen during compression.

Discussion and Comparison: It is with distinct trilete mark, which should be assigned to *Illinites* but it is hard to classify it into species level due to no good preservation.

Turma **Saccites** Erdtman, 1947.

Subturma **Disaccites** Cookson, 1947.

Infraturma **Striatiti** Pant, 1954.

Genus ***Chordasporites*** Klaus, 1960.

Type species: *Chordasporites singulichorda* Klaus, 1960, pl. 33, fig. 45.

Diagnosis (after Klaus, 1960, pp. 157–158; translation): Bisaccate microspores, the central body of which has a distinct exine cord (chorda) or fold on one side, in the case of saccus convergence on the side facing away from it, in the direction of the longitudinal axis, which is usually directly connected to the sac base lines or runs a piece into them as a fold. Sometimes, however, it is so short that it does not reach the bases of the sacs. The exine strand, often in the form of a bulge that looks like it is twisted, is never divided into itself like a closed back (or crown-back laesur) accompanied by bulges, but more or less homogeneous. As the cord splits and pops open. There is no sharp step between the edge of the saccus base and the cord. Often the Sacci converge on the side opposite the exine strand and leave a less or no ornamented, thinner exine point free between them. The strand side of the central body is usually made thicker and structured approximately up to the equator. The cord is occasionally accompanied by narrow, thinned areas of exine on both sides.

Discussion: It is different from *Lueckisporites* on its one cord-like striate.

Affinity: Peltasperm-type pteridosperms (Aggarwal et al. 2018).

Chordasporites spp. (Fig. 11.5-22).

Material examined: Registered specimens: ZK4703_ZK31_282, ZK4703_ZK45_077, ZK4703_ZK51_045.

Occurrence: Lower Triassic Kayitou Formation, commonly in eastern Yunnan.

Description: Some grains here show bisaccate pollens with one cord-like striate over central body, haploxylonoid, body is in shape of circle to transverse oval, sacs are in shape of semicircle. And the sacs is smaller than body. Some folds can be seen during the compression.

Discussion and Comparison: They are featured by its one cord-like striate over body. But they are different from each other by central body size.

Genus ***Lueckisporites*** Potonié and Klaus, 1954.

Type species: *Lueckisporites virkkiae* Potonié and Klaus 1954, Solvay Werke Borth, Germany.

Diagnosis (after Hart, 1965)**:** Disaccate Striatiti, diploxylonoid in outline. The central body is circular or elongated in the 1-a direction, and has a proximal cap showing two longitudinal ribs in polar view. The sacci are more or less semi-circular in shape and have an infra-reticulate structure. The distal roots are less than t-a (sacci) and leave a distal zone that is usually about one-third or less 1-a (C.B.) in dimension.

Discussion: It is similar to *Gardenasporites* but it has the laesurae in intexine and more darker body.

Affinity: Conifers (Balme 1995).

Lueckisporites sp. (Fig. 11.5-25).

Material examined: Registered specimen: ZK4703_ZK52_085.

Occurrence: Lower Triassic Kayitou Formation, commonly in eastern Yunnan.

Description: Bisaccate pollens, diploxylonoid, with two distinct taeniae on proximal surface and one laesurae inside the proximal surface according to its lighter color than other regions. Central body with circular to slightly elongate oval. Sacs with roundly crescent shape. Some folds can be seen on body surface and a little bit darker color for whole body than sacs.

Discussion and Comparison: It is distinct for assigning to *Lueckisporites* by its two taeniae on body surface.

Genus ***Striatopodocarpites*** (Sedova, 1956) Hart, 1964.

Type species: *Striatopodocarpites tojmensis* Sedova 1956, plate XLI, figure 8, North Dvina basin, River Lower Toima, Lower Kazanian Beds.

Diagnosis (Hart, 1965 emended)**:** Disaccate Striatiti, strongly to moderately diploxylonoid in outline. The length of the distal roots of the sacci are less than the diameter of the central body in transverse direction. The central body is circular to slightly oval in outline and possesses a proximal cap divided into more than four longitudinal ribs. The sacci are distinctly larger than the central body in size and greater than semi-circular in shape. The terminal sacci may in rare cases be united to form slight lateral bladders.

Discussion and Comparison: It is different from *Protohaploxypinus* and *Lunatisporites* by its diploxylonoid and relatively thin and dense striates.

Affinity: Ginkgophytes or Pteridosperms (Balme 1995).

Striatopodocarpites sp. (Fig. 11.5-26).

Material examined: Registered specimen: ZK4703_ZK45_078.

Occurrence: Lower Triassic Kayitou Formation, commonly in eastern Yunnan.

Description: Bisaccate pollens, moderately diploxylonoid, with numerous thin striates (about 11) on body surface and centrally with possible laesurae. Body with circular shape and sacs with semicircle shape. Body is smaller than sacs. Sacs with reticulate sculptures.

Discussion and Comparison: It is distinct by numerous thin striates and moderately diploxylonoid outline.

Genus ***Protohaploxypinus*** (Samoilovich, 1953) Hart, 1964.

Type species: *Protohaploxypinus latissimus* (Luber and Valts, 1941) Samoilovich, 1953. Solikamsk region, Western Pre-Urals, U.S.S.R.

Diagnosis (Hart, 1965 emended): Disaccatc Striatiti, haploxylonoid or slightly diploxylonoid in outline. The central body is circular, or oval with a slight t-a or slight I-a elongation. The proximal cap is divided into four or more longitudinal ribs. The sacci are semi-circular, less than semi-circular or slightly greater than semi-circular in shape. The sacci structure is infrareticulate. The length of the distal root is equal to or slightly less than the t-a (C.B.). The distal zone is less than or equal to two-thirds 1-a (C.B.).

Discussion: It is different from *Lunatisporites* on its numerous striates (over 5) rather than a few taeniae.

Affinity: Ginkgophytes or Peltaspermales (Balme 1995).

Protohaploxypinus sp. (Fig. 11.5-27).

Material examined: Registered specimen: ZK4703_ZK52_090.

Occurrence: Lower Triassic Kayitou Formation, commonly in eastern Yunnan.

Description: Bisaccate pollen, haploxylonoid, with over 10 striates on body surface with relatively darker color and along with lighter-color grooves. Body shape is transverse oval to lenticular, sacs are in semicircle and body is larger than sac.

Discussion and Comparison: This grain is distinguished by other genus on its numerous striates with haploxylonoid outline.

References

Balme BA (1995) Fossil in situ spores and pollen grains: an annotated catalogue. Rev Palaeobot Palynol 87:81–323

Balme BE (1970) Palynology of Permian-Tiassic strata in the Salt Range and Surghar, West Pakistan. In: B. Kummel & C. Teichert (eds) Stratigraphic boundary problems: Permian and Triassic of West Pakistan. University of Kansas, Special Publication 4, pp 306–453

Grauvogel-Stamm L, Duringer P (1983) Annalepis zeilleri Fliche 1910 emend., un organe reproducteur de Lycophyte de la Lettenkohle de l'Est de la France. Morphologie, spores in situ et paléoécologie. Geologische Rundschau 72(1):23–51

Ouyang S (1986) Palynology of upper Permian and Lower Triassic strata of Fuyuan district, eastern Yunnan. Science Press, Beijing. 122 pp (in Chinese with English Summary)

Ouyang S, Li ZP (1980) The micro-floral in the Kayitou Bed of Fuyuan, Yunnan and its stratigraphic and paleobotanic significance. In: Nanjing Institute of Geology and Palaeontology, Academica Sinica, (eds) Late Permian coal-bearing strata and palaeontological fauna in western Guizhou and eastern Yunnan. Science Press, Beijing, pp 123–194 (in Chinese).

Bibliography

Broutin J, Yu JX, Shi X, Shu WC, Xue Q (2020) Terrestrial palaeofloral succession across the permian–triassic boundary in the north and south china blocks: a brief review. Paläontologische Zeitschrift, 94(1)

Chen JH, Yu JX, Huang QS, Broutin J, Song QQ, Chen B (2011) New research progress on the paleoflora in the earliest Triassic of western Guizhou and eastern Yunnan, South China. Earth Sci J China Uni Geosci 36(3):500–510 (In Chinese with English abstract)

Galfetti T, Hochuli PA, Brayard A, Bucher H, Weissert H, Vigran JO (2007) Smithian-Spathian boundary event: evidence for global climatic change in the wake of the End-Permian biotic crisis. Geology 35(4):291–294

Hofmann R, Goudemand N, Wasmer M, Bucher H, Hautmann M (2011) New trace fossil evidence for an early recovery signal in the aftermath of the End-Permian Mass Extinction. Palaeogeogr Palaeoclimatol Palaeoecol 310:216–226

Jin YX, Shang QH, Hou JP, Li L, Wang YJ, Zhu ZL, Fei SY (2000) Stratigraphical lexicon of China: Permian. Geological Publishing House, Beijing (In Chinese)

Kidder DL, Worsley TR (2003) Causes and consequences of extreme Permo-Triassic warming to globally equable climate and relation to the Permo-Triassic extinction and recovery. Palaeogeogr Palaeoclimatol Palaeoecol 203:207–237

Krassilov V, Karasev E (2009) Paleofloristic evidence of climate change near and beyond the Permian-Triassic boundary. Palaeogeogr Palaeoclimatol Palaeoecol 284:326–336

Liu LJ, Yao ZQ (2007) Plant megafossils from the permian changhsingian marine deposits of fusui, guangxi, china. Acta Palaeontologica Sinica 46(2):195–212

Liu LL, Yao ZQ (2013) The conifer-remains from the Permian of South China. Acta Palaeontologica Sinica 52(2):182–201

Lozovsky VR (1998) The Permian-Triassic boundary in the continental series of Eurasia. Palaeogeogr Palaeoclimatol Palaeoecol 143:273–283

Lucas SG, Kozur H, Molina-Garza R (1996) Reconsideration of Dalongkou as an auxiliary GSSP for the Permian-Triassic boundary. Albertiana 18:10–11

Meng FS (1996) Floral palaeoecological environment of the Badong Formation in the Yangtze Gorges area. Geology Mineral Resources South China, pp 1–13. (In Chinese with English abstract)

Neregato R, D'Apolito C, Glasspool IJ, Wang S-J, Liu F, Windslow P, Lu J, Shao L, Hilton J (2016) Palynological constriants on the provenance and stratigraphic range of a Lopingian (Late Permian)

inter-extinction floral lagerstatte from the Xuanwei Formation, Guizhou Province, China. Int J Coal Geol 162:139–150
Ouyang S (1991) Transitional palynofloras from basal lower Triassic of China and their ecological implications, with special reference to Paleophyte/Mesophyte problems. Palaeoecology of China 1. Nanjing, Nanjing University Press 168–196
Pal PK, Ghosh AK (1997) Megafloral zonation of Permian-Triassic sequence in the Kamthi Formation, Talcher Coalfield, Orissa. The Palaeobotanists 46:81–87
Shen SZ, Schineider JW, Angiolini CM, Henderson L (2013) The international Permian timescale: March 2013 update. Bull-N M Mus Nat Hist Sci 60:411–416
Song QQ, Yu JX, Feng JP, Huang QS (2015) Palaeobotany of the upper Permian Dalong Formation (marine facies) in south Guizhou. Geological Sci Technol Inform 34(1):63–66 (In Chinese with English Abstract)
Stanley SM (2009) Evidence from ammonoids and conodonts for multiple Early Triassic mass extinctions. Proc Natl Acad Sci 106(36):15264–15267
Sze HC (1953) Notes on some fossil remains from the Shihchienfeng Series in northwestern Shensi. Acta Palaeont Sinica 1(1):11–22 (in Chinese with English summary)
Tian YT, Yu JX, Feng QL (2008) The Discovery of Ophiceras in the Kayitou Formation in Eastern Yunnan Province and Its signification. J Stratigr 32:153–158
Vega JA, Fernández C, Fonturbel T (2005) Throughfall, runoff and soil erosion after prescribed burning in gorse shrubland in Galicia (New Spain). Land Degrad Dev 16(1):37–51
Wang SQ (1978) Fossil Ostracods of the late Permian and early Triassic in western Guizhou and Northeastern Yunnan. Acta Palaeontologica Sinica 17(3):279–308 (in Chinese with English abstract)
Wang YG (1984) Earliest Triassic ammonoid faunas from Jiangsu and Zhejiang and their bearing on the definition of Permo-Triassic boundary. Acta Palaeontologica Sinica 23(3):257–269 (in Chinese with English abstract)
Wang ZQ, Lou YY (1990) Studies on the megaspores of the Mid-Triassic lycopods in Shanxi. China. Acta Bot Sinica 32(5):397–403 (in Chinese with English abstract)
Wang ZQ, Zhang ZP (1997) Gymnosperms on the eve of the terminal Permian mass extinction in North China and their survival strategies. China Science Bulletin 42:2134–2141 (in Chinese with English abstract)
Whittaker RH (1979) Classifying species according to their demographic strategy. Am Nat 113:185–200
Wignall PB (2015) The worst of times: how life on earth survived 80 million years of extinction. Princeton University Press, pp 224
Xiong CH, Wang Q (2011) Permian-Triassic land-plant diversity in South China: was there a mass extinction at the Permian/Triassic boundary? Paleobiology 37(1):157–167
Xu R, Zhu JR, Chen Y, Duan SY, Hu YF, Zhu WQ (1979) China late Triassic Baoding Flora. Science Press 1–130
Yang ZY, Yin HF, Wu SB, Yang FQ, Ding MH, Xu GR (1987) Permian-Triassic boundary stratigraphy and fauna of China. In: Geological memoirs, People's Republic of China ministry of geology and mineral resources, series 2, 6. Geological Publishing House, Beijing, pp 1–379
Yin HF, Ding MH, Zhang KX, Tong JN, Yang FQ, Lai XL (1995) Dongwuan-Indosinian (Late Permian-Middle Triassic) ecostratigraphy of the Yangtze region and its margins. Science Press, Beijing, pp 1–338
Yin HF, Jiang HS, Xia WC, Feng QL, Zhang N, Shen J (2013) The end-Permian regression in South China and its implication on mass extinction. Earth Sci Rev. https://doi.org/10.1016/j.earscirev.2013.06.003
Zhang H, Cao CQ, Liu XL, Mu L, Zheng QF, Liu F, Xiang L, Liu LJ, Shen SZ (2016) The terrestrial end-Permian mass extinction in South China. Palaeogeogr Palaeoclimatol Palaeoecol 448:108–124

GPSR Compliance

The European Union's (EU) General Product Safety Regulation (GPSR) is a set of rules that requires consumer products to be safe and our obligations to ensure this.

If you have any concerns about our products, you can contact us on ProductSafety@springernature.com

In case Publisher is established outside the EU, the EU authorized representative is:

Springer Nature Customer Service Center GmbH
Europaplatz 3
69115 Heidelberg, Germany

Batch number: 10726135

Printed by Printforce, the Netherlands